PerOUS AND COMPLEX FLOW STRUCTURES
IN MODERN TECHNOLOGIES

Springer

New York
Berlin
Heidelberg
Hong Kong
London
Milan
Paris
Tokyo

Adrian Bejan
Ibrahim Dincer
Sylvie Lorente
Antonio F. Miguel
A. Heitor Reis

Porous and Complex Flow Structures in Modern Technologies

With 336 Illustrations

 Springer

Adrian Bejan
Duke University
Dept. of Mechanical Engineering
  and Materials Science
Durham, NC 27708-0300
USA
dalford@duke.edu

Ibrahim Dincer
University of Ontario Institute
  of Technology
School of Manufacturing Engineering
Oshawa, Ontario L1H 7K4
CANADA
Ibrahim.Dincer@uoit.ca

Sylvie Lorente
Laboratoire Matériaux et
  Durabilité des Constructions
Institut National des Sciences Appliquées
Département de Génie Civil
31077 Toulouse
FRANCE
lorente@insa-tlse.fr

Antonio Miguel
Universidade de Évora
Departamento de Fisica
Colegio Luis Verney
7000-671 Évora
PORTUGAL
afm@uevora.pt

Heitor Reis
Universidade de Évora
Departamento de Fisica
Colegio Luis Verney
7000-671 Évora
PORTUGAL
ahr@uevora.pt

Library of Congress Cataloging-in-Publication Data
Porous and complex flow structures in modern technologies / Adrian Bejan ... [et al.].
    p. cm.
  Includes bibliographical references and index.
  ISBN 0-387-20225-0 (alk. paper)
  1. Fluid dynamics. 2. Porous materials--Fluid dynamics. I. Bejan, Adrian, 1948-

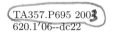
TA357.P695 2003
620.1'06--dc22   2003060456

ISBN 0-387-20225-0   Printed on acid-free paper.

Printed in the United States of America.

9 8 7 6 5 4 3 2 1   SPIN 10949856

Springer-Verlag is a part of *Springer Science+Business Media*

springeronline.com

Contents

Preface

This is a text and reference book on the fundamentals of flows in porous media and the key roles played by these advances in technologies that are important today and in the foreseeable future. The fundamental topic of flows in porous media is the vehicle for bringing together a long list of critically important issues from diverse fields such as energy, civil, biotechnology, chemical, and environmental engineering. This is an interdisciplinary book, because these many technological issues are being brought together with purpose: they are current and important, and are supported by a common scientific structure, the principles and behavior of flows in porous media.

Our first objective in writing this book was to present a new approach, in several respects. Unlike the most recent reference books on porous media, which catalogue in pure and relatively abstract terms the current position of fundamental research on porous media, our book invites the reader to see the real-life needs of today's engineering. The development of engineering science is driven not only by curiosity, but also by needs, objectives, and limitations. This need-based design approach characterizes not only our coverage of flow phenomena in porous media, but also the coverage of the fundamentals of the other disciplines that overlap in this book.

Energy engineering is an important interface because the environmental impact of energy conversion (e.g., power plants) is, in many cases, due to flows through subterranean porous layers. The latter also play the central role in energy exploration (e.g., petroleum, geothermal fluids, methane hydrate deposits). Yet the traditional approach to energy engineering starts from classical thermodynamics—the first and the second laws, and analyses of "thermodynamic systems," which are defined specifically to be distinct from their environments. Our view is considerably more inclusive. Energy engineering also means design, optimization, and the generation of optimal flow structure. Furthermore, realistic models of energy systems demand the combined treatment of installations and their flowing surroundings, more so when the installations are large and their spheres of impact greater. We apply these principles to the analysis and design of energy storage systems and fuel cells.

These examples highlight the connection between exergy, environment, and sustainable development.

Another new direction that is explored in our book is the impact that small-scale devices (compact heat exchangers, electronics) have on the very fundamentals of heat and fluid flow through porous media. Again, current technological needs dictate the development of fundamentals. The traditional treatment of porous media refers mainly to homogeneous and isotropic porous media, Darcy and Darcy-modified flow models, and local thermal equilibrium (one-temperature) heat transfer models. New and considerably more challenging are the models demanded by strikingly coarse porous structures where the representative elemental volume assumption fails, highly conductive structures where the local thermal equilibrium assumption fails, and heterogeneous media composed of porous domains and domains occupied by pure fluids.

Aerosol transport and collection in porous media (e.g., filters) is another area where environmental impact and a variety of modern technologies comes together for the purpose of developing and using new porous-medium models and results. Applications range from nuclear engineering, agricultural products, food technology, and semiconductor manufacturing, to environmental control in hospitals, museums, and large buildings in general. Porous filters are used to protect workers, occupants, and materials.

Biomedical engineering is another major area that bursts with activity, and enriches the understanding of vascularized flow structures as complex and designed (optimized) porous media. Vascularized tissues and the lung can be presented as porous media that function as mass and heat exchangers with flow structures optimized over many length scales, starting with microchannels. The lung can also be described as a porous structure functioning as a periodic aerosol filter. The modeling of heat transfer in living tissues, which is so important to designing heat-treatment and organ preservation techniques, depends greatly on descriptions of porous media with designed structure (e.g., dendritic channels, counterflow pairs).

Food technologies demand porous-media treatments that account for mesopores, water vapor flow, time-dependent flows, phase-change, and the movement of the two-phase interface through the porous structures. These phenomena are important in the drying and storing of grain, as well as in the thermal processing of food products. Energy-intensive processes such as drying and food processing rely not only on porous media but also on principles of energy engineering.

Civil engineering also demands the porous-media backbone of our book. One example is the modeling of environmental impact, such as, the spreading of pollutants. Energy conservation in buildings is also important, for example, the loss of heat through walls with porous inserts and small cavities. The transfer of chemical species and chemical reactions in concrete are critical to understanding and improving the durability and environmental safety of buildings. They govern the penetration of chloride through concrete, and the corrosion of the metallic structure that reinforces the concrete. Another example is the ionic decontamination by electrokinetic processes. The collection and distribution of water (cold or hot), sewage, rainwater, and so on, require increasingly more complex dendritic networks (larger, finer, multiple scales). Urban flows are porous structures that are designed. Such designs have a lot in common with the bioengineering applications mentioned above. Once again, conceptual connections are made not only with civil engineering and porous-media fundamentals, but also with living systems.

Electronics, and the miniaturization of structures that must be cooled intensively, is the technological frontier that, alone, demands the treatment presented in this book. Cooling is the technology that stops the march toward smaller scales, greater processing density, greater complexity, and superior global performance. We are seeing this not only in microscale heat transfer but also in the development of compact heat exchangers. For our treatment of porous media, the new opportunity stems from the need for smaller and smaller flow passages and fins in heat exchangers and electronics cooling. Smaller ducts mean higher heat transfer rates, and laminar flow. In the aggregate, however, macroscopic flow structures (heat exchangers) approach a limit that is much better suited for porous-medium modeling, and for the design of porous structure.

The need for considering the broad picture—the macroscopic system— is great and universal. No matter how successful we are in discovering and understanding small-scale phenomena and processes, we are forced to face the challenge of assembling these invisible elements into palpable devices. The challenge is to construct, that is, to assemble and to optimize while assembling. This challenge is becoming more difficult, because while the smallest scales are becoming smaller, the number of components and the complexity of the useful device (always macroscopic) become correspondingly greater.

This observation deserves emphasis, because it is widely overlooked in discussions of shrinking scales and "nanotechnology." Technology means a lot more than the new physics that may appear on the frontiers of progressively smaller scales. A technology is truly new when it is made useful in the form of devices (macroscopic constructs) that improve our lives. Usefulness demands that we must discover not only new physics, but also the strategy for connecting and packing the smallest-scale elements into devices for use at our macroscopic scales.

We wrote this book together during 2001 through 2003. We had the opportunity to try it as textbook material for summer courses taught at the Ovidius University of Constanţa, Romania, and the University of Évora through the Physics Department and the Évora Geophysics Center, Portugal. The manuscript was typed by Mrs. Linda Hayes. We acknowledge with gratitude the support received from the University of Évora, Ovidius University, King Fahd University of Petroleum and Minerals, Lord Foundation of North Carolina, and Pratt School of Engineering of Duke University. We especially thank our colleagues, Professors Jean Pierre Ollivier (INSA Toulouse), Kristina M. Johnson (Duke), Rui Rosa and Ana Silva (University of Évora),

Donald A. Nield (University of Auckland), José L. Lage (Southern Methodist University), and Doctors Alexandre K. da Silva and Wishsanuruk Wechsatol (Duke University).

Adrian Bejan
Ibrahim Dincer
Sylvie Lorente
Antonio F. Miguel
A. Heitor Reis

1

Porous Media Fundamentals

1.1 Structure

A porous medium consists of a solid structure with void spaces that are in general complicated and distributed throughout the structure. The void spaces can be interconnected or not. The pores are identifiable regions that serve as elements for the void space. The traditional view of porous media was inspired by porous structures found in nature, for example, packed sand saturated with water that seeps through the pores. Natural porous structures have random features, such as irregular pore shapes and sizes, and irregular connections between the pores. Today we are seeing a growing number of technologies that rely on flows through complex and small-scale passages. The structures formed by such passages can be viewed as designed porous media—structures where the pore shapes, sizes, and connections are special and purposeful, not irregular or random. Designed porous media are components of larger systems and installations that meet global objectives and perform functions under constraints.

We consider both types of flow structures; irregular porous media that have been traditionally described based on volume-averaging, and complex flow structures that result from design. Our treatment is organized in the direction from the simple to the complex, and from the old to the new. We start with the classical volume-averaged description of flows through porous media. This approach has generated much of the terminology that is in use today. We review in tutorial manner the basics that support the phenomena and applications that form the core of this book. The review is based on the treatments available in Bejan (1995a) and Nield and Bejan (1999), which may be consulted for greater detail. Additional treatments are provided by Bear (1972), Cheng (1978), Ene and Poliševski (1987), Greenkorn (1983), Bear and Bachmat (1990), Ingham and Pop (1998, 2002), Nakayama (1995), Kaviany (1995), Sahimi (1995), and Pop and Ingham (2001). The subject is also covered in the latest handbooks (Vafai, 2000; Bejan and Kraus, 2003).

Porous media are a highly diverse class of structures. Several classifications have been made. The fluid that fills the pores may be single-phase (gas, liquid), or multiphase. The associated flows are treated as single- or multiphase flows through porous media. The fluid may be a single-component substance or a multicomponent mixture. When the fluid fills the pores completely, the porous medium is saturated.

The length scales of the flow structure also differentiate between types of porous media. Even when the voids are irregular, as in many of the porous media that occur naturally, the pore dimensions may be concentrated in a narrow range, or they may be distributed (smoothly or discretely) over a wide range. The latter are flow structures with multiple scales, which is a defining feature of many designed structures, such as tree-shaped flow distribution networks (Chapter 4) and miniaturized heat exchanges (Chapter 5).

How we treat a flow through a porous structure is largely a question of distance—the distance between the problem-solver and the actual flow structure. When the distance is short, the observer sees only one or two channels, or one or two open or closed cavities. In this case it is possible to use conventional fluid mechanics and convective heat transfer to describe what happens at every point of the fluid- and solid-filled spaces. When the distance is large so that there are many channels and cavities in the problem-solver's field of vision, the complications of the flow paths rule out the conventional approach. In this limit, volume-averaging and global measurements (e.g., permeability, conductivity) are useful in describing the flow, and in simplifying the description. As engineers focus more and more on designed porous media at decreasing pore scales, the problems tend to fall between the extremes noted above. In this intermediate range, the challenge is not only to describe *coarse* porous structures, but also to *optimize* flow elements, and to *assemble* them. The resulting flow structures are *designed* porous media (see, e.g., Chapter 5).

A basic concept in the volume-averaged treatment of flows through porous media is that of *representative elementary volume* (REV). The equations for the conservation of mass, momentum, and energy are averaged over each representative elementary volume, and the averaged properties (velocity, temperature, pressure, etc.) are assigned to the center of the element. The elementary volume is "representative" when the results of averaging do not depend on the size of its volume (Figure 1.1, top). The representative elementary volume is the volume above which the fluctuations of the void volume present in the porous medium are negligible. This is the case when the element contains a sufficient number of pores and solid features (grains), for example, when the length scale of the elementary volume is at least one order of magnitude greater than the pore scale. Said another way, the volume-averaged description of the larger system is adequate when the pores are sufficiently small and numerous so that the macroscopic system can be viewed as an assembly of a large number of representative elementary volumes. The $\Delta x \Delta y W$ volume element, enhanced for the sake of clarity in the lower part of Figure 1.1, would

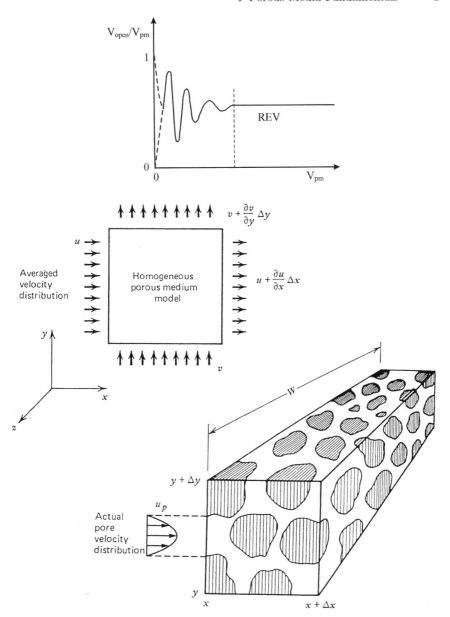

Fig. 1.1. Void volume, porous medium volume, and representative elementary volume REV, and the averaging of the pore velocity distribution in the development of the homogenous porous medium model (Bejan, 1995a; from Bejan, *Convection Heat Transfer*, 2nd ed., Copyright © 1995 Wiley. This material is used by permission of John Wiley & Sons, Inc.).

be "representative" if Δx and Δy were at least one order of magnitude larger than the pore and grain size shown in the figure.

Porous media are formed in nature as a consequence of many physical, chemical, and physiochemical processes. There are four main formation types of porous structures (Rouquerol *et al.*, 1994):

Agglomeration of small particles, the final structure depending on the nature, shape, and size of the primary particles;

Crystallization, leading to regular structures with void spaces (e.g., porous clays, zeolites);

Selective removal of elements of the original structure, leading to pore creation (e.g., by thermal decomposition, chemical reactions, mechanical action, or erosion); and

Organization of living structures, in animals and plants.

Except for the structures originated by crystallization, the void spaces are normally highly irregular in shape and dimension. Following the definition of the International Union of Pure and Applied Chemistry [IUPAC; Rouquerol *et al.* (1994)], pores are cavities, channels, or interstices that are deeper than they are wide. The pore width characterizes the *pore size*. The pore size statistics is usually described by the *pore size distribution* $n(r)$, that is, the number of pores per unit mass with sizes between r and $r + dr$. The pore size distributions of porous solids are usually obtained from mercury porosimetry, or from the adsorption isotherms using the Kelvin equation in the case of mesoporous media (Gregg and Sing, 1982; Ruthven, 1984). The pore size distributions are very useful for characterizing the porous microstructure, and for determining other porous medium properties.

An important quantity that characterizes a porous medium is the *porosity* ϕ. This is defined as the volume fraction of the void space present in the porous medium. The porosity may be determined by several methods, the most common being mercury injection, imbibition, helium porosimetry, scanning electron microscope (SEM) measurements, differential scanning calorimetry (DSC), and adsorption-related techniques (the measurement of microporosity). The materials are usually classified according to their characteristic pore size. *Microporous* materials have pore widths on the order of 2 nm. *Mesoporous* materials have sizes between 2 nm and 50 nm, and *macroporous* materials have pore widths higher than 500 nm [see IUPAC Recommendations (2001)].

Another important geometric property is the *internal surface area* of the porous medium. The methods most commonly used for measuring the internal surface area are the "BET method" [after Brunnauer *et al.* (1938)], which uses gas adsorption isotherms, liquid adsorption calorimetry, and electron microscopy. Some materials, such as super-activated carbons (M-38), have internal surface areas as high as $3800 \, \text{m}^2/\text{g}$.

Many porous media reveal self-similarity at different scales, therefore allowing a fractal description. For instance, Thompson *et al.* (1987) showed that the pore size distributions of sandstones in the range 10^2 to 10^4 nm follow closely the relationship $n(r) \propto r^{2-d_f}$, where $2 < d_f < 3$ is the fractal dimension, relating the number of pores (N) in the volume of the self-similar region (L^3) to the average pore size $\langle r \rangle$; that is, $N \propto (L/\langle r \rangle)^{d_f}$, and $n(r)$ is the number of pores of size r per unit mass. From scanning electron microscope measurements, Katz and Thompson (1985) also found a relationship between the fractal dimension and the porosity ϕ, of the form $\phi = A_c(r_u/r_i)^{d_f-3}$, where $A_c \approx 1$, and r_u and r_s are, respectively, the pore sizes characteristic of the upper and lower limits of the self-similar regions.

The relationship between the internal surface area A_s and the fractal dimension is of the form (Wong, 1988) $A_s \propto r_u^2(r_u/r_i)^{d_f-2}$, where the *specific internal surface area* a_s (m^2/g) scales with $r_u^{-1}(r_u/r_i)^{d_f-2}$. Consequently, by taking into account the above relation between porosity and fractal dimension, it follows that $a_s \propto \phi/r_i$. This means that for a given porosity, that is, when r_u changes proportionally with r_i (cf. the ϕ formula in the preceding paragraph), the specific surface area increases as the characteristic dimension of the lower self-similar region decreases.

1.1.1 Microporous Media

The characteristic pore size of microporous media is on the order of nanometers. Examples of microporous media are some activated carbons, silica gels, carbon molecular sieves, and some crystalline structures such as the zeolites. These materials have a wide range of applications. Activated carbons are used in many purification processes (e.g., water and air purification), solvent recovery, the adsorption processes (water and air purification), the adsorption of gases and vapors, and also as catalysts. Carbon molecular sieves are prepared to have a very narrow range of micropore sizes that enables them to have a high selectivity in the separation of molecules of different size. Silica gels may be used in the chemical industry in separation processes, or as powerful desiccants. Zeolites are open crystal lattices of aluminosilicates that may appear in many configurations (Baerlocher *et al.*, 2001). Because of the crystalline structure, the pore size is very well defined. This property distinguishes zeolites as highly selective molecular sieves. Zeolites have many applications as catalysts, and in separation and purification processes.

The molecules of the fluids inside the pores interact with the pore walls under short range forces that depend on the ratio (r/r_m) between the pore size r and the collision radius r_m of the molecule, which is defined as the square root of the cross-section for collision divided by π (Gregg and Sing, 1982). In microporous materials the distance between the opposite pore walls is on the order of the interaction length. In this way, in very fine pores the molecule/pore potential may be about twice the potential between one molecule and a single pore wall. The enhancement of the potential causes the almost complete filling of the micropores at a relative pressure that is quite low. This aspect distinguishes microporous materials from other porous structures. Figure 1.2 shows the shape of a type I adsorption isotherm of

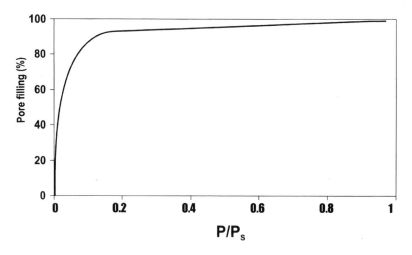

Fig. 1.2. Fluid adsorption in microporous media.

microporous adsorbents, which is one of the six standard isotherms [see Gregg and Sing (1982)]. The start of the plateau corresponds to complete pore filling. The volume corresponding to the liquid phase of the adsorbed fluid represents the total pore volume.

1.1.2 Mesoporous Media

The pore size range of the mesoporous media is 2 to 50 nm. These sizes are found in inorganic xerogels, such as alumina and silica powders, porous glasses, and pillared or nonpillared clays. They are also found in a class of mesostructured materials of the M41S-type that have an ordered pore system (Beck et al., 1992). A great variety of pore structures can be found in mesoporous materials: crystalline (mainly hexagonal), polyhedral, lamellar, pillared, nanotubes (carbons), "hairy tubes," and so on. These materials are used mainly as catalysts and in separation processes.

Gas adsorption in mesoporous materials takes place at low subsaturation, $P/P_s < 0.3$. Pore filling extends to a wide pressure range $0.3 < P/P_s < 1$, until the pores become filled completely just before the saturation pressure (see Figure 1.3). The plateau at the end of the isotherm (close to $P/P_s = 1$) corresponds to complete pore filling, where the isotherms of the same fluid coincide. Other fluids exhibit this behavior when the amount adsorbed is expressed in terms of the equivalent volume of the liquid phase [the Gurvitsch rule, see Gregg and Sing (1982)]. The desorption isotherms of many mesoporous materials do not coincide entirely with the respective

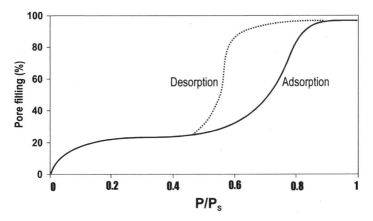

Fig. 1.3. Fluid adsorption and desorption in mesoporous media.

sorption isotherms, showing a hysteresis loop in the part corresponding to pore filling (the dashed curve in Figure 1.3).

1.1.3 Macroporous Media

The characteristic pore sizes of macroporous media are larger than 50 nm, and abound in nature. Soils, fractured rocks, sandstones, wood materials, and various foods are examples of macroporous media. Foods are generally macroporous, but often exhibit multiple porosity scales. This feature is revealed by the pore size distributions, which show two or more peaks (Karathanos *et al.*, 1996). Man-made materials such as thermal insulation materials, silicalite, some ceramics, cements, synthetic resins, and many other synthetic materials are also macroporous. Macroporous materials find application in many processes, for example, filtration of liquids (resins, sand beds), electronic devices (macroporous silicon photonic crystal waveguides), complex textile structures (Reuss *et al.*, 2002), bioceramics (bone substitutes), liquid chromatography, and biotechnology.

1.2 Mass Conservation

The description of flows through porous media begins with the observation that the actual heat, mass, and fluid flow fields are very complicated. Our main interest is not in a detailed local description, but in the global performance of the flow system. Out of this conflict between detail and expeditiousness comes the decision to smooth out (to assume away) the complicated features of the real flow. This decision is analogous to the time-averaging of a turbulent flow field. The effect of volume-averaged pores and grains is analogous to the effect of time-averaged eddy motion.

The volume-averaging method consists of applying the conservation principles to a "gray" medium without the structure of grains and voids. We illustrate this step in this section for mass conservation; for the other conservation statements we list only the resulting equations (Sections 1.4 to 1.6). The transition from a porous structure to a medium without structure is illustrated in Figure 1.1. The larger system that houses the flow has boundary conditions such that the volume-averaged flow is two-dimensional, in the x-y plane. The real flow is three-dimensional everywhere: the flow field in the $\Delta x \Delta y W$ volume element differs from one constant-z plane to another. This is another conceptual link between flows through porous media and turbulent flows: in the latter, the real flow is three-dimensional even if in special cases the time-averaged flow is two-dimensional.

Assume that in the volume element of Figure 1.1 the dimension W is sufficiently larger than either Δx or Δy so that, for the purpose of mass-flow accounting, the important flowrates are in the x and y directions only. Consider first the mass flowrate entering the $\Delta x \Delta y W$ volume element from the left, through the constant-x plane,

$$\dot{m}_x = \rho \int_y^{y+\Delta y} \int_0^W u_p dz dy. \qquad (1.1)$$

Here $u_p(y, z)$ is the nonuniform x-velocity distribution over the void portions of the $x = $ constant plane, and ρ is the density of the fluid. Imagining a control surface $W\Delta y$ sufficiently larger than the pore and solid grain cross-sections, we define the area-averaged velocity in the x-direction,

$$u = \frac{1}{W\Delta y} \int_y^{y+\Delta y} \int_0^W u_p(y, z) dz dy. \qquad (1.2)$$

In other words, the mass flowrate is $\dot{m}_x = \rho u W \Delta y$. The area-averaged velocity in the y-direction is defined in the same way,

$$v = \frac{1}{W\Delta x} \int_x^{x+\Delta x} \int_0^W v_p(x, z) dz dx \qquad (1.3)$$

so that the mass flowrate in the y-direction can be expressed as $\dot{m}_y = \rho v W \Delta x$. Note that in this derivation we treated the density ρ as constant in the $\Delta x \Delta y$ element of the two-dimensional flow: this does not mean that ρ is constant throughout the flow field.

The reward for smoothing out the complications of the real flow and introducing the area-averaged velocities (u, v) is that the averaged flow looks like any other homogeneous fluid flow. Therefore, applying the mass conservation principle to the $\Delta x \Delta y W$ element yields

$$\frac{\partial}{\partial t}(\rho \phi W \Delta x \Delta y) + \frac{\partial \dot{m}_x}{\partial x}\Delta x + \frac{\partial \dot{m}_y}{\partial y}\Delta y = 0, \qquad (1.4)$$

Table 1.1. Properties of Common Porous Materials[a]

Material	Porosity ϕ	Permeability K (cm^2)	Surface per unit volume (cm^{-1})
Agar-agar	0.57–0.66	2×10^{-10}–4.4×10^{-9}	
Black slate powder	0.12–0.34	4.9×10^{-10}–1.2×10^{-9}	7×10^3–8.9×10^3
Brick	0.45	4.8×10^{-11}–2.2×10^{-9}	
Catalyst (Fischer–Tropsch, granules only)			5.6×10^5
Cigarette	0.17–0.49	1.1×10^{-5}	
Cigarette filters	0.02–0.12		
Coal	~ 0.10		
Concrete (ordinary mixes)			
Concrete (bituminous)	0.09–0.34	1×10^{-9}–2.3×10^{-7}	
Copper powder (hot-compacted)		3.3×10^{-6}–1.5×10^{-5}	
Corkboard		2.4×10^{-7}–5.1×10^{-7}	
Fiberglass	0.88–0.93		560–770
Granular crushed rock	0.45		
Hair, on mammals	0.95–0.99		
Hair felt		8.3×10^{-6}–1.2×10^{-5}	
Leather	0.56–0.59	9.5×10^{-10}–1.2×10^{-9}	1.2×10^4–1.6×10^4
Limestone (dolomite)	0.04–0.10	2×10^{-11}–4.5×10^{-10}	
Sand	0.37–0.50	2×10^{-7}–1.8×10^{-6}	150–220
Sandstone ("oil sand")	0.08–0.38	5×10^{-12}–3×10^{-8}	
Silica grains	0.65		
Silica powder	0.37–0.49	1.3×10^{-10}–5.1×10^{-10}	6.8×10^3–8.9×10^3
Soil	0.43–0.54	2.9×10^{-9}–1.4×10^{-7}	
Spherical packings (well shaken)	0.36–0.43		
Wire crimps	0.68–0.76	3.8×10^{-5}–1×10^{-4}	29–40

[a]Based on data compiled by Scheidegger (1974) and Bejan and Lage (1991).

where ϕ is the porosity or void fraction of the medium (the void volume divided by the total volume), and $\phi W \Delta x \Delta y$ is the volume occupied by fluid in the $W \Delta x \Delta y$ element. Examples of porosity values for common porous materials are given in Table 1.1. Most naturally occurring porous media have ϕ values less than 0.6. Equation (1.4) and the assumption that ϕ is independent of time yield

$$\phi \frac{\partial \rho}{\partial t} + \frac{\partial(\rho u)}{\partial x} + \frac{\partial(\rho v)}{\partial y} = 0. \tag{1.5}$$

In general, the mass conservation equation for three-dimensional averaged flow reads

$$\phi \frac{\partial \rho}{\partial t} + \nabla \cdot (\rho \mathbf{v}) = 0, \tag{1.6}$$

where $\mathbf{v}$ is the volume-averaged velocity vector (u, v, w). Note that Equation (1.6) with $\phi = 1$ is the equation for mass conservation in a space with pure fluid. This coincidence is not surprising, because the concept of area-averaged velocity was introduced in order to be able to apply to flows through porous media the pure-fluids mathematical apparatus developed earlier for fluid mechanics.

1.3 Darcy Flow and More Advanced Models

In the fluid mechanics of porous media, just as in turbulence, the place of momentum equations or force balances is occupied by the experimental observations summarized in simple terms called models. Such observations were first reported and modeled by Darcy (1856) who, based on measurement alone, discovered that the area-averaged fluid velocity through a column of porous material is proportional to the pressure gradient maintained along the column. Subsequent studies showed that the area-averaged velocity is, in addition, inversely proportional to the viscosity (μ) of the fluid seeping through the porous material. A detailed review of these advances was provided by Lage (1998).

With reference to the one-dimensional forced flow configuration shown in Figure 1.4, the Darcy flow model is the statement

$$u = \frac{K}{\mu} \left(-\frac{dP}{dx} \right) \quad \text{or} \quad \nabla P = -\frac{\mu}{K} \mathbf{v}. \tag{1.7}$$

The empirical factor K is the permeability of the medium the units of which are (length)^2. Representative values of K for several porous media, and for their typical flow ranges, are listed in Table 1.1. The units of K can be seen by comparing Equation (1.7) with the equation for Hagen–Poiseuille flow of mean velocity U through a tube of diameter D [e.g., Bejan (1995a, p. 99)]

$$U = \frac{D^2}{32\mu} \left(-\frac{dP}{dx} \right). \tag{1.8}$$

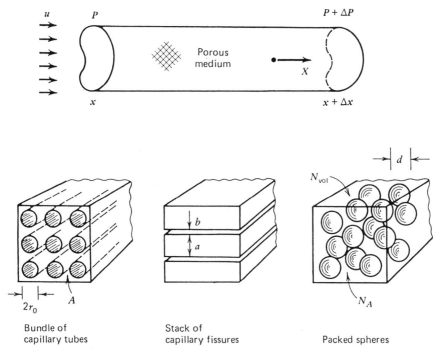

Fig. 1.4. Dercy flow experiment, and three possible models for estimating the permeability of the porous medium (Bejan, 1995a; from Bejan, *Convection Heat Transfer*, 2nd ed., Copyright © 1995 Wiley. This material is used by permission of John Wiley & Sons, Inc.).

It is important to keep in mind that Darcy's model (1.7) is not a balance of forces averaged over a representative elementary volume (Nield and Bejan, 1999). The similarity between Equations (1.7) and (1.8) suggests that the Darcy flow is the macroscopic manifestation of a highly viscous (Newtonian) flow through the pores of the permeable structure, and that $K^{1/2}$ is a length scale representative of the effective pore diameter. By imagining a small-scale network of channels of known geometry, and assuming Hagen–Poiseuille flow through each channel, it is possible to derive Equation (1.7) in such a way that K emerges as a function of the network geometry. According to the hydraulic radius theory of Carman–Kozeny (Dullien, 1992), the permeability of a bed of particles is $K = D_{p2}^2 \phi^3/[180(1 - \phi)^2]$, where D_{p2} is an effective average particle diameter, and 180 is an empirical constant. This equation is accurate when the particles do not deviate much from the spherical shape, and their diameters fall in a narrow range. More general and recent alternatives for the Carman–Kozeny formula are reviewed in Nield and Bejan (1999). Ergun (1952) proposed $K = d^2 \phi^3/[150(1 - \phi)^2]$ as a correlation for the measured permeabilities of columns of packed spheres of diameter d and porosity ϕ.

Using $K^{1/2}$ as a length scale to define the Reynolds number $\mathrm{Re}_K = uK^{1/2}/\nu$ and friction factor

$$f_K = \left(-\frac{dP}{dx}\right)\frac{K^{1/2}}{\rho u^2}, \tag{1.9}$$

we rewrite the Darcy law (1.7) as

$$f_K = \frac{1}{\mathrm{Re}_K}. \tag{1.10}$$

This form is reminiscent of the friction factor for Hagen–Poiseuille flow. Experimental measurements (Ward, 1964) have shown that Equations (1.7) and (1.10) are valid when Re_K is less than the 1 to 10 range, Figure 1.5. If Re_K exceeds 10, inertia effects (or form drag) flatten the $f_K(\mathrm{Re}_K)$ curve in a manner reminiscent of the friction factor curve in turbulent flow over a rough surface,

$$f_K = \frac{1}{\mathrm{Re}_K} + c_F, \tag{1.11}$$

where c_F is an empirical constant originally believed to be approximately equal to 0.55. A more general friction factor model is due to Dupuit's (1863) modification of the Darcy model, later modified by Forchheimer (1901) [see Lage (1998)],

$$-\frac{dP}{dx} = \frac{\mu}{K}u + b\rho|u|u, \tag{1.12}$$

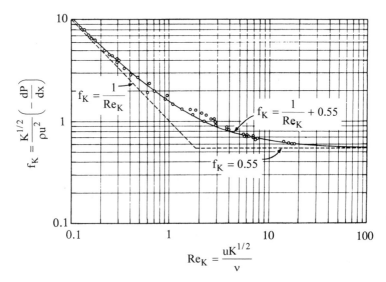

Fig. 1.5. The transition from the Darcy regime to the Forchheimer regime in unidirectional flow through an isothermal porous medium (Ward, 1964).

where b is another empirical constant. In Ergun's (1952) correlation for a column of packed spheres with diameter d, this constant is $b = 1.75(1 - \phi)/(\phi^3 d)$. It is easy to verify that Equation (1.12) accounts for the behavior noted in Equation (1.11) and Figure 1.5. The second term on the right side of Equation (1.12) accounts for fluid inertia, and is known as the Forchheimer term, or the quadratic term. Joseph *et al.* (1982) showed that the appropriate form of (1.12) is, this time, in three dimensions,

$$\nabla P = -\frac{\mu}{K}\mathbf{v} - c_F K^{-1/2}\rho|\mathbf{v}|\mathbf{v}. \tag{1.13}$$

The factor c_F is dimensionless and accounts for inertia effects, or form drag. In Ward's Equation (1.11), $c_F = 0.55$. This factor varies with the type of porous structure, and can be as small as 0.1 for foam metals. The walls that bound a porous bed of particles influence the values of c_F and K. Beavers *et al.* (1973) correlated their data with $c_F = 0.55(1 - 5.5d/D_e)$, where d is the diameter of the packed spheres, and $D_e = 2wh/(w + h)$ is the effective diameter of the bed of height h and width w. Further refinements of the Dupuit–Forchheimer model are reviewed in Nield and Bejan (1999). The use of this model for flows of non-Newtonian fluids through saturated porous media is reviewed in Shenoy (1993).

The transition from Darcy flow (1.7) to Darcy–Forchheimer flow (1.13) occurs when Re_K is of order 10^2. This transition is associated with the occurrence of the first eddies in the fluid flow, for example, the rotating fluid behind an obstacle, or a backward facing step. The order of magnitude $\mathrm{Re}_K \sim 10^2$ is one in a long list of examples that show that the laminar-turbulent transition is associated with a universal *local Reynolds number* of order 10^2 (Bejan, 1995a, p. 280).

To derive $\mathrm{Re}_K \sim 10^2$ from turbulence, assume that the porous structure is made of three-dimensional random fibers that are so sparsely distributed that $\phi \lesssim 1$. According to Koponen *et al.* (1998), in this limit the permeability of the structure is correlated very well by the expression $K = 1.39D^2/[e^{10.1(1-\phi)} - 1]$, where D is the fiber diameter. In this limit the volume-averaged velocity u has the same scale as the velocity of the free stream that bathes every fiber. It is well known that vortex shedding occurs when $\mathrm{Re}_D = uD/\nu \sim 10^2$ [cf. Bejan (1995a, p. 280)]. By eliminating D between the above expressions for K and Re_D, we calculate $\mathrm{Re}_K = uK^{1/2}/\nu$ and find that when eddies begin to appear, the Re_K value is in the range 100 to 300 when ϕ is in the range 0.9 to 0.99.

The transition to turbulence in porous media and the modeling of the turbulent flow continue to attract attention (Masuoka and Takatsu, 1996, 2002; Nield, 1997; Kuwahara *et al.*, 1998; Getachew *et al.*, 2000; Lage *et al.*, 2002; Kim and Kim, 2002). Models of turbulent flow through porous media have been developed by Antohe and Lage (1997), Nakayama and Kuwahara (1999), and Pedras and de Lemos (2001a,b). Another direction of development is the modeling of flows through dense plant canopies and submerged vegetation [e.g., Nepf (1999), Hoffmann (2000), and Hoffmann and van der Meer (2002)]. The deformation of the porous matrix constituted by the stems of plants can be modeled by calculating the bending of thin elastic cylinders in banks immersed in cross-flow [e.g., Fowler and Bejan (1994)]. A related and very active field is the modeling of turbulent airflow through forest canopies (Lai et al., 2000; Siqueira et al., 2000; Katul et al., 2000).

The effect of solid walls has been the target of Brinkman's (1947a,b) modification of Darcy's model,

$$\nabla P = -\frac{\mu}{K}\mathbf{v} + \tilde{\mu}\nabla^2\mathbf{v}. \tag{1.14}$$

The Laplacian term is analogous to the fluid friction term in the corresponding Navier–Stokes equation for a pure fluid. One of its consequences is that it permits the application of no-slip boundary conditions along the solid walls that confine the porous medium. There are two viscosities in Equation (1.14): μ is the viscosity of the fluid, and $\tilde{\mu}/\mu$ depends on the geometry of the porous structure. Bear and Bachmat (1990) showed that for an isotropic porous medium $\tilde{\mu}/\mu = (\phi T^*)^{-1}$, where T^* is the tortuosity of the flow structure.

The tortuosity is a dimensionless parameter that accounts for the fact that the flow path is in general not straight. In the current literature, $\tilde{\mu}$ is often assumed to be equal to μ. The viscosity μ is usually treated as a constant. The tortuosity is known, or can be calculated in designed porous media. In natural porous media it has to be measured. Tortuosity is a relevant concept because the mass fluxes induced by pressure, temperature, or concentration gradients inside porous media are usually smaller when compared with the fluxes driven by the same gradients in free bulk. In porous media, an additional resistance comes from the fact that the mass fluxes are constrained by the pore geometry. Let us consider a one-dimensional mass flux driven by a potential ($\Omega = C, T, P, \mu$) between two parallel planes cutting the porous medium, which are separated by the distance dx. Let dS_i be the path that the particles have to travel through pore i between the two planes. Then, if k_Ω denotes the bulk fluid conductivity associated with the field Ω, the flux along the pore is $j_i = -k_\Omega d\Omega/ds_i$.

Let θ_i be the angle that the pore axis makes with the x-direction (Figure 1.6), and let $T_i^* = 1/\cos\theta_i = ds_i/dx$. Then the flux component normal to the cross-section is $j_{ix} = (-k_\Omega/T_i^{*2})(d\Omega/dx)$. Upscaling is achieved by averaging the fluxes along all the pores in an elemental volume Adx, which leads to the average mass flux in the x-direction,

$$j_x = -\frac{k_\Omega}{T^*}\frac{d\Omega}{dx}, \tag{1.15}$$

where

$$T^* = \left[\sum_i \left(\frac{1}{T_i^{*2}}\right)\right]^{-1} \tag{1.15'}$$

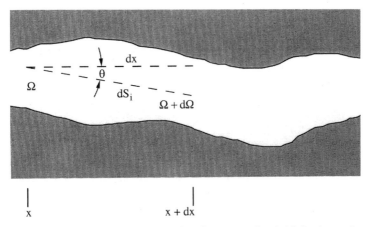

Fig. 1.6. The definition of tortuosity: the change in the field Ω along the pore is smaller (by the factor $\cos\theta$) than the change in the x-direction.

is the tortuosity factor. To summarize, in porous media the usual bulk conductivities are lowered by the factor $1/T^*$. For example, if Ω is the concentration of a certain species C, then k_Ω denotes the corresponding diffusivity coefficient D (Fick's law). In this way, Fick's law describes the average diffusion of the species through the porous medium, as the bulk diffusivity D has been replaced by D/T^*. The tortuosity factor has been related to other geometrical properties such as the fracture width in fractured media (Drazer and Koplik, 2001). It has also been used to account for the presence of extracellular macromolecules in the brain tissue that increase the path length of diffusing particles (Rusakov and Kullmann, 1998).

The effect of temperature on viscosity and convection is documented by Narasimhan and Lage (2001, 2002). Another direction of development is the modeling of moisture content effect on transport properties of porous media (Reis *et al.*, 1994; Miguel, 2000).

Nield and Bejan's (1999) review of the literature based on Equation (1.14) concluded that for many practical situations it is not necessary to include the Laplacian term, and that the Darcy model is sufficient. If no-slip wall conditions must be satisfied, then the Laplacian term is needed but its effect is felt in a very thin boundary layer of thickness $(\tilde{\mu}K/\mu)^{1/2}$, which is of order $K^{1/2}$, that is, comparable with the pore scale. Such a boundary layer is very thin because of the representative elementary volume assumption (Section 1.1), which amounts to $K^{1/2} \ll L$, where L is the macroscopic length scale of the flow system. The full Brinkman model (1.14) is useful in numerical simulations of flow fields with interfaces between fluid-saturated porous media and pure fluids. The modeling of such interfaces is an active area of research; see Beavers and Joseph (1967), Saghir *et al.* (2002), and pp. 15–19 in Nield and Bejan (1999).

More general flow models are being developed in order to facilitate numerical simulations of flows and transport in porous media. One direction is to account for both fluid inertia and wall effects in a combined Forchheimer–Brinkman modification of the Darcy model. Notable in this respect is the model of Vafai and Tien (1981). Another direction is to recognize that in a porous bed packed between solid walls there is a porosity increase close to the walls. This effect is due to the inability of particles to fit close to one another when a solid plane surface confines them (Cheng *et al.*, 1991). In the same wall region, the concept of representative elementary volume is no longer applicable. The effect of porosity variation on the quadratic drag term has been studied by Georgiadis and Catton (1987). More recently, we have seen studies of turbulent flow through porous media, where the effect of the solid structure is to damp the turbulence, for example, Antohe and Lage (1997). The effect of turbulence may be significant when porosities are high. These and other refinements of the flow model are discussed further in Nield and Bejan (1999). A Brinkman–Forchheimer model for forced convection in a parallel-plates channel was used by Kuznetsov (2002). Analytical solutions of the one-dimensional Forchheimer–Brinkman–Darcy equation for a porous medium layer between solid plates can be found in Nield *et al.* (1996), Kuznetsov (1997), Marafie and Vafai (2001), and Miguel and Reis (2004). The last-mentioned includes the transient fluid flow effects and, among other results, indicates that an appropriate periodic forcing of the pressure gradient may increase the effective permeability of the porous layer. Said another way, the effective permeability increases if the flow is forced by intensifying the pressure gradient for a short time and then is let free of forcing for an interval on the order of the relaxation time.

In the presence of a body force per unit volume ρg_x, the Darcy model (1.7) becomes

$$u = \frac{K}{\mu}\left(-\frac{\partial P}{\partial x} + \rho g_x\right). \tag{1.16}$$

The flow through the porous column of Figure 1.4 stops $(u = 0)$ when the externally controlled pressure gradient dP/dx matches the hydrostatic gradient ρg_x. The vectorial generalization of Equation (1.16) is

$$\mathbf{v} = \frac{K}{\mu}(-\nabla P + \rho \mathbf{g}), \tag{1.16'}$$

where $\mathbf{v}(u, v, w)$ is the velocity vector and $\mathbf{g}(g_x, g_y, g_z)$ the body acceleration vector.

In many problems involving only the seepage flow of water through soil, ρ and μ may be regarded as constant. With the y-axis oriented upward against the gravitational acceleration g, the body acceleration vector is $(0, -g, 0)$, and Equation (1.16') becomes

$$\mathbf{v} = -\frac{K}{\mu}\nabla\psi, \qquad \psi(x, y, z) = P + \rho g y. \tag{1.17}$$

Note that under the same conditions (ρ = constant), the mass conservation statement (1.6) becomes

$$\nabla \cdot \mathbf{v} = 0. \tag{1.18}$$

Combining Equations (1.17) and (1.18), we find that many seepage flows in environmental engineering are governed by the Laplace equation

$$\nabla^2 \psi = 0, \tag{1.19}$$

which, in the absence of free surfaces, can be solved in the (x, y, z) space according to, for example, the classical methods of steady-state conduction heat transfer (Carslaw and Jaeger, 1959; Bejan, 1993). For many solutions of Equation (1.19), and for a special transformation designed to handle the free surface (P = constant) boundary condition, the reader is directed to classical books by Muskat (1937) and Yih (1969).

1.4 Energy Conservation

A simple way to illustrate the derivation of the energy conservation equation for a porous medium is based on the one-dimensional heat and fluid flow model of Figure 1.7. The energy equation is the first law of thermodynamics. The figure shows the building block suggested earlier by models such as the capillary tube bundle and the capillary fissures of Figure 1.4. The void space contained in the volume element $A\Delta x$ is $A_p \Delta x$. The volume element is defined such that the ratio $(A_p \Delta x)/(A \Delta x)$ matches the porosity ratio of the porous medium from which the volume element has been isolated, $\phi = A_p \Delta x/(A \Delta x)$.

To derive the energy conservation equation for a porous medium that after volume-averaging is a homogeneous medium, we start with the energy equations for the solid and fluid parts. Later we average these equations over the volume $A\Delta x$. For the solid part, we write the conduction equation [e.g., Bejan (1993, p. 9)]

$$\rho_s c_s \frac{\partial T}{\partial t} = k_s \frac{\partial^2 T}{\partial x^2} + q_s''', \tag{1.20}$$

where $(\rho, c, k)_s$ are the properties of the solid, and q_s''' is the rate of internal heat generation per unit volume of solid material. We assume local thermal equilibrium, that is, one local temperature T for the solid matrix and the pore fluid. The integral of Equation (1.20) over the space occupied by the solid yields

$$\Delta x(A - A_p)\rho_s c_s \frac{\partial T}{\partial t} = \Delta x(A - A_p)k_s \frac{\partial^2 T}{\partial x^2} + \Delta x(A - A_p)q_s'''. \tag{1.21}$$

The energy conservation equation at a point inside the space occupied by fluid is [e.g., Bejan (1995a, p. 14)]

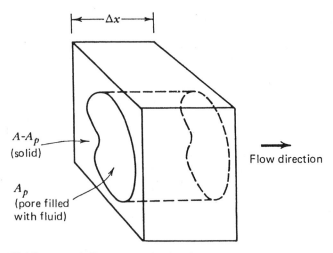

Fig. 1.7. Unidirectional flow example for formulating the energy conservation equation for convection in a porous medium.

$$\rho_f c_{Pf} \left(\frac{\partial T}{\partial t} + u_p \frac{\partial T}{\partial x} \right) = k_f \frac{\partial^2 T}{\partial x^2} + \mu \Phi, \tag{1.22}$$

where u_p and Φ are the velocity inside the pore and the viscous dissipation function. Note that $(\rho, c_P, k)_f$ are fluid properties. The subscript f is used to distinguish the fluid properties only in the energy equation: ρ_f is the same as the fluid density ρ used in the mass conservation equation (1.6) and in the Darcy flow model (1.16). It is assumed that the compressibility term $\beta T DP/Dt$ is negligible in the energy equation for the fluid. It is also assumed that $(c, k)_s$ and $(c_P, k)_f$ are known constants. Note further that T is the temperature of both parts, solid and fluid. This assumption, although adequate for small-pore media such as geothermal reservoirs and fibrous insulation, must be relaxed in the study of nuclear reactor cores and electrical windings where the temperature difference between solid and fluid (coolant) is a very important safety parameter. We consider this case later in Equations (1.36) and (1.37).

Integrating Equation (1.22) over the pore volume $A_p \Delta x$ yields

$$\Delta x A_p \rho_f c_{Pf} \frac{\partial T}{\partial t} + \Delta x A \rho_f c_{Pf} u \frac{\partial T}{\partial x} = \Delta x A_p k_f \frac{\partial^2 T}{\partial x^2} + \Delta x \mu \int_{A_p} \Phi \, dA_p. \tag{1.23}$$

In the second term on the left-hand side of the above equation, we made use of the definition of average fluid velocity, $Au = \int_{A_p} u_p \, dA_p$. The last term on the right-hand side represents the internal heating associated with viscous dissipation. The dissipation term equals the mechanical power needed to extrude the viscous fluid through the pore. This power requirement is equal to the mass flow rate times the externally maintained pressure drop, divided by the fluid density,

$$\Delta x \mu \int_{A_p} \Phi \, dA_p = Au \left(-\frac{\partial P}{\partial x} + \rho_f g_x \right) \Delta x. \tag{1.24}$$

It is easy to prove this identity in the case of known Hagen–Poiseuille flows through pores with simple cross-sections; however, Equation (1.24) is expected to hold for any pore geometry.

Volumetric averaging of the energy conservation statement is achieved by adding Equations (1.21) and (1.23) side by side and dividing by the volume element $A\Delta x$ of the porous structure regarded as a homogeneous medium:

$$[\phi \rho_f c_{Pf} + (1 - \phi)\rho_s c_s]\frac{\partial T}{\partial t} + \rho_f c_{Pf} u \frac{\partial T}{\partial x}$$

$$= [\phi k_f + (1 - \phi)k_s]\frac{\partial^2 T}{\partial x^2} + (1 - \phi)q_s''' + u \left(-\frac{\partial P}{\partial x} + \rho_f g_x \right). \tag{1.25}$$

The thermal conductivity group that emerges in front of $\partial^2 T/\partial x^2$ is a combination of the conductivities of the two constituents,

$$k_1 = \phi k_f + (1 - \phi)k_s. \tag{1.26}$$

This simple expression, however, is the result of the one-dimensional model of Figure 1.7, which is a parallel conduction model. If the solid and fluid spaces are arranged such that heat is conducted with the solid and fluid resistances in series, the thermal conductivity of the porous medium is

$$k_2 = [\phi k_f^{-1} + (1 - \phi)k_s^{-1}]^{-1}. \tag{1.27}$$

In general, the actual k value must be determined experimentally, as the effective thermal conductivity of the porous matrix filled with fluid. When k_1 and k_2 are not much different, an adequate estimate of the effective thermal conductivity is provided by (Nield, 1991)

$$k = k_f^\phi k_s^{1-\phi}. \tag{1.28}$$

The first term on the left side of Equation (1.25) shows that the thermal inertia of the medium depends on the inertias of the solid and the fluid. This complication is accounted for by using the capacity ratio

$$\sigma = \frac{\phi \rho_f c_{Pf} + (1 - \phi)\rho_s c_s}{\rho_f c_{Pf}}. \tag{1.29}$$

The heat generation rate averaged over the volume of the porous medium (q''') decreases as the porosity increases; $q''' = (1 - \phi)q_s'''$. With the new notation defined in Equations (1.26) to (1.29), the energy equation for the homogeneous volume-averaged porous medium with Darcy flow reads

$$\rho_f c_{Pf} \left(\sigma \frac{\partial T}{\partial t} + u \frac{\partial T}{\partial x} \right) = k \frac{\partial^2 T}{\partial x^2} + q''' + \frac{\mu}{K} u^2. \tag{1.30}$$

The corresponding equation for a three-dimensional flow is

$$\rho_f c_{Pf} \left(\sigma \frac{\partial T}{\partial t} + \mathbf{v} \cdot \nabla T \right) = k \nabla^2 T + q''' + \frac{\mu}{K} (\mathbf{v})^2. \tag{1.31}$$

In situations without internal heat generation q''', and when the viscous dissipation effect $(\mu/K)(\mathbf{v})^2$ is negligible, the energy conservation equation assumes the simpler form

$$\sigma \frac{\partial T}{\partial t} + u \frac{\partial T}{\partial x} + v \frac{\partial T}{\partial y} + w \frac{\partial T}{\partial z} = \alpha \left(\frac{\partial^2 T}{\partial x^2} + \frac{\partial^2 T}{\partial y^2} + \frac{\partial^2 T}{\partial z^2} \right). \tag{1.32}$$

The thermal diffusivity of the homogeneous porous medium α is defined as the ratio

$$\alpha = \frac{k}{\rho_f c_{Pf}}. \tag{1.33}$$

Note that k and α are aggregate properties of the fluid-saturated porous medium, whereas $(\rho_f c_{Pf})$ is a property of the fluid only.

Most treatments of heat and fluid flow through porous media rely on Equations (1.6), (1.16′), and (1.32), which hold when the volume-averaged porous medium is homogeneous and isotropic. The porous medium is said to be homogeneous when its solid and fluid-filled pores are distributed evenly in space. The medium is isotropic when properties such as K and k do not depend on the direction of the experiment in which they are measured.

When the medium is anisotropic, the permeability, conductivity, and thermal diffusivity depend on the direction of the measurement, namely (K_x, K_y, K_z), (k_x, k_y, k_z), and $(\alpha_x, \alpha_y, \alpha_z) = (k_x, k_y, k_z)/(\rho_f c_{Pf})$. When the principal directions (Bejan, 1997, p. 691) coincide with the axes x, y, and z, the Darcy model (1.16′) and the energy equation (1.32) are replaced by

$$u = \frac{K_x}{\mu} \left(-\frac{\partial P}{\partial x} + \rho g_x \right), \qquad v = \frac{K_y}{\mu} \left(-\frac{\partial P}{\partial y} + \rho g_y \right),$$

$$w = \frac{K_z}{\mu} \left(-\frac{\partial P}{\partial z} + \rho g_z \right), \tag{1.34}$$

$$\sigma \frac{\partial T}{\partial t} + u \frac{\partial T}{\partial x} + v \frac{\partial T}{\partial y} + w \frac{\partial T}{\partial z} = \alpha_x \frac{\partial^2 T}{\partial x^2} + \alpha_y \frac{\partial^2 T}{\partial y^2} + \alpha_z \frac{\partial^2 T}{\partial z^2}. \tag{1.35}$$

If the porous medium is nonhomogeneous, k is a function of position, and the first term on the right-hand side of Equation (1.31) is replaced by $\nabla \cdot (k \nabla T)$. Heterogeneous porous media with multiphase flows require more complex models, as shown in Kuznetsov and Nield (2001) and Chella et al. (1998).

Another major assumption that stands behind the volume-averaged energy equation (1.32) is that the solid and pore fluid are locally in thermal equilibrium. The single temperature T refers to both solid and fluid at the point (x, y, z) in the volume-averaged medium. There are applications in which the local thermal equilibrium assumption is not adequate, for example, when the solid has a high thermal conductivity relative to the fluid, and when the flow is strong. One example is the porous structure formed by dense fins and even hair and fibers, attached to a solid body and permeated vigorously by fluid (Bejan, 1990a,b). In such cases, a two-temperature model provides a more realistic description of the local thermal nonequilibrium. There are two temperatures at every point: T_s for the solid and T_f for the fluid. There are also two equations,

$$(1 - \phi)\rho_s c_s \frac{\partial T_s}{\partial t} = (1 - \phi)\nabla \cdot (k_s \nabla T_s) + (1 - \phi)q_s'''$$

$$+ ha_{sf}(T_f - T_s), \tag{1.36}$$

$$\phi \rho_f c_{Pf} \frac{\partial T_f}{\partial t} + \rho_f c_{Pf} \mathbf{v} \cdot \nabla T_f = \phi \nabla \cdot (k_f \nabla T_f) + \phi q_f''' + ha_{sf}(T_s - T_f). \tag{1.37}$$

The term $ha_{sf}(T_s - T_f)$, which appears in both equations, accounts for the local heat transfer between the solid and the fluid. The factors h and a_{sf} are the heat transfer coefficient and the heat transfer area per unit volume. Both factors depend on the geometry of the pore structure and the flow conditions. Techniques for evaluating h are developed case by case, and begin with using the principles of convection [e.g., Bejan (1995a)] and a sufficiently simple model of the solid–fluid flow geometry. Examples concerning nonequilibrium flows through beds of particles are reviewed in Nield and Bejan (1999). Another approach is to determine an overall h value experimentally, for example, during the transient heating or cooling of a space filled with the solid-fluid structure of interest. Such time-dependent processes and nonequilibrium modeling are relevant to the energy storage applications discussed in Chapter 3. Porous media without local thermal equilibrium were studied by Nield et al. (2002), Nield and Kuznetsov (2001), Marafie and Vafai (2001), Nakayama et al. (2001), and Alazmi and Vafai (2002).

Thermal dispersion is another effect that may be included in a refined model of heat transfer. Dispersion is an additional fluid mixing effect that increases the effective thermal conductivity of the porous medium. The effect is due to the complicated tortuous path followed by the fluid. The distance between two small fluid packets changes as the pair flows from one pore into the next. This flow mixing phenomenon adds to the molecular diffusion effect that is accounted for in Equations (1.26) to (1.28). It is particularly relevant in applications with vigorous forced convection in packed columns. Methods for estimating the dispersion effect on thermal conductivity were reviewed in Nield and Bejan (1999). Dispersion in forced convection was studied recently by Cheng and Lin (2002).

Natural, or buoyancy-driven, convection is due to the fact that the fluid density is sensitive to temperature changes. In thermodynamics this property is represented by the coefficient of volumetric thermal expansion, or volume expansivity (Bejan, 1997),

$$\beta = -\frac{1}{\rho}\left(\frac{\partial \rho}{\partial T}\right)_P. \tag{1.38}$$

The equation of state $\rho = \rho(T, P)$, which must be used in conjunction with the Darcy model (1.16') and its modifications, is usually approximated linearly:

$$\rho = \rho_0[1 - \beta(T - T_0)]. \tag{1.39}$$

This is part and parcel of the Oberbeck–Boussinesq approximation (Oberbeck, 1879; Boussinesq, 1903), on which most of the natural convection in porous media literature is based. It results from expanding $\rho(T, P)$ as a Taylor series at constant P, around the state (ρ_0, T_0), and retaining only the first term of the series. The coefficient β is evaluated at the reference state or, for better accuracy, at an appropriately defined "film" temperature (Bejan, 1995a).

Convection processes through fluid-saturated porous media are thermodynamically irreversible, partly due to the transfer of heat in the direction of finite temperature gradients, and partly due to the highly viscous flow through the pores. In the simplest description (Bejan, 1995a), the second law of thermodynamics may be invoked in the analysis of the one-dimensional Darcy flow model of Figure 1.7, and the result is the entropy generation rate per unit volume of homogeneous porous medium,

$$\dot{S}'''_{\text{gen}} = \frac{k}{T^2}\left(\frac{\partial T}{\partial x}\right)^2 + \frac{\mu u^2}{KT} \geq 0. \tag{1.40}$$

This expression is restricted to Darcy flow, and is based on the assumption that $q''' = 0$ and $\text{Re}_K < 1$. For a general case of convection in three dimensions, the local rate of entropy generation is

$$\dot{S}'''_{\text{gen}} = \frac{k}{T^2}(\nabla T)^2 + \frac{\mu}{KT}(\mathbf{v})^2 \geq 0. \tag{1.41}$$

Each of the two terms on the right-hand side is nonnegative. Furthermore, it must be stressed that T represents the absolute temperature and, consequently, the viscous irreversibility term may not be negligible, even in cases when the viscous dissipation term can be neglected in the energy equation (1.31). Second-law aspects of heat and fluid flow through porous media are relevant to the thermodynamic optimization of energy systems that employ such flow structures. One example is the operation of gauze-type regenerative heat exchangers for Stirling engines and refrigerators (Organ, 1992, 1997; Finkelstein and Organ, 2001). In addition, the irreversibility of heat and fluid flow through porous media is important in a fundamental sense, as shown in a recent study by Baytas and Pop (2000).

1.5 Heat and Mass Transfer

The treatment of the propagation of a species through a mixture that flows through a porous structure is analogous to the heat transfer process modeled in Section 1.4. Analogous volume-averaging arguments lead to the species conservation equation

$$\phi \frac{\partial C}{\partial t} + \mathbf{v} \cdot \nabla C = \nabla \cdot (D \nabla C) + \dot{m}''', \tag{1.42}$$

where C is the species concentration, and $\dot{m}'''$ is the rate of species generation per unit time and volume. The D coefficient is the mass diffusivity of the species, relative to the porous medium with the fluid mixture in its pores. The diffusivity D is analogous to the porous medium thermal conductivity k. The first term on the right-hand side of Equation (1.42) reflects the use of Fick's law,

$$\mathbf{j} = -D \nabla C, \tag{1.43}$$

where $\mathbf{j}$ is the mass flux vector. Equation (1.43) is analogous to Fourier's law, $\mathbf{q}'' = -k \nabla T$, which was invoked but not mentioned in the derivation of Equation (1.31). Note that D is a function of position when the porous medium is nonhomogeneous. Equation (1.43) is also known as Fick's first law, whereas the mass balance (1.42) is Fick's second law. In Chapter 4, we show how to improve these statements in order to treat ionic transport.

Buoyancy-driven flows may be complicated by the effect of concentration gradients. This may happen when the fluid mixture density depends on the concentration of the species of interest. The equation of state $\rho = \rho(T, P, C)$ can be approximated in the vicinity of the reference state (ρ_0, T_0, C_0) as the linear function

$$\rho = \rho_0[1 - \beta(T - T_0) - \beta_C(C - C_0)], \tag{1.44}$$

where the thermal and concentration expansion coefficients are

$$\beta = -\frac{1}{\rho} \left(\frac{\partial \rho}{\partial T} \right)_{P,C}, \qquad \beta_C = -\frac{1}{\rho} \left(\frac{\partial \rho}{\partial C} \right)_{T,P}. \tag{1.45}$$

The β and β_C effects are taken into account by substituting Equation (1.44) into the flow model [e.g., Equation (1.34)]. Such a treatment is needed when natural convection is driven by a combination of temperature and concentration gradients.

There is also the possibility of the creation of mass fluxes due to temperature gradients (Soret effect), and heat fluxes due to concentration gradients [Dufour effect, Bejan (1997, Chapter 12)]. When heat and species generation is absent, Equations (1.32) and (1.42) are replaced by

$$\sigma \frac{\partial T}{\partial t} + \mathbf{v} \cdot \nabla T = \nabla \cdot (\alpha \nabla T + D_{TC} \nabla C), \tag{1.46}$$

$$\phi \frac{\partial C}{\partial t} + \mathbf{v} \cdot \nabla C = \nabla \cdot (D \nabla C + D_{CT} \nabla T), \tag{1.47}$$

where D_{TC} and D_{CT} are, respectively, the Dufour and Soret coefficients of
the porous medium.

In this section we take a closer and more formal look at the phenomenon
of entropy generation when heat transfer and mass transfer are present. The
internal energy of a rigid porous matrix is distributed by the existing vibrational degrees of freedom (modes) of the constituent particles. In addition to
vibrational modes, in a fluid the internal energy is also distributed by rotational and translational modes. Translational motion endows the fluid with
the capacity of filling and flowing through the pores of the matrix. On the
other hand, the energy of the organized translational motion may be transferred to the rotational and vibrational modes, in the so-called *dissipative
processes*, which account for the generation of entropy.

Porous media are effective entropy generators because their pore structure
is very effective in the destruction of the organized translational motion of
fluids. The translational motion of fluid particles is random, but when the
average particle velocity $\mathbf{v}$ is not zero in an elemental fluid volume dV with
fluid density ρ, we can speak of an organized translational motion (mass flux
or flow), which may be described quantitatively by:

$$\mathbf{j} = \rho\mathbf{v}. \tag{1.48}$$

Mass conservation requires

$$\frac{\partial \rho}{\partial t} + \nabla \cdot \mathbf{j} = 0 \tag{1.49}$$

or the equivalent form

$$\frac{1}{\rho}\frac{d\rho}{dt} + \nabla \cdot \mathbf{v} = 0. \tag{1.50}$$

The vectors $\mathbf{j}$ and $\mathbf{v}$ represent two fields in the fluid domain. If the field $\mathbf{j}$ is
not divergent ($\nabla \cdot \mathbf{j} = 0$), the vector $\mathbf{j}$ represents a *flow*. When $\mathbf{j}$ is divergent
($\nabla \cdot \mathbf{j} \neq 0$), the vector $\mathbf{j}$ describes a *diffusive mass flux*, or a *flow* and a *diffusive
mass flux*. Note also that a nondivergent $\mathbf{v}$ field defines an *incompressible flow*,
whereas a divergent $\mathbf{v}$ field represents a *compressible flow*.

Let s_v and u_v denote the entropy per unit volume and the internal energy
per unit volume. The entropy change in the elemental fluid volume dV_0 is
[e.g., Kondepudi and Prigogine (1998)]:

$$\frac{\partial s_v}{\partial t} = \frac{1}{T}\frac{\partial u_v}{\partial t} - \frac{\mu}{T}\frac{\partial \rho}{\partial t}. \tag{1.51}$$

The subscript v indicates properties per unit volume. The energy present
in the elemental volume dV is the sum of the internal energy u_v and the
mechanical (kinetic and potential) energy $w_v = \rho(v^2/2 + \Phi)$,

$$e_v = u_v + w_v = u_v + \rho\left(\frac{v^2}{2} + \Phi\right). \tag{1.52}$$

Let $\mathbf{j_u}$ and $(w_v/\rho)\mathbf{j}$ be the flux of internal energy and the flux of mechanical energy, respectively. Then the equation for the energy conservation reads:

$$\frac{\partial(u_v + w_v)}{\partial t} + \nabla \cdot \left(\mathbf{j_u} + \left(\frac{w_v}{\rho}\right)\mathbf{j}\right) = 0. \tag{1.53}$$

In the equation for the mechanical energy,

$$\frac{\partial w_v}{\partial t} + \nabla \cdot \left(\frac{w_v}{\rho}\mathbf{j}\right) = \dot{w}_{v,d}, \tag{1.54}$$

the term $\dot{w}_{v,d}$ accounts for mechanical energy generation ($\dot{w}_{v,d} > 0$), or energy depletion ($\dot{w}_{v,d} < 0$). Generation of mechanical energy means depletion of internal energy (and vice versa) because the total energy is conserved. Combining the equation for mass conservation (1.49), the equation for energy conservation (1.53), and the equation for mechanical energy (1.54), Equation (1.51) becomes

$$\frac{\partial s_v}{\partial t} + \nabla \cdot \left[\frac{1}{T}(\mathbf{j_u} - \mu\mathbf{j})\right] = \mathbf{j_u} \cdot \nabla\left(\frac{1}{T}\right) - \mathbf{j} \cdot \nabla\left(\frac{\mu}{T}\right) - \frac{\dot{w}_{v,d}}{T}. \tag{1.55}$$

Equation (1.55) represents the entropy "balance" in the elemental volume dV. This is not truly a balance because entropy is generated, not conserved. The first term in Equation (1.55) represents the local entropy variation. The second term represents the divergence of the entropy flux

$$\mathbf{j_s} = \frac{1}{T}(\mathbf{j_u} - \mu\mathbf{j}). \tag{1.56}$$

The group on the right side of Equation (1.55) represents the entropy generation, or destruction per unit volume,

$$\dot{S}'''_{\text{gen}} = \mathbf{j_u} \cdot \nabla\left(\frac{1}{T}\right) - \mathbf{j} \cdot \nabla\left(\frac{\mu}{T}\right) - \frac{\dot{w}_{v,d}}{T}. \tag{1.57}$$

The heat flux is the energy associated with the entropy flux ($T\mathbf{j_s}$) minus the part associated with the mass flux,

$$\mathbf{q}'' = T\mathbf{j_s} - Ts\mathbf{j}. \tag{1.58}$$

Combining Equations (1.56) and (1.58), noting that $\rho s = s_v$, and that by the equation for the internal energy per unit volume $\rho\mu = u_v - Ts_v + P$ [e.g., Kondepudi and Prigogine (1998)], Equation (1.57) becomes

$$\dot{S}'''_{\text{gen}} = \left(\mathbf{q}'' + \frac{u_v + P}{\rho}\mathbf{j}\right) \cdot \nabla\left(\frac{1}{T}\right) - \mathbf{j} \cdot \nabla\left(\frac{\mu}{T}\right) - \frac{\dot{w}_{v,d}}{T}. \tag{1.59}$$

The Gibbs–Duhem relation in entropy representation,

$$u_v d\left(\frac{1}{T}\right) + d\left(\frac{P}{T}\right) - \rho d\left(\frac{\mu}{T}\right) = 0 \tag{1.60}$$

allows us to rewrite Equation (1.59) as

$$\dot{S}'''_{\text{gen}} = -\frac{\mathbf{q}''}{T^2} \cdot \nabla T - \frac{\mathbf{j}}{\rho T} \cdot \nabla P - \frac{\dot{w}_{v,d}}{T}. \tag{1.61}$$

In conclusion, entropy may be generated in a fluid by heat fluxes the direction of which is opposite to that of the temperature gradients, by mass fluxes (or flows) oriented against the pressure gradient, by the destruction of mechanical energy ($\dot{w}_{v,d} < 0$), or by a combination of all these effects.

The second law of thermodynamics states that $\dot{S}'''_{\text{gen}} \geq 0$. Therefore, if only pressure gradients are present in a fluid, a mass flux is spontaneously set up in order to generate entropy. Similarly, temperature gradients induce heat fluxes that also produce entropy. Conversely, Equation (1.61) shows that entropy is destroyed in the elemental volume if the heat flux and the temperature gradient (or the mass flux and the pressure gradient) have the same direction. The second law of thermodynamics forbids this, unless an external entity is present. In that case, the external device generates more entropy than what is being destroyed in the elemental volume.

Equation (1.61) accounts for dissipative processes in many situations. One example is fluid flow, where the balance of momentum is described by the Navier–Stokes equation:

$$\frac{\partial \mathbf{v}}{\partial t} + \mathbf{v} \cdot \nabla \mathbf{v} = -\nabla \Phi - \frac{1}{\rho} \nabla P + \left(\nu' + \frac{1}{3}\nu \right) \nabla \nabla \cdot \mathbf{v} + \nu \nabla^2 \mathbf{v}, \tag{1.62}$$

where ν represents the usual viscosity—the shear viscosity resulting from the random translatory motion of particles—and ν' stands for the bulk viscosity which is associated with the fluid internal relaxation phenomena [for details, see, e.g., Woods (1986)]. In most cases, the Navier–Stokes equation is written in the reduced form

$$\frac{\partial \mathbf{v}}{\partial t} + \mathbf{v} \cdot \nabla \mathbf{v} = -\nabla \Phi - \frac{1}{\rho} \nabla P + \nu \nabla^2 \mathbf{v}. \tag{1.62'}$$

Performing the scalar product with $\mathbf{j}$ for each term in Equation (1.62), and assuming that the potential Φ does not depend on time, we obtain

$$\frac{\partial w_v}{\partial t} + \nabla \cdot \left(\frac{w_v}{\rho} \mathbf{j} \right) = \dot{w}_{v,d} = -\mathbf{j} \cdot \left[\frac{1}{\rho} \nabla P - \left(\nu' + \frac{1}{3} \right) \nabla \nabla \cdot \mathbf{v} - \nu \nabla^2 \mathbf{v} \right]. \tag{1.63}$$

Here we can identify the terms corresponding to processes that destroy (or create) mechanical energy. Substituting $\dot{w}_{v,d}$ of Equation (1.63) into Equation (1.61), we see the processes that generate entropy in fluid flow:

$$\dot{S}'''_{\text{gen}} = -\frac{\mathbf{q}''}{T^2} \cdot \nabla T - \frac{\mathbf{j}}{T} \cdot \left[\left(\nu' + \frac{1}{3}\nu \right) \nabla (\nabla \cdot \mathbf{v}) - \nu \nabla^2 \mathbf{v} \right]. \tag{1.64}$$

These processes are heat flow and viscous dissipation. In porous media, because of short-range forces, fluid particles attach themselves to the pore walls and bring the bulk velocity equal to zero at the wall. The trapped particles retard the motion of neighboring particles and the like, and this results in a distribution of velocity (a profile) near the wall. In this way, the kinetic energy associated with translational motion is converted into internal energy associated with the vibrational and rotational modes. The end result is the generation of entropy.

It is interesting to point out that some flows exhibit a perfect balance between the creation and the destruction of the mechanical energy, so $\dot{w}_{v,d} = 0$. For example, in the Hagen–Poiseuille flow the velocity field is given by [cf. Equation (1.63)]

$$\frac{1}{\rho}\nabla P = \left(\nu' + \frac{1}{3}\nu\right)\nabla\nabla\cdot v + \nu\nabla^2\mathbf{v}. \tag{1.65}$$

Nevertheless, such flows generate entropy, as shown by combining Equations (1.64) and (1.65), or by combining Equation (1.61) with $\dot{w}_{v,d} = 0$,

$$\dot{S}'''_{\text{gen}} = -\frac{\mathbf{q}''}{T^2}\cdot\nabla T - \frac{\mathbf{j}}{\rho T}\cdot\nabla P. \tag{1.66}$$

Finally, reversible processes (zero entropy generation) are described by the equation obtained from Equation (1.61) with $\dot{S}'''_{\text{gen}} = 0$,

$$\frac{\mathbf{q}''}{T}\cdot\nabla T + \frac{\mathbf{j}}{\rho}\cdot\nabla P + \dot{w}_{v,d} = 0. \tag{1.67}$$

Let us consider processes in which the mechanical energy is constant. Then, by using Equation (1.66), the heat and the mass fluxes are related through linear combinations of the existing gradients [e.g., see Kondepudi and Prigogine (1998)]:

$$\mathbf{q}'' = \frac{L_{qq}}{T^2}\nabla T + \frac{L_{qm}}{T}\nabla P, \tag{1.68}$$

$$\mathbf{j_m} = \frac{L_{mq}}{T^2}\nabla T + \frac{L_{mm}}{T}\nabla P, \tag{1.69}$$

where L_{qq}, L_{qm}, L_{mq}, and L_{mm} are the phenomenological coefficients. In view of Equation (1.61) and the second law of thermodynamics, the matrix $[L_{ij}]$ has to be positive definite, so that $\dot{S}'''_{\text{gen}} > 0$,

$$L_{qq} > 0, \quad L_{mm} > 0, \quad (L_{qm} + L_{mq})^2 < 4L_{qq}L_{mm}. \tag{1.70}$$

Because of the Onsager reciprocity relations,

$$L_{qm} = L_{mq}, \tag{1.71}$$

only three of the four coefficients remain to be determined from phenomenological relations.

Equations (1.68) and (1.69) are the general expressions for the coupled
heat and mass fluxes. They show that heat and mass fluxes may be driven
by the temperature and/or pressure gradients. In every case, the phenomenological coefficients must be determined by taking into account the specific
features of the flow configuration. In most cases, the coefficients L_{ij} may be
determined from known phenomenological relationships, such as Darcy's law,
Fourier's law, and so on. In some cases, they can also be determined based
on dimensional analysis. The following are two simple cases involving porous
media.

1.5.1 Fluid Flow

In the case of isothermal fluid flow through a porous medium [$\nabla T = 0$ in
Equation (1.66)], we may think that the area-averaged flow rate $\langle \mathbf{j} \rangle$ of the
porous medium is related to several properties: a property of the solid matrix
such as the sum of the cross-sectional area of the pores (Σa_p) that are present
in the unit cross-sectional area of the porous medium, a fluid property such as
the kinematic viscosity ν, and the pressure gradient ∇P as the driving force.
These four quantities are expressed in terms of three fundamental dimensions:
length L, mass M, and time t. Then the Buckingham pi theorem assures
us that only one dimensionless group (G_D) can be formed with the four
quantities,

$$\langle j \rangle (\Sigma a_p)^a (\nu)^b (\nabla P)^c = G_D. \tag{1.72}$$

The exponents a, b, and c are unknown. Moreover, G_D must be constant. In
terms of the fundamental dimensions, Equation (1.72) requires

$$(ML^{-2}t^{-1})(L^2)^a (L^2 t^{-1})^b (ML^{-2}t^{-2})^c = (MLt)^0, \tag{1.73}$$

which means that a, b, and c must satisfy

$$1 + c = 0, \qquad -2 + 2a + 2b - 2c = 0, \qquad -1 - b - 2c = 0. \tag{1.74}$$

The solution is $a = -1$, $b = 1$, and $c = -1$; in other words,

$$\langle j \rangle = G_D \frac{\Sigma a_p}{\nu} \nabla P, \tag{1.75}$$

which is the analytical form of Darcy's law. By comparing this form with
Equation (1.7) we identify $-G_D(\Sigma a_p)$ as the permeability of the porous
medium. Finally, Equations (1.69) and (1.75) reveal the coefficient

$$L_{mm} = G_D T \Sigma a_p / \nu, \tag{1.76}$$

where the dimensionless factor G_D is constant for every porous medium.

1.5.2 Heat Flow

Assume that the isobaric heat flux $\mathbf{q}''$ depends on a solid-matrix property such as $(k_M \Sigma a_M + k_f \Sigma a_p)$, where (k_M, k_f) and $(\Sigma a_M, \Sigma a_p)$ are the thermal conductivities and cross-sectional areas of the matrix and fluid (pores), respectively. The heat flux may also depend on the sum $(\Sigma a_M + \Sigma a_p)$ and the driving force $\nabla T/T$. Dimensional analysis and the Buckingham pi theorem lead to

$$\mathbf{q}'' = -\langle k \rangle \nabla T, \qquad \langle k \rangle = -G_F \frac{k_M \Sigma a_M + k_f \Sigma a_p}{\Sigma a_M + \Sigma a_p}, \qquad (1.77)$$

where $G_F < 0$ is the appropriate dimensionless constant. Equation (1.77) is Fourier's law. The factor $\langle k \rangle$ is the area-averaged thermal conductivity of the matrix with the fluid in the pores. Finally, Equation (1.68) shows that

$$L_{qq} = -\langle k \rangle T^2. \qquad (1.78)$$

The coupling coefficients $L_{qm} = L_{mq}$ are not as easy to identify, and generally are determined from phenomenological relationships. Equations (1.66), (1.68), and (1.69) permit the calculation of the entropy generated by heat and mass fluxes at constant mechanical energy.

In this first chapter we reviewed the simplest and most essential features of modeling flows through porous media. This review serves as an introduction for the results obtained for flows in some of the most basic and frequent configurations, which are presented in the next chapter. Additional modeling features are reviewed in Nield and Bejan (1999), for example, the generation of species in chemical reactions [the term $\dot{m}'''$ in Equation (1.42)], and the complications associated with the presence of multiphase flow. Chemically reactive flows are described by Nakayama and Kuwahara (2000), Chao *et al.* (1994, 1996), and Pop *et al.* (2002). Another modeling direction is pursued in later chapters, in applications involving flows through coarse porous media and designed complex flow structures.

The special case of isothermal mass diffusion refers to the part of the mass flux that is driven by the gradient of concentration, Equation (1.43). We return to this subject in Section 4.8.

2

Flows in Porous Media

2.1 Use Simple Methods First

We now turn our attention to the results that constitute the core of modern research on convective heat and mass transfer through porous media. Our objective is not only to organize the compact presentation of these results, but also to explain their origin. We want to show the student how to anticipate these results and the results for related problems. This is why we begin with methodology. We emphasize the freedom that educators and researchers have in choosing methods to solve problems, present the results, and put results into practice.

The field of convection in porous media is an excellent candidate for stressing this important message. It is mature enough, and at the same time it is rich: its results cover a wide spectrum of problems and applications in thermal engineering, physics, geophysics, bioengineering, civil engineering, and many other fields. These fields are united by several key phenomena, some of which are selected for review in this chapter. The opportunity that the maturity of our field offers is this: after a few decades of development, we find that more than one method is available for attacking a certain problem. Older methods, such as analysis and experiments (direct laboratory measurements), are as eligible to be used as the newer methods based on computational analysis. The point is that the available methods *compete* for the researcher's attention. The researcher has the freedom to choose the method that suits him or her (Bejan, 1995a).

Methods are literally competitive because each is an example of a tradeoff between cost (effort) and accuracy. The simpler methods require less effort and produce less accurate results than the more complicated methods. The researcher enters the marketplace of methodology with a personal profile: talent, time to work, interest in details, and users of the results of the contemplated work (e.g., customers, students). The match of researcher, problem, and method is not the result of chance. It is an optimization decision (structure, configuration) in the sense of *constructal* theory and design (Bejan, 2000). This configuration maximizes the performance (benefit) for all parties concerned. The intellectual work that goes on in research and education is a conglomerate of mind–problem–method matches of the kind illustrated in this book. The so-called "knowledge industry" thrives on the optimization of matches and connections—it thrives on the development (growth) and optimization of structure.

Among the methods that have emerged in fluid mechanics and heat transfer, scale analysis (scaling) is one of the simplest and most cost effective (Bejan, 1984, 1995a); see Sections 2.1, 2.5, and 4.4. Scaling is now used by many, yet the need to explain its rules and promise remains. To accomplish this in a compact and effective format is the first objective of this chapter. Scale analysis is so cost effective that it is beneficial as a first step (preliminary, prerequisite) even in situations where the appropriate method is more laborious and the sought results are more accurate. The results of scale analysis serve as a guide. They tell the researcher what to expect before the use of a more laborious method, what the ultimate (accurate) results mean, and how to report them in dimensionless form. The engineering advice to "try the simplest first" fits perfectly in the optimization of research.

The second objective of this chapter is to explain the rules and the promise of another simple method: the intersection of asymptotes (Section 2.11). This method was born by accident, in the search for a quick solution to a homework problem, namely, the optimal spacing between parallel plates with natural convection (Bejan, 1984, problem 11, p. 157). What led to the quick solution is an idea of more permanent and general value: when one is challenged to describe a complicated system or phenomenon (e.g., a flow structure), it is helpful to describe the phenomenon in the simpler extremes (asymptotes) in which it may manifest itself. The complicated phenomenon lies somewhere between the extremes, and its behavior may be viewed as the result of the competition (clash, collision) between the extremes. This idea has helped us in many areas. For example, the highly complicated relations between the thermodynamic properties of a real substance (e.g., steam tables) make more sense when viewed as the intersection of two extremes of thermodynamic behavior: the incompressible substance model and the ideal gas model.

Although newer than scaling, the intersection of asymptotes method is now used frequently in thermal design and optimization (e.g., Sadeghipour and Razi, 2001). Optimization of global performance is an integral part of the physics of flow structures: flows choose certain patterns (shapes, structures, regimes) as compromises between the available extremes. Flows design for themselves paths for easiest access (Bejan, 2000). To illustrate this constructal characteristic of natural flows, in Section 2.11 we take a look at the classical phenomenon of natural convection in a porous layer saturated with fluid and heated from below (Horton and Rogers, 1945; Lapwood, 1948). This phenomenon has been investigated with increasingly accurate analytical, numerical, and experimental tools, as shown by recent reviews [e.g., Nield and Bejan (1999)]. Just like scaling, the intersection of asymptotes method provides a surprisingly direct alternative, a short cut to the most important characteristics of the flow.

2.2 Scale Analysis of Forced Convection Boundary Layers

Scaling is a method for determining answers to concrete problems, such as the heat transfer rate in a configuration that is described completely. The results are correct and accurate, but only in an order of magnitude sense. The following examples exhibit an accuracy better than within a factor of 2 or 1/2. The analysis is based on the complete problem statement, that is, the conservation equations and all the initial and boundary conditions. Partial differential equations are replaced by global algebraic statements, which are approximate. Scale analysis is a problem-solving method—a method of solution that should not be confused with dimensional analysis.

A simple class of flows that can be described based on scale analysis is boundary layers. Figure 2.1 shows the thermal boundary layer in the vicinity of an isothermal solid wall (T_0) embedded in a saturated porous medium

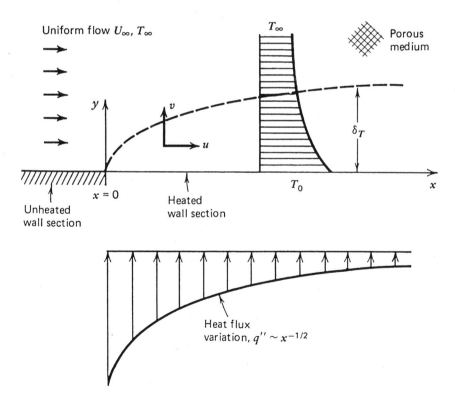

Fig. 2.1. Thermal boundary layer near an isothermal wall with parallel flow.

with uniform flow parallel to the wall (U_∞, T_∞). For simplicity, we assume temperature-independent properties, so that the temperature field is independent of the flow field. The flow field is known: it is the uniform flow U_∞, which in Darcy or Darcy–Forchheimer flow (cf. Section 1.3) is driven by a uniform and constant pressure gradient (dP/dx). Unknown are the temperature distribution in the vicinity of the wall and the heat transfer between the wall and the porous medium.

The analysis refers to the boundary layer regime, which is based on the assumption that the region in which the thermal effect of the wall is felt is *slender*,

$$\delta_T \ll x. \tag{2.1}$$

The boundary layer region has the length x and thickness δ_T. The latter is defined as the distance y where the temperature is practically the same as the approaching temperature (T_∞), or where $\partial T/\partial y \cong 0$. The thermal boundary layer thickness δ_T is the unknown geometric feature that is also the solution to the heat transfer problem. The wall heat flux is given by

$$q'' = k\left(-\frac{\partial T}{\partial y}\right)_{y=0}, \tag{2.2}$$

where k is the effective thermal conductivity of the porous medium saturated with fluid. The scale of the temperature gradient appearing in Equation (2.2) is given by

$$\left(-\frac{\partial T}{\partial y}\right)_{y=0} \sim -\frac{T_\infty - T_0}{\delta_T - 0} = \frac{\Delta T}{\delta_T}, \tag{2.3}$$

where ΔT is the overall temperature difference that drives q''. In conclusion, the heat flux scale is given by

$$q'' \sim k\frac{\Delta T}{\delta_T}, \tag{2.4}$$

which means that in order to estimate q'' we must first estimate δ_T.

The required equation for δ_T is provided by the energy equation, namely, Equation (1.32),

$$u\frac{\partial T}{\partial x} + v\frac{\partial T}{\partial y} = \alpha\frac{\partial^2 T}{\partial y^2}, \tag{2.5}$$

where $\alpha = k/(\rho c_p)_f$ is the thermal diffusivity of the saturated porous medium. The volume-averaged velocity components of the uniform flow are $u = U_\infty$ and $v = 0$, such that Equation (2.5) becomes

$$U_\infty\frac{\partial T}{\partial x} = \alpha\frac{\partial^2 T}{\partial y^2}. \tag{2.6}$$

To determine the order of magnitude of $\partial^2 T/\partial y^2$, we use the same technique as in Equation (2.3): we evaluate the change in $\partial T/\partial y$ from $y = 0$ to $y \sim \delta_T$,

$$\frac{\partial^2 T}{\partial y^2} = \frac{\partial}{\partial y}\left(\frac{\partial T}{\partial y}\right) \sim \frac{(\partial T/\partial y)_{\delta_T} - (\partial T/\partial y)_0}{\delta_T - 0} = \frac{0 + \Delta T/\delta_T}{\delta_T} = \frac{\Delta T}{\delta_T^2}. \quad (2.7)$$

Similarly, for $\partial T/\partial x$ we look at the change in T along the system, from $x = 0$ to x, at a constant y sufficiently close to the wall:

$$\frac{\partial T}{\partial x} \sim \frac{(T)_x - (T)_{x=0}}{x - 0} = \frac{T_0 - T_\infty}{x} = \frac{\Delta T}{x}. \quad (2.8)$$

Together, Equations (2.6) to (2.8) produce

$$U_\infty \frac{\Delta T}{x} \sim \alpha \frac{\Delta T}{\delta_T^2}, \quad (2.9)$$

which is the approximate algebraic statement that replaces the partial differential Equation (2.6). The boundary layer thickness follows from Equation (2.9),

$$\delta_T \sim \left(\frac{\alpha x}{U_\infty}\right)^{1/2} \quad (2.10)$$

and so does the conclusion that δ_T increases as $x^{1/2}$, as shown in Figure 2.1. The heat flux decreases as $x^{-1/2}$, [cf. Equation (2.4)],

$$q'' \sim k\Delta T \left(\frac{U_\infty}{\alpha x}\right)^{1/2}. \quad (2.11)$$

The dimensionless version of this heat transfer rate is given by

$$\mathrm{Nu}_x \sim \mathrm{Pe}_x^{1/2}, \quad (2.12)$$

where the Nusselt and Péclet numbers are given by

$$\mathrm{Nu}_x = \frac{q''x}{k\Delta T} \qquad \mathrm{Pe}_x = \frac{U_\infty x}{\alpha}. \quad (2.13)$$

These estimates are valid when the heart of boundary layer theory, the slenderness assumption (1), is respected, and this translates into the requirement that $\mathrm{Pe}_x \gg 1$, or that U_∞ and/or x must be sufficiently large.

How approximate is this heat transfer solution? The exact solution to the same thermal boundary layer problem is available in closed form, after solving Equation (2.6) in similarity formulation. The details of this analysis may be found in Bejan (1995a). The similarity solution for the local heat flux is given by

$$\mathrm{Nu}_x = 0.564\mathrm{Pe}_x^{1/2}. \quad (2.14)$$

This agrees with the scaling estimate (2.12). The heat flux averaged over a wall of length L,

$$\overline{q''} = \frac{1}{L} \int_0^L q'' \, dx \qquad (2.15)$$

can be estimated based on Equation (2.14),

$$\overline{\mathrm{Nu}} = 1.128 \mathrm{Pe}_L, \qquad (2.16)$$

where corresponding Nusselt and Péclet number definitions are given by

$$\overline{\mathrm{Nu}} = \frac{\overline{q''}L}{k\Delta T}, \qquad \mathrm{Pe}_L = \frac{U_\infty L}{\alpha}. \qquad (2.17)$$

Equation (2.16) shows that the scaling estimate (2.12) is again accurate within a factor of order 1. Furthermore, scale analysis makes no distinction between local flux and wall-averaged heat flux: both have the same scale, and the correct scale is delivered by scale analysis, Equation (2.11).

The analysis that produced Equation (2.11) did not require the wall temperature be uniform, $\Delta T = $ constant. We made this assumption only later, when we formulated Nu_x and $\overline{\mathrm{Nu}}$. Equation (2.11) is the general and correct relation between the heat flux scale (q'') and wall excess temperature scale (ΔT) in the boundary layer configuration. Equation (2.11) can be used in situations other than the isothermal-wall case of Figure 2.1. For example, when the wall heat flux is uniform, Equation (2.11) delivers the scale and character of the wall temperature distribution, $\Delta T = T_0(x) - T_\infty$. In local Nusselt number formulation, the scaling result is

$$\mathrm{Nu}_x = \frac{q''x}{k[T_0(x) - T_\infty]} \sim \mathrm{Pe}_x^{1/2}. \qquad (2.18)$$

The corresponding local and overall Nusselt numbers derived from the similarity solution to the same problem are

$$\mathrm{Nu}_x = \frac{q''x}{k[T_0(x) - T_\infty]} = 0.886 \mathrm{Pe}_x^{1/2} \qquad (2.19)$$

$$\overline{\mathrm{Nu}} = \frac{q''L}{k(\bar{T} - T_\infty)} = 1.329 \mathrm{Pe}_L^{1/2}, \qquad (2.20)$$

where $\bar{T}$ is $T(x)$ averaged from $x = 0$ to $x = L$. Once again, the scaling result (2.18) anticipates within 12 and 33% the similarity solution, Equations (2.19) and (2.20). The trends are identical, and correct. To appreciate how simple, direct, and cost effective scale analysis is, the reader should try to solve the problems using other methods.

2.3 Sphere and Cylinder with Forced Convection

We continue with several examples of forced convection around other bodies embedded in porous media. Figure 2.2a shows the thermal boundary layer region around a sphere, or around a circular cylinder that is perpendicular to a uniform flow with volume averaged velocity u. The sphere or cylinder radius is r_0, and the surface temperature is T_w. The distributions of heat flux around the sphere and cylinder in Darcy flow were determined in Cheng (1982). With reference to the angular coordinate θ, the local peripheral Nusselt numbers for the sphere are

$$\mathrm{Nu}_\theta = 0.564 \left(\frac{ur_0\theta}{\alpha}\right)^{1/2} \left(\frac{3}{2}\theta\right)^{1/2} \sin^2\theta \left(\frac{1}{3}\cos^3\theta - \cos\theta + \frac{2}{3}\right)^{-1/2} \quad (2.21)$$

and for the cylinder:

$$\mathrm{Nu}_\theta = 0.564 \left(\frac{ur_0\theta}{\alpha}\right)^{1/2} (2\theta)^{1/2} \sin\theta \, (1 - \cos\theta)^{-1/2}. \quad (2.22)$$

The Péclet number is based on the swept arc $r_0\theta$, namely, $\mathrm{Pe}_\theta = ur_0\theta/\alpha$. The local Nusselt number is defined as

$$\mathrm{Nu}_\theta = \frac{q''}{T_w - T_\infty} \frac{r_0\theta}{k}. \quad (2.23)$$

Equations (2.21) and (2.22) are valid when the boundary layers are distinct (thin), that is, when the boundary layer thickness $r_0 \mathrm{Pe}_\theta^{-1/2}$ is smaller than the radius r_0. This requirement can also be written as $\mathrm{Pe}_\theta^{1/2} \gg 1$, or $\mathrm{Nu}_\theta \gg 1$.

The analogy between the thermal boundary layers of the cylinder and the sphere (Figure 2.2a) and that of the flat wall (Figure 2.1) is illustrated further by Nield and Bejan's (1999) correlation of the heat transfer results for these

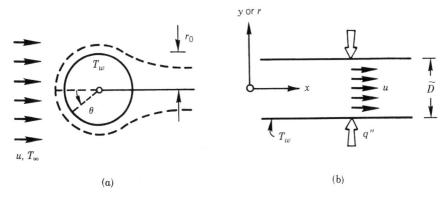

(a)   (b)

Fig. 2.2. Forced convection in porous media: (a) boundary layer around a sphere or cylinder; (b) channel filled with a porous medium.

three configurations. The heat flux averaged over the area of the cylinder and sphere, $\overline{q}''$, can be calculated by averaging the local heat flux q'' expressed by Equations (2.21) through (2.23). The results for the sphere are

$$\overline{\mathrm{Nu}}_D = 1.128\mathrm{Pe}_D^{1/2} \qquad (2.24)$$

and for the cylinder

$$\overline{\mathrm{Nu}}_D = 1.015\mathrm{Pe}_D^{1/2}. \qquad (2.25)$$

In these expressions, the Nusselt and Péclet numbers are based on the diameter $D = 2r_0$,

$$\overline{\mathrm{Nu}}_D = \frac{\overline{q}''D}{(T_w - T_\infty)k}, \qquad \mathrm{Pe}_D = \frac{uD}{\alpha}. \qquad (2.26)$$

In summary, the overall Nusselt number is nearly equal to $\mathrm{Pe}_D^{1/2}$, in accordance with the scale analysis presented in Section 2.2. The purpose of the present section was to review the exact results that are available for the sphere and cylinder, and to direct the reader to the appropriate sources in the literature.

2.4 Channels with Porous Media and Forced Convection

Consider now the forced convection heat transfer in a channel or duct packed with a uniform and isotropic porous material as in Figure 2.2b. In the Darcy flow regime the longitudinal volume-averaged velocity u is uniform over the channel cross-section. When the temperature field is fully developed, the relationship between the wall heat flux q'' and the local temperature difference $(T_w - T_b)$ is analogous to the relationship for fully developed heat transfer to slug flow through a channel without a porous matrix (Bejan, 1995a). The temperature T_b is the mean or bulk temperature of the stream that flows through the channel,

$$T_b = \frac{1}{A} \int_A T \, dA, \qquad (2.27)$$

in which A is the area of the channel cross-section. In general, when the velocity u is not uniform over the channel cross-section, T_b is weighted with the velocity,

$$T_b = \frac{1}{\bar{u}A} \int_A uT \, dA, \qquad (2.27')$$

where $\bar{u}$ is the velocity averaged over A. In cases where the confining wall is a tube with internal diameter D, the relation for fully developed heat transfer can be expressed as a constant Nusselt number (Rohsenow and Choi, 1961):

$$\mathrm{Nu}_D = \frac{q''(x)}{T_w - T_b(x)} \frac{D}{k} = 5.78 \quad (\text{tube}, T_w = \text{uniform}), \qquad (2.28)$$

$$\mathrm{Nu}_D = \frac{q''}{T_w(x) - T_b(x)} \frac{D}{k} = 8 \quad (\text{tube}, q'' = \text{uniform}). \qquad (2.29)$$

When the porous matrix is sandwiched between two parallel plates with the spacing D, the corresponding Nusselt numbers are (Rohsenow and Hartnett, 1973)

$$\mathrm{Nu}_D = \frac{q''(x)}{T_w - T_b(x)} \frac{D}{k} = 4.93 \quad (\text{parallel plates}, T_w = \text{uniform}), \quad (2.30)$$

$$\mathrm{Nu}_D = \frac{q''}{T_w(x) - T_b(x)} \frac{D}{k} = 6 \quad (\text{parallel plates}, q'' = \text{uniform}). \quad (2.31)$$

The forced convection results of Equations (2.28) through (2.31) are valid when the temperature profile across the channel is fully developed (sufficiently far from the entrance $x = 0$). The entrance length, or the length needed for the temperature profile to become fully developed, can be estimated by noting that the thermal boundary layer thickness scales as $(\alpha x/u)^{1/2}$. Setting $(\alpha x/u)^{1/2} \sim D$, the thermal entrance length $x_T \sim D^2 u/\alpha$ is obtained. Inside the entrance region $0 < x < x_T$, the heat transfer is impeded by the forced convection thermal boundary layers that line the channel walls, and can be calculated approximately using the results of Section 2.2.

One important application of the results for a channel packed with a porous material is in the area of heat transfer augmentation. As shown by Nield and Bejan (1999), the Nusselt numbers for fully developed heat transfer in a channel without a porous matrix are given by expressions similar to Equations (2.28) through (2.31) except that the saturated porous medium conductivity k is replaced by the thermal conductivity of the fluid alone, k_f. The relative heat transfer augmentation effect is indicated approximately by the ratio

$$\frac{h_x(\text{with porous matrix})}{h_x(\text{without porous matrix})} \sim \frac{k}{k_f} \qquad (2.32)$$

in which h_x is the local heat transfer coefficient $q''/(T_w - T_b)$. In conclusion, a significant heat transfer augmentation effect can be achieved by using a high-conductivity matrix material, so that k is considerably greater than k_f. This effect is accompanied by a significant increase in fluid flow resistance.

Key results for forced convection in porous media have been developed based on constructal theory for tree networks of cracks (Bejan, 2000), time-dependent heating, annular channels, stepwise changes in wall temperature, local thermal nonequilibrium, and other external flows (such as over a cone or wedge). The concepts of heatfunctions and heatlines were introduced for the purpose of visualizing the true path of the flow of energy through a convective medium (Bejan, 1984, 1995a). The heatfunction accounts simultaneously for the transfer of heat by conduction and convection at every point in the medium. The heatlines are a generalization of the flux lines used routinely in the field of conduction. The concept of heatfunction is a spatial generalization of the concept of the Nusselt number, that is, a way of indicating the magnitude of the heat transfer rate through any unit surface drawn through any point on the convective medium. The heatline method was extended to several configurations of convection through fluid saturated porous media (Morega and Bejan, 1994).

2.5 Scale Analysis of Natural Convection Boundary Layers

We return to the method of scale analysis, and consider the natural convection boundary layer near a vertical impermeable wall embedded in a saturated porous medium at a different temperature. The boundary conditions are indicated in Figure 2.3, where the gravitational acceleration points in the negative y-direction. The equations for mass conservation and Darcy flow,

$$\frac{\partial u}{\partial x} + \frac{\partial v}{\partial y} = 0, \tag{2.33}$$

$$u = -\frac{K}{\mu}\frac{\partial P}{\partial x}, \qquad v = -\frac{K}{\mu}\left(\frac{\partial P}{\partial y} + \rho g\right), \tag{2.34}$$

can be rewritten as a single equation

$$\frac{\partial^2 \psi}{\partial x^2} + \frac{\partial^2 \psi}{\partial y^2} = -\frac{Kg\beta}{\nu}\frac{\partial T}{\partial x}, \tag{2.35}$$

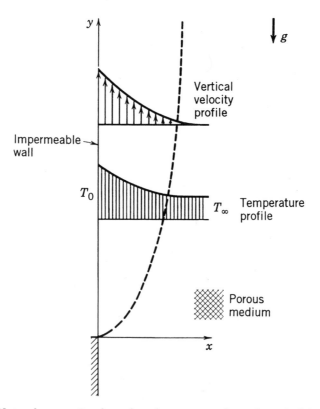

Fig. 2.3. Natural convection boundary layer near a heated vertical impermeable wall.

where $\psi(x, y)$ is the streamfunction for volume-averaged flow, $u = \partial\psi/\partial y$ and $v = -\partial\psi/\partial x$. In Equation (2.35) we used the Boussinesq approximation $\rho = \rho_0[1 - \beta(T - T_0)]$, which effects the coupling of the velocity field to the temperature field. The energy conservation equation for boundary layer flow is given by [cf. Equation (2.5)]

$$\frac{\partial\psi}{\partial y}\frac{\partial T}{\partial x} - \frac{\partial\psi}{\partial x}\frac{\partial T}{\partial y} = \alpha\frac{\partial^2 T}{\partial x^2}. \tag{2.36}$$

The problem is to determine the heat transfer rate between the wall and the medium, $q'' \sim k\Delta T/\delta_T$, where $\Delta T = T_0 - T_\infty$, and δ_T is the thickness of the boundary layer region ($\delta_T \ll y$). By invoking the temperature boundary conditions sketched in Figure 2.3, and using the rules of scale analysis outlined in Section 2.2, we find that the terms of Equations (2.35) and (2.36) are represented by the following scales,

$$\frac{\psi}{\delta_T^2}, \quad \frac{\psi}{y^2} \sim \frac{Kg\beta\Delta T}{\nu\delta_T}, \tag{2.37}$$

$$\frac{\psi\Delta T}{y\delta_T}, \quad \frac{\psi\Delta T}{\delta_T y} \sim \alpha\frac{\Delta T}{\delta_T^2}. \tag{2.38}$$

On the left side of Equation (2.37) we retain the first scale, because $\psi/\delta_T^2 > \psi/y^2$. On the left side of Equation (2.38), the two scales are represented by the same order of magnitude, $\psi\Delta T/(y\delta_T)$. Together, Equations (2.37) and (2.38) are sufficient for determining the two unknown scales, δ_T and ψ,

$$\frac{\delta_T}{y} \sim \mathrm{Ra}_y^{-1/2}, \quad \psi \sim \alpha\mathrm{Ra}_y^{1/2}, \tag{2.39}$$

where Ra_y is the Darcy-modified Rayleigh number, $\mathrm{Ra}_y = Kg\beta y\,\Delta T/(\alpha\nu)$. From the ψ scale we conclude that the vertical velocity scale is $v \sim \psi/\delta_T$, or $v \sim (\alpha/y)\mathrm{Ra}_y \sim Kg\beta\,\Delta T/\nu$. From the δ_T scale we deduce the heat flux, or the Nusselt number,

$$\mathrm{Nu}_y = \frac{q''y}{k\Delta T} \sim \mathrm{Ra}_y^{1/2}. \tag{2.40}$$

This scale agrees with the similarity solution to the problem of the boundary layer along an isothermal wall of temperature T_0 (Cheng and Minkowycz, 1977),

$$\mathrm{Nu}_y = \frac{q''y}{(T_0 - T_\infty)k} = 0.444\mathrm{Ra}_y^{1/2}, \tag{2.41}$$

$$\overline{\mathrm{Nu}} = \frac{\overline{q''}H}{(T_0 - T_\infty)k} = 0.888\mathrm{Ra}_H^{1/2}. \tag{2.42}$$

If the wall has uniform heat flux, then the local wall temperature and the boundary layer thickness must vary such that

$$q'' \sim k \frac{T_0(y) - T_\infty}{\delta_T} = \text{constant.} \qquad (2.43)$$

Combining this with the δ_T scale (2.39), we conclude that

$$\frac{\delta_T}{y} \sim \text{Ra}_{*y}^{-1/3}, \qquad (2.44)$$

where Ra_{*y} is the Darcy-modified Rayleigh number based on heat flux, $\text{Ra}_{*y} = Kg\beta y^2 q''/(\alpha\nu k)$. The local heat transfer rate must therefore scale as

$$\text{Nu}_y = \frac{q''}{T_0(y) - T_\infty} \frac{y}{k} \sim \text{Ra}_{*y}^{1/3}. \qquad (2.45)$$

The numerical solution to the similarity for formulation of this problem is given by (Bejan, 1995a)

$$\text{Nu}_y = \frac{q''}{T_0(y) - T_\infty} \frac{y}{k} = 0.772 \text{Ra}_{*y}^{1/3}, \qquad (2.46)$$

$$\overline{\text{Nu}} = \frac{q''}{\bar{T}_0 - T_\infty} \frac{H}{k} = 1.03 \text{Ra}_{*H}^{1/3}, \qquad (2.47)$$

where H is the wall height. In conclusion, Equations (2.45) to (2.47) show that the exact results are anticipated within 23% by the results of scale analysis. More examples of boundary layer natural convection in porous media are reviewed in the next section.

2.6 Thermal Stratification and Vertical Partitions

When the porous medium of Figure 2.3 is finite in the x- and y-directions, the discharge of the heated vertical stream into the rest of the medium leads in time to thermal stratification. This problem was considered in Bejan (1984). As shown in Figure 2.4, the original wall excess temperature is $T_0 - T_{\infty,0}$, and the porous medium is stratified according to the positive vertical gradient $\gamma = dT_\infty/dy$. The local temperature difference $T_0 - T_\infty(y)$ decreases as y increases, which is why a monotonic decrease in the total heat transfer rate as γ increases should be expected. This trend is confirmed by the right-hand part of the figure, which shows the integral solution developed for this configuration. The overall Nusselt number, Rayleigh number, and stratification parameter are defined as

$$\overline{\text{Nu}}_H = \frac{\overline{q''}H}{k(T_0 - T_{\infty,0})}, \qquad \text{Ra}_H = \frac{Kg\beta H}{\alpha\nu}(T_0 - T_{\infty,0}), \qquad b = \frac{\gamma H}{T_0 - T_{\infty,0}}.$$

$$\qquad (2.48)$$

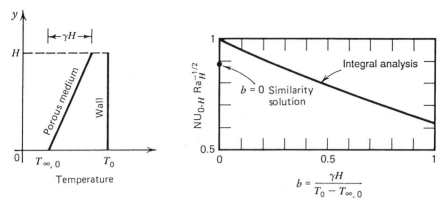

Fig. 2.4. Heat transfer solution for a vertical isothermal wall facing a linearly stratified porous medium saturated with fluid (Bejan, 1984; from Bejan, *Convection Heat Transfer*, 2nd ed., Copyright © 1995 Wiley. This material is used by permission of John Wiley & Sons, Inc.).

The accuracy of this integral solution can be assessed by comparing its $b = 0$ limit

$$\frac{\overline{\mathrm{Nu}_H}}{\mathrm{Ra}_H^{1/2}} = 1 \qquad (b = 0) \tag{2.49}$$

with the similarity solution for an isothermal wall adjacent to an isothermal porous medium, Equation (2.42). The integral solution overestimates the global heat transfer rate by only 12.6% (Figure 2.4).

The breakdown of the Darcy flow model in vertical boundary layer natural convection was the subject of several studies (Plumb and Huenefeld, 1981; Bejan and Poulikakos, 1984; Nield and Joseph, 1985). Assuming the Dupuit–Forchheimer modification of the Darcy flow model, at local pore Reynolds numbers greater than 10, the local Nusselt number for the vertical wall configuration of Figure 2.3 approaches the following limits (Bejan and Poulikakos, 1984),

$$\mathrm{Nu}_y = 0.494\mathrm{Ra}_{\infty,y}^{1/4} \quad \text{for the isothermal wall,} \tag{2.50}$$

$$\mathrm{Nu}_y = 0.804\mathrm{Ra}_{\infty,y}^{*1/5} \quad \text{for the constant heat flux wall,} \tag{2.51}$$

where $\mathrm{Ra}_{\infty,y} = g\beta y^2(T_w - T_\infty)/(b\alpha^2)$ and $\mathrm{Ra}_{\infty,y}^* = g\beta y^3 q''/(kb\alpha^2)$, and $b[m^{-1}]$ is defined in Equation (1.12). Equations (2.50) and (2.51) are valid provided $G \ll 1$, where $G = (\nu/K)[bg\beta(T_w - T_\infty)]^{-1/2}$. In the intermediate range between the Darcy limit and the inertia-dominated limit (or form drag limit), that is, in the range where G is of order one, numerical results for a vertical isothermal wall (Bejan and Poulikakos, 1984) are correlated within 2% by the closed-form expression (Bejan, 1987)

$$\mathrm{Nu}_y = [(0.494)^n + (0.444G^{-1/2})^n]^{1/n}\mathrm{Ra}_{\infty,y}^{1/4}, \tag{2.52}$$

where $n = -3$. The heat transfer results summarized in this section also apply to configurations where the vertical wall is inclined (slightly) to the vertical. In such cases, the gravitational acceleration that appears in the definition of the Rayleigh-type numbers in this section must be replaced by the gravitational acceleration component that acts along the nearly vertical wall.

When a vertical wall divides two porous media, and a temperature difference exists between the two systems, there is a pair of conjugate boundary layers, one on each side of the wall, with neither the temperature nor the heat flux specified on the wall but rather to be found as part of the solution to the problem (Figure 2.5a) (Bejan and Anderson, 1981). The overall Nusselt number results for this configuration are correlated within 1% by the expression

$$\overline{\mathrm{Nu}}_H = 0.382(1 + 0.615\omega)^{-0.875}\mathrm{Ra}_H^{1/2}, \tag{2.53}$$

where $\overline{\mathrm{Nu}}_H = \overline{q''}H/[(T_{\infty,H} - T_{\infty,L})k]$, and where $\overline{q''}$ is the heat flux averaged over the entire height H. In addition, $\mathrm{Ra}_H = Kg\beta H(T_{\infty,H} - T_{\infty,L})/(\alpha\nu)$.

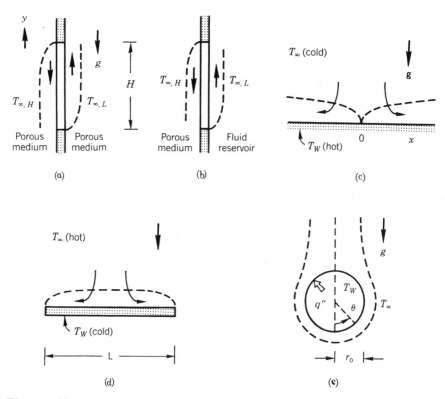

Fig. 2.5. Natural convection boundary layers in porous media: (a) vertical partition with porous media on both sides; (b) vertical wall separating a porous medium and a fluid reservoir; (c) hot surface facing upward; (d) cold surface facing downward; (e) sphere or cylinder embedded in a porous medium.

The wall thickness parameter ω is

$$\omega = \frac{W}{H}\frac{k}{k_w}\mathrm{Ra}_H^{1/2}, \qquad (2.54)$$

where W and H are the width and height of the wall cross-section, k and k_w are the conductivities of the porous medium and wall material, respectively, and Ra_H is the Rayleigh number based on H and the temperature difference between the two systems. The progress on conjugate boundary layers was reviewed by Kimura $et\ al.$ (1997).

In thermal insulation and architectural applications, the porous media on both sides of the vertical partition of Figure 2.5a may be thermally stratified. If the stratification on both sides is the same and linear (e.g., Figure 2.4), so that the vertical temperature gradient far enough from the wall is $dT/dy = b_1(T_{\infty,H} - T_{\infty,L})/H$, where b_1 is a constant, and if the partition is thin enough so that $\omega \approx 0$, then it is found that the overall Nusselt number increases substantially with the degree of stratification (Bejan and Anderson, 1983). In the range $0 < b_1 < 1.5$, these findings are summarized by the correlation

$$\overline{\mathrm{Nu}}_H = 0.382(1 + 0.662b_1 - 0.073b_1^2)\mathrm{Ra}_H^{1/2}. \qquad (2.55)$$

Another configuration of engineering interest is sketched in Figure 2.5b: a vertical impermeable surface separates a porous medium of temperature $T_{\infty,H}$ from a fluid reservoir of temperature $T_{\infty,L}$ (Bejan and Anderson, 1983). When both sides of the interface are lined by boundary layers, the overall Nusselt number is

$$\overline{\mathrm{Nu}}_H = [(0.638)^{-1} + (0.888B)^{-1}]^{-1}\mathrm{Ra}_{H,f}^{1/4}, \qquad (2.56)$$

where $\overline{\mathrm{Nu}}_H = \overline{q''}H/[(T_{\infty,H} - T_{\infty,L})k]$ and $B = k\mathrm{Ra}_H^{1/2}/(k_f\mathrm{Ra}_{H,f}^{1/4})$. The parameter k_f is the fluid-side thermal conductivity, and the fluid-side Rayleigh number $\mathrm{Ra}_{H,f} = g(\beta/\alpha\nu)_f H^3(T_{\infty,H} - T_{\infty,L})$. Equation (2.56) is valid in the regime where both boundary layers are distinct, $\mathrm{Ra}_H^{1/2} \gg 1$ and $\mathrm{Ra}_{H,f}^{1/4} \gg 1$. It is also assumed that the fluid on the right side of the partition in Figure 2.5b has a Prandtl number of order 1 or greater. For a numerical model and simulation of a porous-fluid interface, see Costa $et\ al.$ (2004). Additional solutions for boundary layer convection in the vicinity of vertical partitions in porous media are reviewed in Nield and Bejan (1999). The heat flow near the interface between a porous medium with natural convection and a conducting solid was illustrated with $heatlines$ by Costa (2003).

2.7 Horizontal Walls with Natural Convection

With reference to Figure 2.5c, boundary layers form in the vicinity of a heated horizontal surface that faces upward (Cheng and Chang, 1976). Measuring x

horizontally away from the vertical plane of symmetry of the flow, the local Nusselt number for an isothermal wall is

$$\mathrm{Nu}_x = 0.42\mathrm{Ra}_x^{1/3}, \tag{2.57}$$

where $\mathrm{Nu}_x = q''x/[k(T_w - T_\infty)]$ and $\mathrm{Ra}_x = Kg\beta x(T_w - T_\infty)/(\alpha\nu)$. The local Nusselt number for a horizontal wall heated with uniform flux is

$$\mathrm{Nu}_x = 0.859\mathrm{Ra}_x^{*1/4}, \tag{2.58}$$

where $\mathrm{Ra}_x^* = Kg\beta x^2 q''/(k\alpha\nu)$. Equations (2.57) and (2.58) are valid in the boundary layer regime, $\mathrm{Ra}_x^{1/3} \gg 1$ and $\mathrm{Ra}_x^{*1/4} \gg 1$, respectively. They also apply to porous media bounded from above by a cold surface; this new configuration is obtained by rotating Figure 2.5c by 180°.

The upward-facing cold plate of Figure 2.5d was studied in Kimura *et al.* (1985). The overall Nusselt number in this configuration is

$$\mathrm{Nu} = 1.47\mathrm{Ra}_L^{1/3}, \tag{2.59}$$

where $\mathrm{Nu} = q'/[k(T_\infty - T_w)]$ and $\mathrm{Ra}_L = Kg\beta L(T_\infty - T_w)/(\alpha\nu)$, and where q' is the overall heat transfer rate through the upward-facing cold plate of length L. Equation (2.59) holds if $\mathrm{Ra}_L \gg 1$, and applies equally to hot horizontal plates facing downward in an isothermal porous medium. Note the exponent $1/3$, which is in contrast to the exponent $1/2$ for the vertical wall in Equation (2.42).

2.8 Sphere and Horizontal Cylinder with Natural Convection

With reference to the coordinate system shown in the circular cross-section sketched in Figure 2.5e, the local Nusselt numbers for boundary layer convection around an impermeable sphere or a horizontal cylinder embedded in an infinite porous medium are, in order (Cheng, 1982),

$$\mathrm{Nu}_\theta = 0.444\mathrm{Ra}_\theta^{1/2}\left(\frac{3}{2}\theta\right)^{1/2}\sin^2\theta\left(\frac{1}{3}\cos^3\theta - \cos\theta + \frac{2}{3}\right)^{-1/2}, \tag{2.60}$$

$$\mathrm{Nu}_\theta = 0.444\mathrm{Ra}_\theta^{1/2}(2\theta)^{1/2}\sin\theta(1 - \cos\theta)^{-1/2}, \tag{2.61}$$

where $\mathrm{Nu}_\theta = q''r_0\theta/[k(T_w - T_\infty)]$ and $\mathrm{Ra}_\theta = Kg\beta\theta r_0(T_w - T_\infty)/(\alpha\nu)$. These steady-state results are valid provided the boundary layer region is slender enough, that is, if $\mathrm{Nu}_\theta \gg 1$. The overall Nusselt numbers for the sphere and horizontal cylinder are, respectively (Nield and Bejan, 1999),

$$\overline{\mathrm{Nu}}_D = 0.362\mathrm{Ra}_D^{1/2}, \tag{2.62}$$

$$\overline{\mathrm{Nu}}_D = 0.565\mathrm{Ra}_D^{1/2}, \tag{2.63}$$

where $\overline{\mathrm{Nu}}_D = \overline{q''}D/[k(T_w - T_\infty)]$ and $\mathrm{Ra}_D = Kg\beta D(T_w - T_\infty)/(\alpha\nu)$. Solutions for convection at low and intermediate Rayleigh numbers are summarized in Nield and Bejan (1999).

In the review conducted in this chapter we considered separately the fundamentals of forced convection and natural convection. There are many practical situations in which these two flow mechanisms occur together. The resulting class of flows is called mixed convection, and is reviewed in Nield and Bejan (1999). A recent study of mixed convection was performed by Magyari et al. (2001). There are many other effects that complicate the modeling of convection in porous media. The effect of a magnetic field and heat generation was considered by Chamkha and Quadri (2001). The modeling of coupled heat and mass transfer was discussed by Mendes et al. (2002). Dissipation effects were discussed by Nield (2000).

2.9 Enclosures Heated from the Side

The most basic configuration of a porous layer heated in the horizontal direction is sketched in Figure 2.6. In Darcy flow, the heat and fluid flow driven by buoyancy depend on two parameters: the geometric aspect ratio H/L, and the Rayleigh number based on height, $\mathrm{Ra}_H = Kg\beta H(T_h - T_c)/(\alpha\nu)$. There exist four heat transfer regimes (Bejan, 1984, 1995a), that is, four ways to calculate the overall heat transfer rate $q' = \int_0^H q'' dy$. These are summarized in Figure 2.6:

Regime I.   The pure conduction regime, defined by $\mathrm{Ra}_H \ll 1$. In this regime, q' is approximately equal to the pure conduction estimate $kH(T_h - T_c)/L$.

Regime II.   The conduction-dominated regime in tall layers, defined by $H/L \gg 1$ and $(L/H)\mathrm{Ra}_H^{1/2} \ll 1$. In this regime, the heat transfer rate scales as $q' \geq kH(T_h - T_c)/L$.

Regime III.   The convection-dominated regime (or high Rayleigh number regime), defined by $\mathrm{Ra}_H^{-1/2} < H/L < \mathrm{Ra}_H^{1/2}$. In this regime, q' scales as $k(T_h - T_c)\mathrm{Ra}_H^{1/2}$.

Regime IV.   The convection-dominated regime in shallow layers, defined by $H/L \ll 1$ and $(H/L)\mathrm{Ra}_H^{1/2} \ll 1$. Here the heat transfer rate scales as $q \leq k(T_h - T_c)\mathrm{Ra}_H^{1/2}$.

Considerable analytical, numerical, and experimental work has been done to estimate more accurately the overall heat transfer rate q' or the overall Nusselt number,

$$\mathrm{Nu} = \frac{q'}{kH(T_h - T_c)/L}. \tag{2.64}$$

Note that unlike the single-wall configuration of Section 2.5, in confined layers of thickness L the Nusselt number is defined as the ratio of the actual heat transfer rate to the pure conduction heat transfer rate. An analytical solution

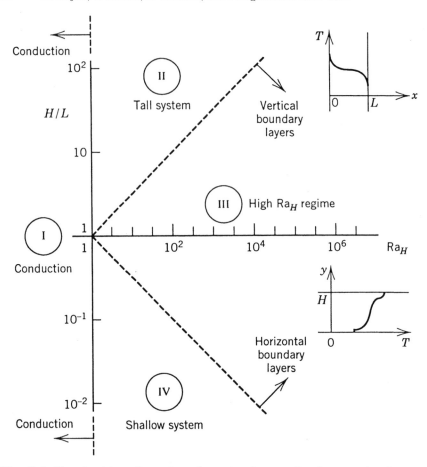

Fig. 2.6. Four heat transfer regimes for natural convection in an enclosed rectangular porous layer heated from the side (Bejan, 1984; from Bejan, *Convection Heat Transfer*, 2nd ed., Copyright © 1995 Wiley. This material is used by permission of John Wiley & Sons, Inc.).

that covers the four heat transfer regimes smoothly is (Bejan and Tien, 1978)

$$\mathrm{Nu} = K_1 + \frac{1}{120} K_1^3 \left(\mathrm{Ra}_H \frac{H}{L} \right)^2 , \qquad (2.65)$$

where $K_1(H/L, \mathrm{Ra}_H)$ is obtained by solving the system

$$\frac{1}{120} \delta_e \mathrm{Ra}_H^2 K_1^3 \left(\frac{H}{L} \right)^3 = 1 - K_1 = \frac{1}{2} K_1 \frac{H}{L} \left(\frac{1}{\delta_e} - \delta_e \right). \qquad (2.66)$$

This result is displayed in chart form in Figure 2.7 along with numerical results from Hickox and Gartling (1981). The asymptotic values of this solution are

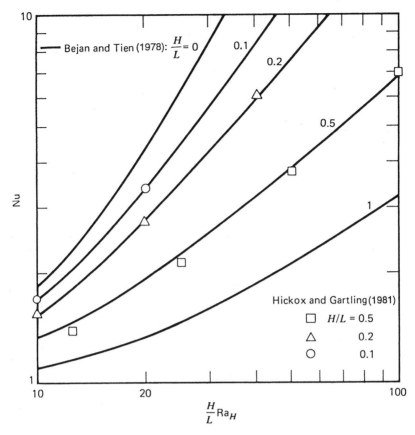

Fig. 2.7. The total heat transfer rate through an enclosed porous layer heated from the side (Bejan and Tien, 1978).

$$\text{Nu} \sim 0.508 \frac{L}{H} \text{Ra}_H^{1/2} \quad \text{as } \text{Ra}_H \to \infty \tag{2.67}$$

$$\text{Nu} \sim 1 + \frac{1}{120} \left(\text{Ra}_H \frac{H}{L} \right)^2 \quad \text{as } \frac{H}{L} \to 0. \tag{2.68}$$

The heat transfer in the convection-dominated Regime III is well represented by Equation (2.67) or by alternate solutions developed solely for Regime III, for example (Weber, 1975), $\text{Nu} = 0.577(L/H)\text{Ra}_H^{1/2}$. More refined estimates for Regime III were developed in Bejan (1979a) and Simpkins and Blythe (1980), where the proportionality factor between Nu and $(L/H)\text{Ra}_H^{1/2}$ is replaced by a function of both H/L and Ra_H; see Figure 2.8. For expedient engineering calculations of heat transfer dominated by convection, Figure 2.7 is recommended for shallow layers ($H/L < 1$), and Figure 2.8 for square and tall layers ($H/L \gtrsim 1$) in the boundary layer regime, $\text{Ra}_H^{-1/2} < H/L < \text{Ra}_H^{1/2}$.

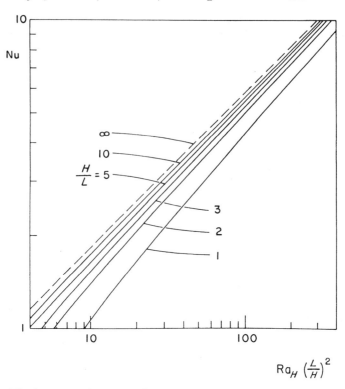

Fig. 2.8. The heat transfer rate in Regime III through a porous layer heated from the side (Bejan, 1979a).

In the field of thermal insulation engineering, a more appropriate model for heat transfer in the configuration of Figure 2.6 is the case where the heat flux q'' is distributed uniformly along the two vertical sides of the porous layer. In the high Rayleigh number regime (Regime III), the overall heat transfer rate is given by (Bejan, 1983a)

$$\mathrm{Nu} = \frac{1}{2}\left(\frac{L}{H}\right)^{4/5}\mathrm{Ra}_H^{*2/5}, \tag{2.69}$$

where $\mathrm{Ra}_H^* = Kg\beta H^2 q''/(\alpha\nu k)$. The overall Nusselt number is defined as in Equation (2.64), where $T_h - T_c$ is the height-averaged temperature difference between the two sides of the rectangular cross-section. Equation (2.69) holds in the high Rayleigh number regime $\mathrm{Ra}_H^{*-1/3} < H/L < \mathrm{Ra}_H^{*1/3}$.

The progress reviewed so far is based on models that assume local thermal equilibrium. Natural convection without local thermal equilibrium was studied based on a two-temperature model by Rees and Pop (2000).

Impermeable partitions (flow obstructions) inserted in the confined porous medium can have a dramatic effect on the overall heat transfer rate across the

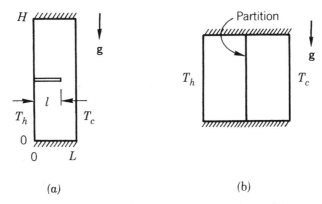

(a)   (b)

Fig. 2.9. Rectangular enclosure filled with porous medium: (a) partial horizontal partition; (b) full vertical partition.

enclosure (Bejan, 1983b). With reference to the two-dimensional geometry of Figure 2.9a, in the convection-dominated Regimes III and IV the overall heat transfer rate decreases steadily as the length l of the horizontal partition approaches L, that is, as the partition divides the porous layer into two shorter layers. The horizontal partition has practically no effect in Regimes I and II where the overall heat transfer rate is dominated by conduction. If the partition is oriented vertically (Figure 2.9b), then in the convection-dominated regime the overall heat transfer rate is approximately 40% of what it would have been in the same porous medium without the internal partition.

The nonuniformity of permeability and thermal diffusivity can have a dominating effect on the overall heat transfer rate (Poulikakos and Bejan, 1983a). In cases where the properties vary so that the porous layer can be modeled as a sandwich of vertical sublayers of different permeability and diffusivity (Figure 2.10a), an important parameter is the ratio of the peripheral sublayer thickness (d_1) to the thermal boundary-layer thickness $(\delta_{T,1})$ based on the

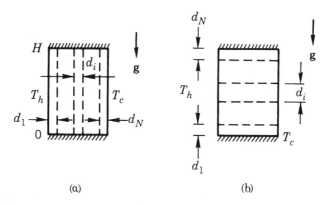

(a)   (b)

Fig. 2.10. Rectangular enclosure filled with several porous layers: (a) vertical layers; (b) horizontal layers.

properties of the d_1 sublayer (note that $\delta_{T,1}$ scales as $H\,\mathrm{Ra}_{H,1}^{-1/2}$, where the Rayleigh number $\mathrm{Ra}_{H,1} = K_1 g\beta H(T_h - T_c)/(\alpha_1\nu)$ and where the subscript l represents the properties of the d_1 sublayer). If $d_1 > \delta_{T,1}$, then the heat transfer through the left side of the porous system of Figure 2.10a is impeded by a thermal resistance of order $\delta_{T,1}/(k_1 H)$. If the sublayer situated next to the right wall (d_N) has exactly the same properties as the d_1 sublayer, and if $\delta_{T,1} < (d_1, d_N)$, then the overall heat transfer rate in the convection-dominated regime can be estimated using Equation (2.67) in which both Nu and Ra_H are based on the properties of the peripheral layers.

When the porous-medium inhomogeneity may be modeled as a sandwich of N horizontal sublayers (Figure 2.10b), the scale of the overall Nusselt number in the convection-dominated regime can be evaluated as (Poulikakos and Bejan, 1983a)

$$\mathrm{Nu} \sim 2^{-3/2}\mathrm{Ra}_{H,1}^{1/2}\frac{L}{H}\sum_{i=1}^{N}\frac{k_i}{k_1}\left(\frac{K_i d_i \alpha_1}{K_1 d_1 \alpha_i}\right)^{1/2}, \qquad (2.70)$$

where both Nu and $\mathrm{Ra}_{H,1}$ are based on the properties of the d_1 sublayer (Figure 2.10b). The correlation of Equation (2.70) was tested via numerical experiments in two-layer systems.

The convection in a porous medium confined in a horizontal cylinder with disk-shaped ends at different temperatures (Figure 2.11a) has features similar to the configuration of Figure 2.6. A parametric solution for the horizontal cylinder problem is reported in Bejan and Tien (1978). The corresponding phenomenon in a porous medium in the shape of a horizontal cylinder with annular cross-section (Figure 2.11b) is documented in Bejan and Tien (1979).

An important geometric configuration in thermal insulation engineering is the horizontal cylindrical annular space filled with fibrous or granular insulation (Figure 2.11c). In this configuration the heat transfer is oriented radially between the concentric cylindrical surfaces of radii r_i and r_o. Experimental

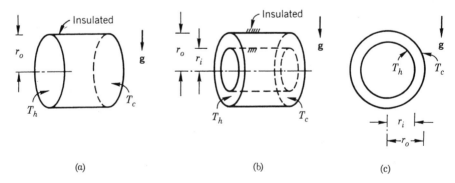

(a)   (b)   (c)

Fig. 2.11. Enclosures filled with porous media: (a) horizontal cylindrical enclosure; (b) horizontal cylindrical annulus with axial heat flow; (c) horizontal cylindrical annular enclosure, or spherical annulus, with radial heat flow.

measurements and numerical solutions for the overall heat transfer in the horizontal cylindrical annulus were reported in Caltagirone (1976) and Burns and Tien (1979). These results were correlated based on scale analysis in the range $1.19 \leq r_o/r_i \leq 4$ (Bejan, 1987),

$$\mathrm{Nu} = \frac{q'_{\mathrm{actual}}}{q'_{\mathrm{conduction}}} \approx 0.44 \mathrm{Ra}_{r_i}^{1/2} \frac{\ln{(r_o/r_i)}}{1 + 0.916(r_i/r_o)^{1/2}}, \tag{2.71}$$

where $\mathrm{Ra}_{r_i} = Kg\beta r_i(T_h - T_c)/(\alpha\nu)$ and $q'_{\mathrm{conduction}} = 2\pi k(T_h - T_c)/\ln(r_o/r_i)$. This correlation is valid in the convection-dominated limit, $\mathrm{Nu} \gg 1$.

Porous media confined to the space formed between two concentric spheres are also an important component in thermal insulation engineering. Figure 2.11c can be interpreted as a vertical cross-section through the concentric-sphere arrangement. Numerical heat transfer solutions for discrete values of Rayleigh number and radius ratio are reported graphically in Burns and Tien (1979). Using the method of scale analysis, the data that correspond to the convection-dominated regime ($\mathrm{Nu} \gtrsim 1.5$) have been correlated within 2% by the scaling-correct expression (Bejan, 1987)

$$\mathrm{Nu} = \frac{q_{\mathrm{actual}}}{q_{\mathrm{conduction}}} = 0.756 \mathrm{Ra}_{r_i}^{1/2} \frac{1 - r_i/r_o}{1 + 1.422(r_i/r_o)^{3/2}}, \tag{2.72}$$

where $\mathrm{Ra}_{r_i} = Kg\beta r_i(T_h - T_c)/(\alpha\nu)$ and $q_{\mathrm{conduction}} = 4\pi k(T_h - T_c)/(r_i^{-1} - r_o^{-1})$.

Natural convection through an annular porous insulation oriented vertically was investigated numerically (Havstad and Burns, 1982) and experimentally (Prasad et al., 1985). These and other results are reviewed in Nield and Bejan (1999).

Natural convection in enclosures with heating from the side continues to attract interest. Heat and mass transfer (double diffusive convection) was studied by Mohamad and Bennacer (2002), Bera and Khalili (2002), Asbik et al. (2002), Benhadji and Vasseur (2001), Kalla et al. (2001), Bansod et al. (2000, 2002), and Rathish Kumar et al. (2002). An enclosure with a vertical fluid layer sandwiched between two porous layers was studied numerically by Bennacer et al. (2003). Enclosed porous media with heat generation were analyzed in Dhanasekaran et al. (2002) and Kim et al. (2001). The effect of variable porosity was documented in Marcondes et al. (2001). Natural convection in a partly porous cavity was described by Mercier et al. (2002). The effect of anisotropy was studied experimentally by Kimura et al. (2002).

2.10 Enclosures Heated from Below

The most basic configuration of a confined porous layer heated in the vertical direction is shown in Figure 2.12a. The most important difference between

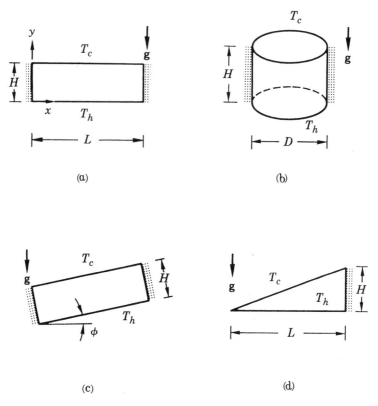

Fig. 2.12. Enclosed porous layers heated from below: (a) rectangular enclosure; (b) vertical cylindrical enclosure; (c) inclined rectangular enclosure; (d) wedge-shaped enclosure.

heat transfer in this configuration and heat transfer in confined layers heated from the side is that in Figure 2.12a convection occurs only when the imposed temperature difference or heating rate exceeds a certain finite value. Recall that in configurations such as Figure 2.6 convection is present even in the limit of vanishingly small temperature differences.

Assume that the fluid saturating the porous medium of Figure 2.12a expands upon heating ($\beta > 0$). By analogy with the phenomenon of Bénard convection in a pure fluid, in the convection regime the flow consists of finite-size cells that become more slender and multiply discretely as the destabilizing temperature difference $T_h - T_c$ increases. If $T_h - T_c$ does not exceed the critical value necessary for the onset of convection, the heat transfer mechanism through the layer of thickness H is that of pure thermal conduction. If the porous layer is heated from above (i.e., if T_h and T_c change places in Figure 2.12a), then the fluid remains stably stratified and the heat transfer is again due to pure thermal conduction: $q' = kL(T_h - T_c)/H$.

The onset of convection in an infinitely long porous layer heated from below was examined on the basis of linearized hydrodynamic stability analysis (Horton and Rogers, 1945; Lapwood, 1948; Nield and Bejan, 1999; Tyvand, 2002; Mamou, 2002). This subject continues to attract attention at the most fundamental level [e.g., Bilgen and Mbaye (2001) Rees (2002), and Banu and Rees (2002)]. For fluid layers confined between impermeable and isothermal horizontal walls, it was found that convection is possible if the Rayleigh number based on height exceeds the critical value

$$\mathrm{Ra}_H = \frac{Kg\beta H(T_h - T_c)}{\alpha\nu} = 4\pi^2 = 39.48. \tag{2.73}$$

A much simpler analysis based on constructal theory (Nelson and Bejan, 1998) predicted the critical Rayleigh number $12\pi = 37.70$, which approaches the hydrodynamic stability result within 5%. This analysis is summarized in Section 2.11. For a history of the early theoretical and experimental work on the onset of Bénard convection in porous media, and for a rigorous generalization of the stability analysis to convection driven by combined buoyancy effects, the reader is directed to Nield (1968), where it is shown that the critical Rayleigh number for the onset of convection in infinitely shallow layers depends to a certain extent on the heat and fluid flow conditions imposed along the two horizontal boundaries.

Of practical interest in heat transfer engineering is the heat transfer rate at Rayleigh numbers that are higher than critical. There has been a considerable amount of analytical, numerical, and experimental work devoted to this issue. Reviews of these advances may be found in Nield and Bejan (1999) and Cheng (1978). Constructal theory anticipates the entire curve relating heat transfer to Rayleigh number (Nelson and Bejan, 1998).

The scale analysis of the convection regime with Darcy flow (Bejan, 1984) indicates that the Nusselt number should increase linearly with the Rayleigh number, whence the relationship

$$\mathrm{Nu} \approx \frac{1}{40}\mathrm{Ra}_H \quad \text{for } \mathrm{Ra}_H > 40. \tag{2.74}$$

This linear relationship is confirmed by numerical heat transfer calculations at large Rayleigh numbers in Darcy flow (Kimura et al., 1986). The experimental data compiled in Cheng (1978) show that the scaling law (2.74) serves as an upper bound for some of the high-Ra_H experimental data available in the literature.

Most of the data show that in the convection regime Nu increases as Ra_H^n, where n becomes progressively smaller than 1 as Ra_H increases. This behavior is anticipated by the constructal theory solution (Nelson and Bejan, 1998); see Section 2.11. The exponent $n \sim 1/2$ revealed by data at high Rayleigh numbers was anticipated based on a scale analysis of convection rolls in the

Forchheimer regime (Bejan, 1995a):

$$\frac{\text{Nu}}{\text{Pr}_p} \sim \left(\frac{\text{Ra}_H}{\text{Pr}_p}\right)^{1/2} \quad (\text{Ra}_H > \text{Pr}_p), \tag{2.75}$$

where Pr_p is the "porous medium Prandtl number" for the Forchheimer regime (Bejan, 1995a),

$$\text{Pr}_p = \frac{H\nu}{bK\alpha} \tag{2.76}$$

and $b[m^{-1}]$ is defined in Equation (1.12). In this formulation Nu is a function of two groups, Ra_H and Pr_p, in which Pr_p accounts for the transition from Darcy to Forchheimer flow (Figure 2.13). In this formulation the Darcy flow result (2.74) becomes

$$\frac{\text{Nu}}{\text{Pr}_p} \sim \frac{1}{40}\frac{\text{Ra}_H}{\text{Pr}_p} \quad (40 < \text{Ra}_H < \text{Pr}_p). \tag{2.77}$$

The experimental data for convection in the entire regime spanned by the asymptotes given by Equations (2.75) and (2.77) are correlated by

$$\text{Nu} = \left\{\left(\frac{\text{Ra}_H}{40}\right)^n + [c(\text{Ra}_H\text{Pr}_p)^{1/2}]^n\right\}^{1/n}, \tag{2.78}$$

where $n = -1.65$ and $c = 1896$ are determined empirically based on measurements reported by many independent sources. The correlations of Equations (2.74) through (2.78) refer to layers with length/height ratios considerably greater than one. They apply when the length (lateral dimension L, perpendicular to gravity) of the confined system is greater than the horizontal length scale of a single convective cell, that is, greater than $H\text{Ra}_H^{-1/2}$, according to the scale analysis of Bejan (1984).

These principles become partial effects in real-life systems that demand considerably more complex models. One important class for thermal and structural engineering are the cavernous walls and multiscale regenerators built with terra-cotta bricks. The terra-cotta material is porous, and moisture and heat diffuse together across it (Vasile et al., 1998). The caverns can be designed to have many shapes, and they play host to a combination of natural convection and radiation (Lorente et al., 1994, 1996). Furthermore, the thermal performance of the cavernous structure is in competition with the mechanical stiffness: from this competition emerges the optimal size and number of caverns (Lorente and Bejan, 2002). This subject is treated in some detail in Section 8.1.

Natural-convection studies have also been reported for porous layers confined in rectangular parallelepipeds heated from below, horizontal circular cylinders, and horizontal annular cylinders. The general conclusion is that the lateral walls have a convection-suppression effect. For example, in a circular

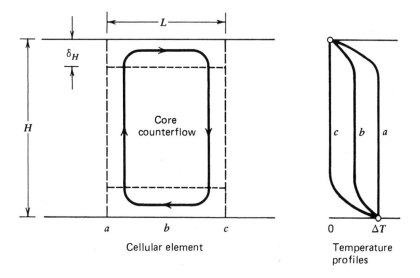

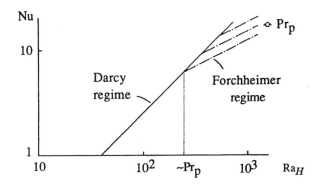

Fig. 2.13. Convective heat transfer in a porous layer heated from below, in the Darcy and Forchheimer regimes (Bejan, 1995a; from Bejan, *Convection Heat Transfer*, 2nd ed., Copyright © 1995 Wiley. This material is used by permission of John Wiley & Sons, Inc.).

cylinder of diameter D and height H (Figure 2.12b), in the limit $D \ll H$ the critical condition for the onset of convection is (Bau and Torrance, 1982)

$$\mathrm{Ra}_H = 13.56 \left(\frac{H}{D}\right)^2 . \tag{2.79}$$

In porous layers inclined from the horizontal position at an angle ϕ (Figure 2.12c), convection sets in at Rayleigh numbers that satisfy the criterion (Combarnous and Bories, 1975)

$$\mathrm{Ra}_H > \frac{39.48}{\cos \phi}, \qquad (2.80)$$

where it is assumed that the boundaries are isothermal and impermeable. The average heat transfer rate at high Rayleigh numbers can be estimated by

$$\mathrm{Nu} = 1 + \sum_{s=1}^{\infty} k_s \left(1 - \frac{4\pi^2 s^2}{\mathrm{Ra}_H \cos \phi} \right), \qquad (2.81)$$

where $k_s = 0$ if $\mathrm{Ra}_H \cos \phi < 4\pi^2 s^2$, and $k_s = 2$ if $\mathrm{Ra}_H \cos \phi \geq 4\pi^2 s^2$.

In a porous medium confined in a wedge-shaped (or attic-shaped) space cooled from above (Figure 2.12d), convection consisting of a single counter-clockwise cell exists even in the limit $\mathrm{Ra}_H \to 0$, because in this direction the imposed heating is not purely vertical. The same observation holds for Figure 2.12c. Numerical solutions of transient high Rayleigh number convection in wedge-shaped layers show the presence of a Bénard-type instability at high enough Rayleigh numbers (Poulikakos and Bejan, 1983b). When $H/L = 0.2$, the instability occurs above $\mathrm{Ra}_H \cong 620$. It was found that this critical Rayleigh number increases as H/L increases.

Nuclear-safety issues have motivated the study of natural convection in horizontal saturated porous layers (Figure 2.12a) heated volumetrically at a rate q'''. Boundary conditions and observations regarding the onset of convection and overall Nusselt numbers are presented in Nield and Bejan (1999). It is found that convection sets in at internal Rayleigh numbers Ra_I in the range 33 to 46 (Kulacki and Freeman, 1979), where

$$\mathrm{Ra}_I = \frac{g\beta H^3 K q'''}{2k\alpha\nu_f} \qquad (2.82)$$

and the subscript f indicates properties of the fluid alone. Top and bottom surface temperature experimental measurements in the convection-dominated regime ($10^3 < \mathrm{Ra}_I < 10^4$) are adequately represented by (Buretta and Berman, 1976)

$$\frac{q''' H^2}{2k(T_h - T_c)} \approx 0.116 \mathrm{Ra}_I^{0.573}, \qquad (2.83)$$

where T_h and T_c are the resulting bottom and top temperatures when q''' is distributed throughout the layer of Figure 2.12a.

2.11 The Method of Intersecting the Asymptotes

In this section we take a closer look at the phenomenon of convection in a porous layer heated from below. Our objective is to show that most of the features of the flow can be determined based on a simple method: the intersection of asymptotes (Nelson and Bejan, 1998). This method was originally used for the optimization of spacings for compact cooling channels for electronics (Bejan, 1984), as we show in Sections 5.2 and 5.3. See also Lewins (2003).

Assume that the system of Figure 2.14 is a porous layer saturated with fluid and that, if present, the flow is two-dimensional and in the Darcy regime. The height H is fixed, and the horizontal dimensions of the layer are infinite in both directions. The fluid has nearly constant properties such that its density–temperature relation is described well by the Boussinesq linearization. The volume-averaged equations that govern the conservation of mass, momentum, and energy are

$$\frac{\partial u}{\partial x} + \frac{\partial v}{\partial y} = 0, \tag{2.84}$$

$$\frac{\partial u}{\partial y} - \frac{\partial v}{\partial x} = -\frac{Kg\beta}{\nu}\frac{\partial T}{\partial x}, \tag{2.85}$$

$$u\frac{\partial T}{\partial x} + v\frac{\partial T}{\partial y} = \alpha\left(\frac{\partial^2 T}{\partial x^2} + \frac{\partial^2 T}{\partial y^2}\right). \tag{2.86}$$

The horizontal length scale of the flow pattern $(2L_r)$, or the geometric aspect ratio of one roll, is unknown. The method consists of analyzing two extreme flow configurations—many counterflows versus few plumes—and intersecting these asymptotes for the purpose of maximizing the global thermal conductance of the flow system [cf. constructal theory, Bejan (2000)].

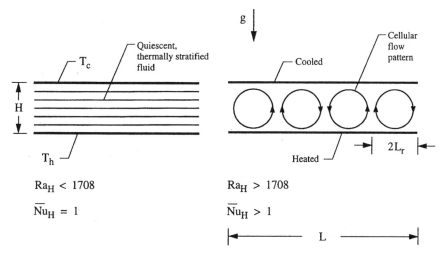

Fig. 2.14. Horizontal porous layer saturated with fluid and heated from below (Nelson and Bejan, 1998).

2.11.1 The Many Counterflows Regime

In the limit $L_r \to 0$ each roll is a very slender vertical counterflow, as shown in Figure 2.15. Because of symmetry, the outer planes of this structure ($x = \pm L_r$) are adiabatic: they represent the center planes of the streams that travel over the distance H. The scale analysis of the $H \times (2L_r)$ region indicates that in the $L_r/H \to 0$ limit the horizontal velocity component u vanishes. This scale analysis is not shown because it is well known as the defining statement of fully developed flow [e.g., Bejan (1995a, p. 97)]. Equations (2.85) and (2.86) reduce to

$$\frac{\partial v}{\partial x} = \frac{Kg\beta}{\nu}\frac{\partial T}{\partial x}, \tag{2.87}$$

$$v\frac{\partial T}{\partial y} = \alpha\frac{\partial^2 T}{\partial x^2}, \tag{2.88}$$

which can be solved exactly for v and T. The boundary conditions are $\partial T/\partial x = 0$ at $x = \pm L_r$, and the requirement that the extreme (corner) temperatures of the counterflow region are dictated by the top and bottom walls, $T(-L_r, H) = T_c$ and $T(L_r, 0) = T_h$. The solution is given by

$$v(x) = \frac{\alpha}{2H}\left[\mathrm{Ra}_H - \left(\frac{\pi H}{2L_r}\right)^2\right]\sin\left(\frac{\pi x}{2L_r}\right), \tag{2.89}$$

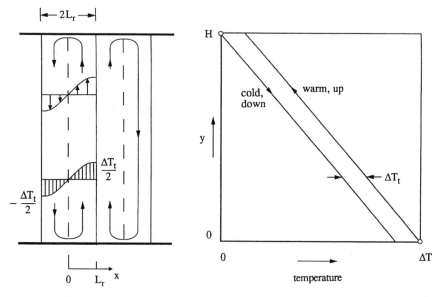

Fig. 2.15. The extreme in which the flow consists of many vertical and slender counterflows (Nelson and Bejan, 1998).

$$T(x, y) = \frac{\nu}{Kg\beta}v(x) + \frac{\nu}{Kg\beta}\left(2\frac{y}{H} - 1\right)\frac{\alpha}{2H}\left[\text{Ra}_H - \left(\frac{\pi H}{2L_r}\right)^2\right]$$

$$+(T_h - T_c)\left(1 - \frac{y}{H}\right), \tag{2.90}$$

where the porous-medium Rayleigh number $\text{Ra}_H = Kg\beta H(T_h - T_c)/(\alpha\nu)$ is a specified constant. The right side of Figure 2.15 shows the temperature distribution along the vertical boundaries of the flow region ($x = \pm L_r$): the vertical temperature gradient $\partial T/\partial y$ is independent of altitude. The transversal (horizontal) temperature difference (ΔT_t) is also a constant,

$$\Delta T_t = T(x = L_r) - T(x = -L_r) = \frac{\nu}{Kg\beta}\frac{\alpha}{H}\left[\text{Ra}_H - \left(\frac{\pi H}{2L_r}\right)^2\right]. \tag{2.91}$$

The counterflow convects heat upward at the rate q', which can be calculated using Equations (2.89) and (2.90):

$$q' = \int_{-L}^{L} (\rho c_P)_f vT\,dx. \tag{2.92}$$

The average heat flux convected in the vertical direction, $q'' = q'/(2L_r)$, can be expressed as an overall thermal conductance

$$\frac{q''}{\Delta T} = \frac{k}{8H\text{Ra}_H}\left[\text{Ra}_H - \left(\frac{\pi H}{2L_r}\right)^2\right]^2. \tag{2.93}$$

This result is valid provided the vertical temperature gradient does not exceed the externally imposed gradient, $(-\partial T/\partial y) < \Delta T/H$. This condition translates into

$$\frac{L_r}{H} > \frac{\pi}{2}\text{Ra}_H^{-1/2}, \tag{2.94}$$

which in combination with the assumed limit $L_r/H \to 0$ means that the domain of validity of Equation (2.93) widens when Ra_H increases. In this domain the thermal conductance $q''/\Delta T$ decreases monotonically as L_r decreases [cf. Figure 2.16].

2.11.2 The Few Plumes Regime

As L_r increases, the number of rolls decreases and the vertical counterflow is replaced by a horizontal counterflow in which the thermal resistance between T_h and T_c is dominated by two horizontal boundary layers, as in Figure 2.17. Let δ be the scale of the thickness of the horizontal boundary layer. The thermal conductance $q''/\Delta T$ can be deduced from the heat transfer solution for

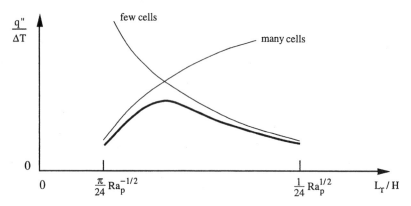

Fig. 2.16. The geometric maximization of the thermal conductance of a fluid-saturated porous layer heated from below (Nelson and Bejan, 1998).

natural convection boundary layer flow over a hot isothermal horizontal surface facing upward, or a cold surface facing downward. The similarity solution for the horizontal surface with power-law temperature variation (Cheng and Chang, 1976) can be used to develop an analytical result, as we show at the end of this section. Plume flows have also been described by Shu and Pop (1997).

A simpler analytical solution can be developed in a few steps using the integral method. Consider the slender flow region $\delta \times (2L_r)$, where $\delta \ll 2L_r$, and integrate Equations (2.84) to (2.86) from $y = 0$ to $y \to \infty$, that is, into the region just above the boundary layer. The surface temperature is T_h, and the temperature outside the boundary layer is T_∞ (constant). The origin $x = 0$ is set at the tip of the wall section of length $2L_r$. The integrals of Equations (2.84) and (2.86) yield

$$\frac{d}{dx} \int_0^\infty u(T - T_\infty)dy = -\alpha \left(\frac{\partial T}{\partial y}\right)_{y=0}. \tag{2.95}$$

The integral of Equation (2.85), in which we neglect $\partial v/\partial x$ in accordance with boundary layer theory, leads to

$$u_0(x) = \frac{Kg\beta}{\nu} \frac{d}{dx} \int_0^\infty T \, dy, \tag{2.96}$$

where u_0 is the velocity along the surface, $u_0 = u(x, 0)$. Reasonable shapes for the u and T profiles are the exponentials

$$\frac{u(x, y)}{u_0(x)} = \exp\left[-\frac{y}{\delta(x)}\right] = \frac{T(x, y) - T_\infty}{T_h - T_\infty} \tag{2.97}$$

2 Flows in Porous Media 63

which transform Equations (2.95) and (2.96) into

$$\frac{d}{dx}(u_0\delta) = \frac{2\alpha}{\delta},$$

(2.98)

$$u_0 = \frac{Kg\beta}{\nu}(T_h - T_\infty)\frac{d\delta}{dx}.$$

(2.99)

These equations can be solved for $u_0(x)$ and $\delta(x)$,

$$\delta(x) = \left[\frac{9\alpha\nu}{Kg\beta(T_h - T_\infty)}\right]^{1/3} x^{2/3}.$$

(2.100)

The solution for $u_0(x)$ is of the type $u_0 \sim x^{-1/3}$, which means that the horizontal velocities are large at the start of the boundary layer, and decrease as x increases. This is consistent with the geometry of the $H \times 2L_r$ roll sketched in Figure 2.17, where the flow generated by one horizontal boundary layer turns the corner and flows vertically as a relatively narrow plume (narrow relative to $2L_r$), to start with high velocity (u_0) a new boundary layer along the opposite horizontal wall.

The thermal resistance of the geometry of Figure 2.17 is determined by estimating the local heat flux $k(T_h - T_\infty)/\delta(x)$ and averaging it over the total length $2L_r$:

$$q'' = \left(\frac{3}{4}\right)^{1/3} \frac{k\Delta T}{H} \left(\frac{T_h - T_\infty}{\Delta T}\right)^{4/3} \mathrm{Ra}_H^{1/3} \left(\frac{H}{L_r}\right)^{2/3}.$$

(2.101)

The symmetry of the sandwich of boundary layers requires $T_h - T_\infty = (1/2)\Delta T$, such that

$$\frac{q''}{\Delta T} = \frac{3^{1/3}k}{4H} \mathrm{Ra}_H^{1/3} \left(\frac{H}{L_r}\right)^{2/3}.$$

(2.102)

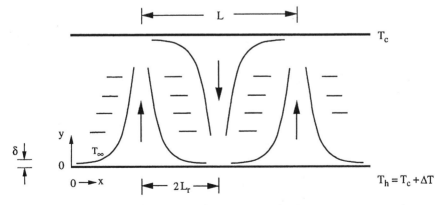

Fig. 2.17. The extreme in which the flow consists of a few isolated plumes (Nelson and Bejan, 1998).

The goodness of this result can be tested against the similarity solution for a hot horizontal surface that faces upward in a porous medium and has an excess temperature that increases as x^λ. The only difference is that the role that was played by $(T_h - T_\infty)$ in the preceding analysis is now played by the excess temperature averaged over the surface length $2L_r$. If we use $\lambda = 1/2$, which corresponds to uniform heat flux, then it can be shown that the solution of Cheng and Chang (1976) leads to the same formula as Equation (2.102), except that the factor $3^{1/3} = 1.442$ is replaced by $0.816(3/2)^{4/3} = 1.401$. Equation (2.102) is valid when the specified Ra_H is such that the horizontal boundary layers do not touch. We write this geometric condition as $\delta(x = 2L_r) < H/2$ and, using Equation (2.100), we obtain

$$\frac{L_r}{H} < \frac{1}{24}\mathrm{Ra}_H^{1/2}. \tag{2.103}$$

Since in this analysis L_r/H was assumed to be very large, we conclude that the L_r/H domain in which Equation (2.102) is valid becomes wider as the specified Ra_H increases. The important feature of the "few rolls" limit is that the thermal conductance decreases as the horizontal dimension L_r increases. This second asymptotic trend has been added to Figure 2.16.

2.11.3 The Intersection of Asymptotes

Figure 2.16 presents a bird's-eye view of the effect of flow shape on thermal conductance. Even though we did not draw $q''/\Delta T$ completely as a function of L_r, the two asymptotes tell us that the thermal conductance is maximum at an optimal L_r value that is close to their intersection. There is a family of such curves, one curve for each Ra_H. The $q''/\Delta T$ peak of the curve rises, and the L_r domain of validity around the peak becomes wider as Ra_H increases. Looking in the direction of small Ra_H values we see that the domain vanishes (and the cellular flow disappears) when the following requirement is violated,

$$\frac{1}{24}H\mathrm{Ra}_H^{1/2} - \frac{\pi}{2}H\mathrm{Ra}_H^{-1/2} \geq 0. \tag{2.104}$$

This inequality means that the flow exists when $\mathrm{Ra}_H \geq 12\pi = 37.70$. This conclusion is extraordinary: it agrees with the stability criterion for the onset of two-dimensional convection, Equation (2.73), namely, $\mathrm{Ra}_H > 4\pi^2 = 39.5$, which was derived based on a lengthier analysis and the assumption that a flow structure exists, the initial disturbances (Horton and Rogers, 1945; Lapwood, 1948).

We obtain the optimal shape of the flow $2L_{r,\mathrm{opt}}/H$, by intersecting the asymptotes (2.93) and (2.102):

$$\pi^2\left(\frac{H}{2L_{r,\mathrm{opt}}}\mathrm{Ra}_H^{-1/2}\right)^2 + 2^{5/6}3^{1/6}\left(\frac{H}{2L_{r,\mathrm{opt}}}\mathrm{Ra}_H^{-1}\right)^{1/3} = 1. \tag{2.105}$$

Over most of the Ra_H domain where Equation (2.104) is valid, Equation (2.105) is well approximated by its high Ra_H asymptote:

$$\frac{2L_{r,\mathrm{opt}}}{H} \cong \pi \mathrm{Ra}_H^{-1/2}. \tag{2.106}$$

The maximum thermal conductance is obtained by substituting the $L_{r,\mathrm{opt}}$ value in either Equation (2.102) or Equation (2.93). This estimate is an upper bound, because the intersection is above the peak of the curve. In the high-Ra_H limit (2.106) this upper bound assumes the analytical form

$$\left(\frac{q''}{\Delta T}\right)_{\mathrm{max}} \frac{H}{k} \lesssim \frac{3^{1/3}}{2^{4/3}\pi^{2/3}} \mathrm{Ra}_H^{2/3}. \tag{2.107}$$

Towards lower Ra_H values the slope of the $(q''/\Delta T)_{\mathrm{max}}$ curve increases such that the exponent of Ra_H approaches 1. This behavior is in excellent agreement with the large volume of experimental data collected for Bénard convection in saturated porous media (Cheng, 1978). The less-than-1 exponent of Ra_H in the empirical $\mathrm{Nu}(\mathrm{Ra}_H)$ curve, and the fact that this exponent decreases as Ra_H increases, has attracted considerable attention from theoreticians during the last two decades (Nield and Bejan, 1999).

In this section and Sections 2.2 and 2.5, we outlined the basic rule for two methods of solution for problems of convection in porous media: scale analysis and the intersection of asymptotes. These are two of the simplest methods that are available. They yield concrete results for engineering problems such as heat transfer rates, flow rates, velocities, temperature differences, and time intervals. They distinguish themselves from other methods because they offer a high return on investment: because they are so simple, they deserve to be tried first, as preliminaries, even in problems where more exact results are needed. Simple methods identify the proper dimensionless formulation for presenting more exact (and more expensive) results developed based on more complicated methods (analytical, numerical, and/or experimental).

Other simple methods are available, for example, the integral method (Karman–Pohlhausen), Equations (2.95) to (2.100), and similarity formulations [e.g., Bejan (1995a)]. A word of caution goes with the use of all "simple" methods. More complicated problems with nonsimilar and singular solutions may require more advanced treatments from the start.

The intersection of asymptotes method relied on an additional principle that applies throughout the physics of flow systems: the constructal law, the generation of flow geometry for the maximization of access for currents in systems far from equilibrium. In Figure 2.16, we invoked this principle when we minimized the global thermal resistance encountered by the flow of heat across the horizontal porous layer. The intersection of the two asymptotes is an approximation of the flow geometry that minimizes the global thermal resistance. The same "constructal principle" has been used to predict flow geometry and transitions between flow regimes in a great variety of configurations (Bejan, 2000).

The most important conclusion is that by learning simple methods, and using them correctly, the young researcher learns two important lessons. One is "try the simplest first." Simple methods are valuable throughout thermal and fluid sciences, not only in porous media. The other lesson is the open competition among methodologies in the search for engineering answers to fundamental questions. A researcher with a personal mathematics background and, most important, with a personal supply of curiosity and time can and should judiciously evaluate the worthiness of any of these methodologies relative to his or her ability and taste (Bejan, 1995a, p. 55).

3

Energy Engineering

3.1 Thermodynamics Fundamentals: Entropy Generation or Exergy Destruction

The energy crisis of the 1970s and the continuing emphasis on efficiency (the conservation of fuel resources) has led to a complete overhaul of the way in which power systems are analyzed and improved thermodynamically. The new methodology is *exergy analysis* and its optimization component is known as thermodynamic optimization, or *entropy generation minimization* (EGM). This new approach is based on the simultaneous application of the first and second laws in analysis and design. In the 1990s it has become the premier method of thermodynamic analysis in engineering education (Moran, 1989; Bejan, 1982, 1996a,b, 1997; Feidt, 1987; Sieniutycz and Salamon, 1990; Kotas, 1995; Moran and Shapiro, 1995; Radcenco, 1994; Shiner, 1996; Bejan *et al.*, 1996) and it is now sweeping every aspect of engineering practice (Stecco and Moran, 1990, 1992; Valero and Tsatsaronis, 1992; Bejan and Mamut, 1999; Bejan *et al.*, 1999). It is particularly well suited for computer-assisted design and optimization (Sciubba and Melli, 1998; Sciubba, 1999a,b).

In this chapter we begin with a review of the fundamentals of the method and a few simple examples of the ways in which the method can be used for system optimization. The chapter continues with an in-depth treatment of several sectors of energy engineering where porous flow structures play critical roles.

First, we must distinguish between exergy and energy in order to avoid any confusion with the traditional energy-based methods of thermal system analysis and design. Energy flows into and out of a system along paths of mass flow, heat transfer, and work (e.g., shafts, piston rods). Energy is conserved, not destroyed: this is the statement made by the first law of thermodynamics.

Exergy is an entirely different concept. It represents quantitatively the "useful" energy, or the ability to do or receive work—the work content—of the great variety of streams (mass, heat, work) that flow through the system. The first attribute of the property "exergy" is that it makes it possible to compare on a common basis different interactions (inputs, outputs, work, heat).

Another benefit is that by accounting for all the exergy streams of the system it is possible to determine the extent to which the system destroys exergy. The destroyed exergy is proportional to the generated entropy. Exergy is always destroyed, partially or totally: this is the statement made by the second law of thermodynamics. The destroyed exergy or the generated entropy is responsible for the less-than-theoretical thermodynamic efficiency of the system.

By performing exergy accounting in smaller and smaller subsystems, we are able to draw a map of how the destruction of exergy is distributed over the engineering system of interest. In this way we are able to pinpoint the components and mechanisms (processes) that destroy exergy the most. This is a real advantage as we strive for higher efficiencies (always by finite means, always against constraints), because it tells us from the start how to allocate hardware, engineering effort, and resources.

In engineering thermodynamics today, emphasis is placed on identifying the mechanisms and system components that are responsible for thermo-dynamic losses, the sizes of these losses (exergy analysis), minimizing the losses subject to the global constraints of the system (entropy generation minimization), and minimizing the total costs associated with building and operating the energy system (thermoeconomics). To review the fundamentals of thermoeconomics is not the purpose of this chapter. A description of ther-moeconomics and its applications, as an evolutionary development beyond exergy analysis and entropy generation minimization, is available in Bejan *et al.* (1996).

The method of thermodynamic optimization or entropy generation min-imization is a field of activity at the interface of heat transfer, engineering thermodynamics, and fluid mechanics. The position of the field is illustrated in Figure 3.1. The method relies on the simultaneous application of principles of heat and mass transfer, fluid mechanics, and engineering thermodynamics, in the pursuit of realistic models of processes, devices, and installations. By realistic models we mean models that account for the inherent irreversibility of engineering systems and processes. Thermodynamic optimization may be used by itself (without cost minimization) in the preliminary stages of design in order to identify trends and the existence of optimization opportunities. The optima and structural characteristics identified based on thermodynamic optimization can be made more realistic through subsequent refinements based on global cost minimization by using the method of thermoeconomics.

Here is why in thermodynamic optimization we must rely on heat transfer and fluid mechanics, not just thermodynamics. Consider the most general system-environment configuration, namely, a system that operates in the unsteady state, Figure 3.2. Its instantaneous inventories of mass, energy, and entropy are M, E, and S. The system experiences the net work transfer rate $\dot{W}$, heat transfer rates $(\dot{Q}_0, \dot{Q}_1, \ldots, \dot{Q}_n)$ with $n+1$ temperature reservoirs

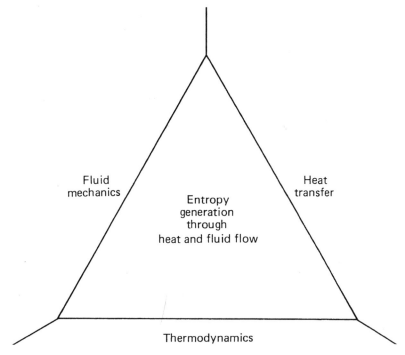

Fluid
mechanics

Heat
transfer

Entropy
generation
through
heat and fluid flow

Thermodynamics

Fig. 3.1. The interdisciplinary field covered by the method of thermodynamic optimization or entropy generation minimization (Bejan, 1982).

$(T_0, T_1, \ldots, T_n)$, and mass flow rates $(\dot{m}_{\mathrm{in}}, \dot{m}_{\mathrm{out}})$ through any number of inlet and outlet ports. Noteworthy in this array of interactions is the heat transfer rate between the system and the environmental (atmospheric) temperature reservoir $\dot{Q}_0$, on which we focus shortly.

The thermodynamics of the system consists of accounting for the first and second laws [e.g., Moran and Shapiro (1995) and Bejan (1997)],

$$\frac{dE}{dt} = \sum_{i=0}^{n} \dot{Q}_i - \dot{W} + \sum_{\mathrm{in}} \dot{m}h - \sum_{\mathrm{out}} \dot{m}h, \qquad (3.1)$$

$$\dot{S}_{\mathrm{gen}} = \frac{dS}{dt} - \sum_{i=0}^{n} \frac{\dot{Q}_i}{T_i} - \sum_{\mathrm{in}} \dot{m}s + \sum_{\mathrm{out}} \dot{m}s \geq 0, \qquad (3.2)$$

where h is shorthand for the sum of specific enthalpy, kinetic energy, and potential energy of a particular stream at the boundary. In Equation (3.2) the total entropy generation rate $\dot{S}_{\mathrm{gen}}$ is simply a definition (notation) for the entire quantity on the left-hand side of the inequality sign. We show that it is advantageous to decrease $\dot{S}_{\mathrm{gen}}$; this can be accomplished by changing at least one of the quantities (properties, interactions) specified along the system boundary.

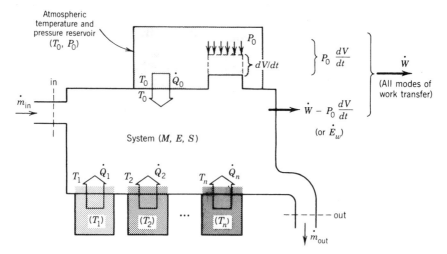

Fig. 3.2. General definition of a flow system (an open thermodynamic system) in communication with the atmosphere (Bejan, 1997; from Bejan, *Advanced Engineering Thermodynamics*, 2nd ed., Copyright © 1997 Wiley. This material is used by permission of John Wiley & Sons, Inc.).

We select the environmental heat transfer $\dot{Q}_0$ as the interaction that is allowed to float as $\dot{S}_{gen}$ varies. Historically, this choice was inspired (and justified) by applications to power plants and refrigeration plants, because the rejection of heat to the atmosphere was of little consequence in the overall cost analysis of the design. Eliminating $\dot{Q}_0$ between Equations (3.1) and (3.2) we obtain

$$\dot{W} = -\frac{d}{dt}(E - T_0 S) + \sum_{i=1}^{n}\left(1 - \frac{T_0}{T_i}\right)\dot{Q}_i + \sum_{in}\dot{m}(h - T_0 s)$$

$$- \sum_{out}\dot{m}(h - T_0 s) - T_0\dot{S}_{gen}. \tag{3.3}$$

The power output or input in the limit of reversible operation ($\dot{S}_{gen} = 0$) is

$$\dot{W}_{rev} = -\frac{d}{dt}(E - T_0 S) + \sum_{i=1}^{n}\left(1 - \frac{T_0}{T_i}\right)\dot{Q}_i$$

$$+ \sum_{in}\dot{m}(h - T_0 s) - \sum_{out}\dot{m}(h - T_0 s). \tag{3.4}$$

In engineering thermodynamics each of the terms on the right-hand side of Equation (3.4) is recognized as an exergy of one type or another (see Section 3.2), and the calculation of $\dot{W}_{rev}$ is known as exergy analysis. Subtracting Equation (3.3) from (3.4) we arrive at the Gouy–Stodola theorem,

$$\dot{W}_{rev} - \dot{W} = T_0\dot{S}_{gen}, \tag{3.5}$$

where $\dot{W}_{\text{rev}}$ is fixed because all the heat and mass flows (other than $\dot{Q}_0$) are fixed.

Pure thermodynamics (e.g., exergy analysis) ends, and the method of entropy generation minimization begins with Equation (3.5). The lost power $(\dot{W}_{\text{rev}} - \dot{W})$ is always positive, regardless of whether the system is a power producer (e.g., power plant) or a power user (e.g., refrigeration plant). To minimize lost power when $\dot{W}_{\text{rev}}$ is fixed is the same as maximizing power output in a power plant, and minimizing power input in a refrigeration plant. This operation is also equivalent to minimizing the total rate of entropy generation.

The critically new aspect of the EGM method—the aspect that makes the use of thermodynamics insufficient and distinguishes EGM from pure exergy analysis—is the minimization of the calculated entropy generation rate. Optimization and *design* (the generation of flow *geometry*) are the difference. Flow structures and designed porous media are the focus of the present book. To minimize the irreversibility of a proposed configuration, the analyst must use the relations between temperature differences and heat transfer rates, and between pressure differences and mass flow rates. The analyst must express the thermodynamic nonideality of the design ($\dot{S}_{\text{gen}}$) as a function of the topology and physical characteristics of the system, namely, finite dimensions, shapes, materials, finite speeds, and finite-time intervals of operation. For this the analyst must rely on heat transfer and fluid mechanics principles, in addition to thermodynamics. Only by varying the physical characteristics of the system, by morphing the architecture, can the analyst bring the design closer to the operation characterized by minimum entropy generation subject to size and time constraints. We illustrate this technique by means of a few very basic models in Section 3.3.

3.2 Exergy Analysis

There is a rich nomenclature and mathematical apparatus associated with defining and calculating the exergies of various entities. Most of these exergy names are attached to the four types of terms shown on the right side of Equation (3.4). This nomenclature must be used with care, especially now as the method is applied for the first time to areas where energy-based methods are still the norm. The key feature is this: exergy is the maximum work that can be extracted (or the minimum work that is required) from an entity (e.g., stream, amount of matter) as the entity passes from a given state to one of equilibrium with the environment. As such, exergy is a measure of the departure of the given state from the environmental state—the larger the departure, the greater the potential for doing work.

To illustrate the calculation of exergy, consider the following examples in which the environmental state is represented by the atmospheric temperature T_0 and pressure P_0. If the entity is a closed system (fixed mass, Figure 3.3) at an initial state represented by the energy E, entropy S, and volume V, then

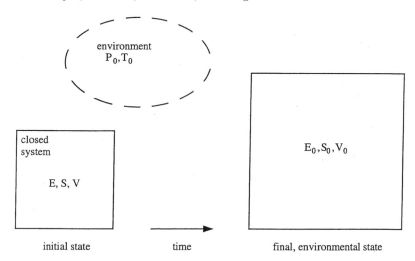

Fig. 3.3. Closed system and process en route to thermal and mechanical equilibrium with the environment.

its exergy Ξ (expressed in joules) relative to the environment is

$$\Xi = E - E_0 - T_0(S - S_0) + P_0(V - V_0). \tag{3.6}$$

In this expression Ξ is known as the nonflow exergy of the given mass (Moran, 1989; Bejan, 1997), and the subscript 0 indicates the system properties in the state of thermal and mechanical equilibrium with the environment. The environmental state is also known as the restricted dead state: "dead" because once in this state the system cannot deliver any more work relative to the environment. It is "restricted" because in this state the system is in thermal and mechanical equilibrium with the environment, but not in chemical equilibrium. Equation (3.6) is general in the sense that the internal construction and materials (e.g., single phase vs. multiphase) of the given mass are not specified. Equation (3.6) can be generalized further for cases where the chemical composition of the given mass may change en route to chemical equilibrium with the environment (Moran, 1989; Bejan, 1997).

Note that the nonflow exergy Ξ has its origin in the first term identified on the right side of Equation (3.3) or (3.4). The system "accumulates" the quantity $(E - T_0S)$ as potential work and, in going from left to right in Figure 3.3, this work potential decreases from $(E - T_0S)$ to $(E_0 - T_0S_0)$. The difference, which is $E - E_0 - T_0(S - S_0)$, represents all the work that could be produced during the process. From this quantity we must subtract the work fraction done by the system against the atmosphere $P_0(V_0 - V)$. The resulting expression is Equation (3.6).

As a second exergy calculation example, consider the steady-flow system with a stream of mass flow rate $\dot{m}$, where the given (inlet) state is represented by the specific enthalpy h, entropy s, kinetic energy $(1/2)V^2$, and gravitational potential energy gz, where z is the altitude of the inlet. The specific flow exergy is expressed in J/kg, and is evaluated relative to the environment (T_0, P_0):

$$e_x = \left(h + \frac{1}{2}V^2 + gz\right) - \left(h + \frac{1}{2}V^2 + gz\right)_0 - T_0(s - s_0). \qquad (3.7)$$

As shown in Figure 3.4, the subscript 0 indicates the properties of the stream that reached thermal and mechanical equilibrium with the environment. In other words, e_x is the change in the value of the group $h + (1/2)V^2 + gz - T_0s$, in going from the inlet to the outlet. The exergy flow rate of this stream is the product $\dot{m}\,e_x$. More general versions of Equation (3.7) are available for streams that can also exchange chemical species with the environment (Moran, 1989; Bejan, 1988).

The streams of flow exergy were identified already in the third and fourth terms on the right-hand side of Equations (3.3) and (3.4), where h was used as shorthand for the sum $(h + (1/2)V^2 + gz)$. Continuing with this shorthand notation, we note that the group $(h - T_0s)$ represents the specific flow availability of the stream. The difference between the flow availability at the indicated state (inlet, Figure 3.4) and the flow availability of the same stream at the environmental state (outlet, Figure 3.4) is the flow exergy, namely, $e_x = (h - T_0s) - (h_0 - T_0s_0)$, which is the same as Equation (3.7).

A third example is the exergy content of heat transfer. If the heat transfer rate $\dot{Q}$ enters the system by crossing a boundary of local temperature T, then its exergy stream relative to the environment (T_0), which is associated with $\dot{Q}$, is

$$\dot{E}_Q = \dot{Q}\left(1 - \frac{T_0}{T}\right). \qquad (3.8)$$

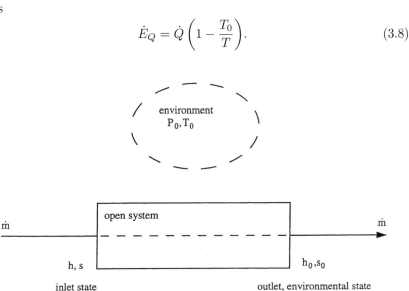

Fig. 3.4. Open system, steady state, and one stream en route to thermal and mechanical equilibrium with the environment.

The heat-transfer exergy flow rate $\dot{E}_Q$ is zero when $T = T_0$, that is, as in the case of the heat rejected to the ambient by power and refrigeration plants. In Figure 3.2 an exergy stream is associated with each of the heat inputs $\dot{Q}_1, \ldots, \dot{Q}_n$, whereas the $\dot{Q}_0$ stream carries no exergy $\dot{E}_{Q_0} = \dot{Q}_0(1 - T_0/T_0) = 0$. The exergy flows associated with heat transfer are accounted for by the second term on the right side of Equations (3.3) and (3.4).

In Figure 3.2 and Equation (3.8) T_0 is the temperature of those regions of the environment that are sufficiently close to the system but not affected by the discharge. The purpose of this modeling decision is to place inside the defined thermodynamic system all the irreversibilities associated with the internal and external effects of the physical installation that resides inside the system. The system is comprised of the installation and the surrounding regions that are affected (e.g., heated) by the installation. In an actual power or refrigeration plant, the rejected heat current $\dot{Q}$ leaves the installation and enters the neighboring environmental fluid (air, water) at a temperature somewhat higher than T_0. Further down the line, the same heat current $\dot{Q}$ reaches the true environment at T_0 (i.e., it crosses the T_0 boundary, Figure 3.2). The interaction between energy systems and their surrounding fluids is illustrated by the environmental flows documented in Sections 4.1 to 4.4.

The spatial distribution of exergy destruction is visible in the example of Figure 3.5. The top figure shows the traditional energy-flow analysis of a simple Rankine cycle power plant. The heat input is $\dot{Q}_H$, and the net power output is $\dot{W}_t - \dot{W}_p = \eta_{II}\dot{E}_{Q_H}$. The fraction η_{II} is the second-law or rational efficiency of the power plant and is a relative measure of the combined imperfections of the plant. The calculated widths of the exergy destruction streams indicate a ranking of the components as candidates for thermodynamic optimization. Exergy analysis can be performed inside each component in order to determine the particular features (e.g., combustion, fouling, heat transfer, pressure drop) that dominate the irreversibility of that component. Finally, the success of the thermodynamic improvements that are implemented can only be evaluated by repeating the exergy analysis and registering the changes in exergy destruction and second-law efficiency.

Exergy principles are being applied to a wide variety of thermal/chemical processes. Avoidable destructions of exergy represent the waste of exergy sources such as oil, natural gas, and coal. By devising ways to avoid the destruction of exergy, better use can be made of fuels. Exergy analysis determines the location, type, and true magnitude of the waste of fuel resources, and plays a central role in developing strategies for more effective fuel use.

Most of the new work and opportunities for advances are in developing strategies for the optimal allocation (configuration, topology) of resources. This work is known as exergy destruction minimization, irreversibility (entropy generation) minimization, or thermodynamic optimization. It is often subjected to overall constraints such as finite sizes, finite times, material

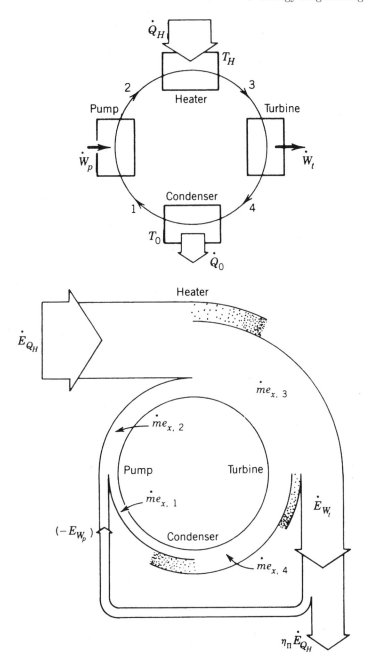

Fig. 3.5. The conversion and partial destruction of exergy in a power plant based on the simple Rankine cycle. Top: the traditional notation and energy-interaction diagram; bottom: the exergy wheel diagram (Bejan, 1997; from Bejan, *Advanced Engineering Thermodynamics*, 2nd ed., Copyright © 1997 Wiley. This material is used by permission of John Wiley & Sons, Inc.).

types, and shapes. Recent reviews of the literature (Bejan, 1996a,b) show that thermodynamic optimization is making fast progress in cryogenics, heat transfer engineering, energy storage systems, solar power plants, fossil-fuel power plants, and refrigeration plants (Feidt, 1998). The thermal energy storage topics discussed in the following sections illustrate the opportunities for devising strategies of optimal allocation.

3.3 Thermal Energy Storage

Thermal energy storage (TES) generally involves a temporary storage of high- or low-temperature thermal energy for later use. Examples are the storage of solar energy for overnight heating, summer heat for winter use, winter ice for space cooling in summer, and heat or coolness generated electrically during off-peak hours for use during subsequent peak demand hours. TES is an excellent candidate to offset this mismatch between availability and demand. As an advanced energy technology, TES has attracted increasing interest for a large number of thermal applications such as space heating, hot water, cooling, and air-conditioning. TES systems have the potential for increasing effective use of thermal energy equipment and for facilitating large-scale switching.

We focus in detail on sensible and latent heat storage. The effectiveness of sensible heat storage depends on the specific heat of the storage material and, if volume is an important constraint, on its density. Sensible storage systems commonly use rocks or water as the storage medium, the thermal energy being stored by increasing the temperature of the storage medium. Latent heat storage systems store energy in phase-change materials, with the thermal energy stored when the material changes phase, usually from solid to liquid. The specific heat of solidification/fusion or vaporization, and the temperature at which the phase change occurs are of design importance. Storage by sensible heat and latent heat may occur in the same storage material. Examples of storage materials are given in Table 3.1. TES may also be due to the reversible scission or reforming of chemical bonds.

TES technology has been used in various forms and specific applications. Most common is the use of sensible heat (oils, molten salts) or latent heat

Table 3.1. Examples of materials for sensible heat and latent heat energy storage

Sensible heat		Latent heat
Short term	Long term (annual)	Short term
Rock beds	Rock beds	Inorganic materials
Earth beds	Earth beds	Organic materials
Water tanks	Large water tanks	Fatty acids
	Aquifers	Aromatics
	Solar ponds	

(ice, phase-change material) storage for refrigeration and/or space heating and cooling needs. As an example of the cost savings and increased efficiency achievable through the use of TES, consider the following case. In some climates it is necessary to provide heating in winter and cooling in summer. Typically these services are provided by using energy to drive heaters and air-conditioners. With TES, it is possible to store heat during summer months for use in winter, while winter storage provides cooling in the summer. This is an example of seasonal storage and its use to help meet the energy needs caused by seasonal fluctuations in temperature. Such a scheme requires a great deal of storage capacity because of the long storage time scales.

The same principle can be applied on a smaller scale to smooth out daily temperature variations. For example, solar energy can be used to heat tiles on a floor during the day. At night, as the ambient temperature falls, the tiles release their stored heat to slow the temperature decrease in the room. Another example of a TES application is the use of thermal storage to take advantage of off-peak electricity tariffs. Chiller units can be run at night when the cost of electricity is relatively low. These units are used to cool down a thermal storage, which then provides cooling for air-conditioning throughout the day. In this way electricity costs are reduced, chiller efficiency is increased because of the lower nighttime ambient temperatures, and the peak electricity demand for electrical-supply utilities is reduced.

A complete storage process involves at least three steps: charging, storing, and discharging. A simple storage cycle is illustrated in Figure 3.6, where the three steps are distinct. In practical systems, some of the steps may occur simultaneously (e.g., charging and storing), and each step may occur more than once during one storage cycle.

A wide variety of storage materials exists, depending on the temperature range and application. For sensible heat storage, water is a common choice because among its other positive attributes, it has one of the highest specific heats of any liquid at ambient temperatures. Although the specific heat of water is not as high as that of many solids, it has the advantage of being a liquid, which can easily be pumped to transport thermal energy. The flow of liquid water also facilitates high heat transfer rates. Solids have the advantage of higher specific heat capacities, which allow for more compact storage units. When higher temperatures are involved (e.g., preheating furnace air supplies), solids become the preferred sensible heat store. In such cases, refractories are used as storage materials. If the storage medium needs to be pumped, liquid metals are often used.

The most common example of latent heat storage is the conversion of water to ice. Cooling systems incorporating ice storage have a distinct size advantage over equivalent-capacity chilled-water units because of the relatively large amount of energy that is stored through the phase change. The small size is the major advantage of latent heat thermal storage. In aerospace applications, lithium fluoride salts are used to store heat in the zero-gravity environment of the space shuttle. Another interesting development is the use

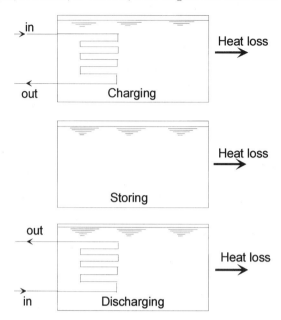

Fig. 3.6. The three stages of a simple heat storage process: charging period (top), storing period (center), and discharging period (bottom), where Q_1 represents the heat loss during each period (Dincer and Rosen, 2002).

of phase-change materials in wall paneling and clothing (see Section 8.2). These panels incorporate compounds that undergo solid-to-solid structural phase changes. With the right choice of material, the phase change occurs at ambient temperature. In this way, the phase-change materials act as high-density heat sinks and sources that resist changes in ambient room temperature.

During energy storage based on reversible endothermic chemical reactions, chemical bonds are broken so that significant amounts of energy can be stored per unit mass of storage material. Although not currently viable, a variety of reactions is being explored. These include catalytic reactions, such as the steam reforming reaction with methane and the decomposition of sulfur trioxide, and thermal dissociation reactions involving metal oxides and metal hydrides. These reactions are expected to be useful in high-temperature nuclear cycles and solar-energy systems, and as topping cycles for industrial boilers. At present low-temperature reactions ($<300°$C) have not been proven to be promising. Their application to TES can be an effective way of reducing costs and increasing efficiency. Endothermic chemical reactions are also used for temperature control during fire and accidents.

The use of TES can contribute significantly to meeting the needs of our society for more efficient, environmentally benign energy use in building heating and cooling, aerospace power, and utility applications. Possible benefits are reduced energy costs, reduced energy consumption, improved indoor air quality, increased flexibility of operation, and reduced initial and maintenance costs. Dincer *et al.* (1997a) pointed out additional advantages of TES: reduced equipment size, more efficient and effective utilization of equipment, conservation of fossil fuels (by facilitating more efficient energy use and/or fuel substitution), and reduced pollutant emissions (e.g., CO_2 and CFCs).

Although TES processes are used in a wide variety of applications, they are all designed to operate on a cyclical basis (usually daily, occasionally seasonally). They are attractive because they fulfill one or more of the following objectives.

Increase generation capacity: Demand for heating, cooling, or power is seldom constant over time, and the excess generation available during low-demand periods can be used to charge a TES in order to increase the effective generation capacity during high-demand periods. This process allows a smaller production unit to be installed (or to add capacity without purchasing additional units) and results in a higher load factor for the units.

Enable better operation of cogeneration plants: Combined heat and power, or cogeneration, plants are generally operated to meet the demands of the connected thermal load, which often results in excess electric generation during periods of low electric use. By incorporating TES, the plant need not be operated to follow a load. Rather, it can be deployed in more advantageous ways.

Shift energy purchases to low-cost periods: This is the demand-side application of the first objective listed, and allows energy consumers subject to time-of-day pricing to shift energy purchases from high-cost to low-cost periods.

Increase system reliability: Any form of energy storage, from the uninterruptible power supply of a small personal computer to a large pumped storage project, normally increases system reliability.

Integration with other functions: In applications where onsite water storage is needed for fire protection, it may be feasible to incorporate thermal storage into a common storage tank. Likewise, the apparatus designed to solve power-quality problems may be adaptable to energy-storage purposes as well.

The time period of storage is an important factor. Diurnal storage systems have certain advantages: capital investment and energy losses are usually low, and units are smaller and can easily be manufactured offsite. The sizing of daily storage for each application is not nearly as critical as it is for larger annual storage. Annual storage, however, may become economical only in multidwelling or industrial park designs, and often requires expensive energy distribution systems and novel institutional arrangements related to ownership and financing. In solar TES applications, the optimal energy storage duration is usually the one that offers the final (delivered) energy at minimum cost, when integrated with the collector field and as a backup into a final application.

3.4 Sensible Heat Storage

The amount of energy input to TES by a sensible heat device is proportional to the difference among the final and initial storage temperatures, the mass of the storage material, and its heat capacity. Each material has its own advantages and disadvantages. For example, water has approximately twice the specific heat of rock and soil. The high heat capacity of water ($\sim$4.2 kJ/kg K) often makes water tanks a logical choice for TES systems that operate in a temperature range needed for building heating or cooling. The relatively low heat capacity of rocks and ceramics ($\sim$0.84 kJ/kg K) is offset by the large temperature changes possible with these materials and by their relatively high density (Tomlinson and Kannberg, 1990).

A sensible TES consists of a storage medium, a container, and input/output devices. Containers must confine the storage material and prevent losses of thermal energy. Thermal stratification, that is, the maintenance of a thermal gradient across storage, is desirable. Maintaining stratification is much simpler in solid storage media than in fluids.

The ability to store sensible heat for a given material depends strongly on the value of the heat capacity ρc_p. Water has a high value and is inexpensive but, being liquid, it must be contained in a better-quality container than a solid. Some common TES materials and their properties are presented in Table 3.2. To be useful in a TES application, the material must be inexpensive and have a good thermal capacity. Another important parameter in sensible TES is the rate at which heat can be released and extracted. This characteristic is a function of thermal diffusivity. For this reason, iron shot

Table 3.2. Thermal capacities at 20°C of some common thermal energy storage materials

Material	Density (kg/m^3)	Specific heat (J/kg K)	Volumetric thermal capacity ($10^6 \text{ J/m}^3\text{K}$)
Clay	1458	879	1.28
Brick	1800	837	1.51
Sandstone	2200	712	1.57
Wood	700	2390	1.67
Concrete	2000	880	1.76
Glass	2710	837	2.27
Aluminum	2710	896	2.43
Iron	7900	452	3.57
Steel	7840	465	3.68
Gravelly earth	2050	1840	3.77
Magnetite	5177	752	3.89
Water	988	4182	4.17

Source: Adapted from Norton (1992).

is an excellent thermal storage medium, having both high heat capacity and high thermal conductance.

For high-temperature sensible TES (i.e., up to several 100°C) iron and iron oxide have thermal properties (per unit volume of storage) that are comparable to those of water. The cost is moderate for oxide pellets and metal balls. Because iron and its oxide have similar thermal characteristics, the slow oxidization of the metal in a high-temperature liquid or air system does not degrade the performance of the TES.

Rock is a good material for sensible TES from the standpoint of cost, but its volumetric thermal capacity is only half that of water. Studies have shown that rock storage bins are practical, their main advantage being that they can easily be used for heat storage at temperatures above 100°C (Dincer, 1997a,b).

Figure 3.7 shows a thermally stratified storage tank, highlighting the positions of the inlet and outlet for both well and poorly designed cases. Also shown is the thermally effective quantity of water. Because the tank stores thermal energy for periods of at least several hours, heat loss/gain occurs as the tank interacts with the environment. The heat-retaining performance of a tank is an important factor in its design.

The sensible heat system most commonly employed at present uses water as the storage medium. An effective TES tank utilizing water as the storage medium satisfies the following general requirements.

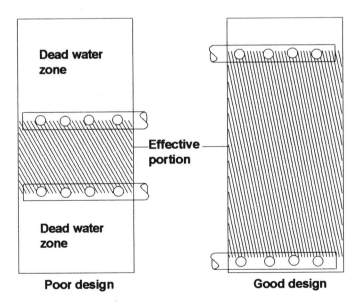

Fig. 3.7. The positions of the inlet and outlet, and the effective quantity of water for a thermally stratified thermal energy storage system (adapted from Shimizu and Fujita, 1985).

(1) The tank should be stratified, that is, hold separate volumes of water at different temperatures. Mixing of the volumes should be minimal, even during charging and discharging periods;

(2) The effective storage capacity should minimize the amount of dead water volume in the tank (see Figure 3.7); and

(3) The heat loss/gain from the tank should be minimized.

Many types of TES tanks have been developed to satisfy these requirements. The principal types are listed in Table 3.3.

A thermally naturally stratified storage tank has no internal partitions. Warm water has low density and floats to the top of the tank, whereas cooler water with higher density sinks to the bottom. The storage volume with this type of system is reduced relative to other systems because the dead water volume is relatively low and the energy efficiency is relatively high. When designing a thermally stratified thermal storage tank, the following criteria can guide the design process.

Geometrical considerations: A deep water-storage container is desirable for the purpose of improving thermal stratification. The water inlet and outlet should be installed in a manner that produces a uniform flow of water to avoid mixing. To minimize the dead water volume, the outlet and inlet connections should be located as close as possible to the top and bottom of the storage volume. The surface area in contact with the storage water should be minimized. The insulating and waterproofing characteristics of the tank should be designed to meet appropriate specifications.

Operating considerations: The temperature difference between the upper and lower parts of the tank should be large, at least 5 to 10°C. Controls can be used to maintain fixed water temperatures in the upper and lower parts of the tank. The velocity of the water flowing into and out of the tank should be low.

Rock storage bins in home air-heating systems are practical (Dincer and Dost, 1996). Rock is an inexpensive TES material, but its volumetric thermal capacity is much less than that of water. The advantage of rock over water is that it can easily be used for TES above 100°C. Rock-bed and water/rock storage types can be utilized in many ways, such as in conjunction with heat pumps to improve efficiency of heat recovery or with more elaborate heat exchangers.

Three configurations of storage tanks for sensible TES applications are shown in Figure 3.8. The most common TES system has a water-filled container in direct contact with the solar collector and the house heating system (Figure 3.8a). Cool water from the bottom of the tank is circulated to the collector for solar heating and then returned to the top of the tank. Warm water from the top of the tank is circulated directly through baseboard radiators or radiant heating panels inside the rooms. Figure 3.8b shows another system, which consists of a copper coiled-finned tube immersed in the tank

Table 3.3. Types and features of stratified thermal energy storage tanks

Type	Schematic representation of cross-section	Efficiency	Remarks
Continuous multitank type		Medium	Underground beam space can be used effectively. Insulation is difficult to install.
Improved dipped weir type		Medium to High	Construction is difficult.
Thermally stratified type		High	Best suited for large-size tank built above ground.
Movable diaphragm type		High	Diaphragm material is problematical. Not easily adapted to tanks with internal pillars and beams.
Multitank water renewing type		High	Underground beam space can be utilized to some extent. Heat loss is large.

Source: Adapted from Shimizu and Fujita (1985).

of solar-heated water. Rocks are the most widely used storage medium for air collectors.

One attractive storage method that uses both water and rocks as storage media is Harry Thomason's method (Figure 3.8c). Heated water from the solar

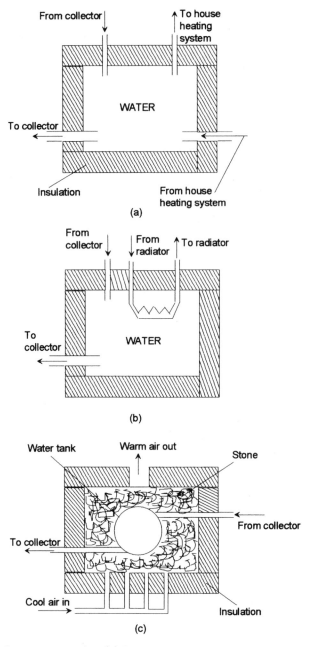

Fig. 3.8. Solar storage tanks: (a) heat storage tank tied directly to the collector and the house heating system; (b) sensible TES system using a heat exchanger to extract solar heat from a storage tank; and (c) Harry Thomason's technique using both water and stone as storage media.

collector enters a water tank at the top, sinks as it cools, and finally leaves at the tank bottom as it is recirculated to the collectors. The water tank is surrounded by river rock through which air is circulated to carry the heat into the house. The entire rock and water tank assembly is contained within insulated walls. The advantage of this system is that the high heat capacity of the water and the extensive area of the rocks container lead to efficient transfer of heat to air.

We end this section with an example to show that there is an opportunity to optimize the history of the storage process. The storage system (the left side of Figure 3.9) contains a batch of liquid (m, c). The liquid is held in an insulated vessel. The hot-gas stream of flow rate $\dot{m}$ enters the system through one port and is gradually cooled as it flows through a heat exchanger immersed in the liquid bath. The spent gas is discharged directly into the atmosphere. As time passes, the bath temperature T and the gas outlet temperature T_{out} approach the hot-gas inlet temperature T_∞.

If we model the hot gas (e.g., steam, products of combustion) as an ideal gas with constant specific heat c_P, the temperature history of the storage system is expressed in closed form by the equations

$$\frac{T(t) - T_0}{T_\infty - T_0} = 1 - \exp(-y\theta), \tag{3.9}$$

$$\frac{T_{\text{out}}(t) - T_0}{T_\infty - T_0} = 1 - y\exp(-y\theta), \tag{3.10}$$

where y and the dimensionless time θ are defined as

$$y = 1 - \exp(-N_{\text{tu}}), \qquad N_{\text{tu}} = \frac{\bar{h}_b A_b}{\dot{m} c_P}, \qquad \theta = \frac{\dot{m} c_P}{mc}t. \tag{3.11}$$

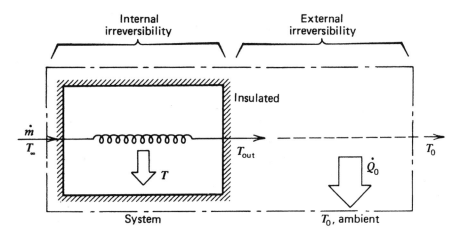

Fig. 3.9. Two sources of irreversibility in the heating (charging) stroke of a sensible heat storage process (Bejan, 1982).

In these equations, A_b is the total heat-exchanger surface area separating the stream from the liquid bath, $\bar{h}_b$ is the mean (in time and space) heat transfer coefficient based on A_b, and N_{tu} is the number of heat transfer units. Built into the model is the assumption that the liquid bath is well mixed, that is, that the liquid temperature (T) is a function of the time (t) only. As expected, both T and T_{out} approach T_∞ asymptotically. They approach T_∞ faster when N_{tu} is higher.

Turning our attention to the irreversibility of the energy-storage process, Figure 3.9 shows that the irreversibility is due to distinct parts of the apparatus: first, there is the finite-$\Delta T(t)$ irreversibility associated with the heat transfer between the hot stream and the cold liquid bath; and second, the stream exhausted into the atmosphere is eventually cooled down to T_0, again by heat transfer across a finite temperature difference. Neglected in the present model is the irreversibility due to the pressure drop across the heat exchanger traveled by the stream $\dot{m}$.

The combined effect of the competing irreversibilities noted in Figure 3.9 is a characteristic of all sensible heat storage systems. Because of it, only a fraction of the exergy content of the hot stream can be stored in the liquid bath. In order to see this, consider the instantaneous rate of entropy generation in the overall system delineated in Figure 3.9,

$$\dot{S}_{gen} = \dot{m}c_P \ln \frac{T_0}{T_\infty} + \frac{\dot{Q}_0}{T_0} + \frac{d}{dt}(mc \ln T), \qquad (3.12)$$

where $\dot{Q}_0 = \dot{m}c_P(T_{out} - T_0)$. The entropy generated during the entire charging-time interval $0 - t$ is important, and can be put in dimensionless form as

$$\frac{1}{mc}\int_0^t \dot{S}_{gen} dt = \theta \left(\ln \frac{T_0}{T_\infty} + \tau\right) + \ln(1 + \tau\eta_I) - \tau\eta_I, \qquad (3.13)$$

where $\tau = (T_\infty - T_0)/T_0$, and the first-law efficiency η_I is shorthand for the right side of Equation (3.9).

According to the Gouy-Stodola theorem, Equation (3.5), $\int_0^t \dot{S}_{gen} dt$ represents the destroyed exergy, the bite taken by irreversibilities out of the total exergy supply brought into the system by the hot stream:

$$E_x = t\dot{E}_x = t\dot{m}c_P \ln[T_\infty - T_0 - T_0 \ln(T_\infty/T_0)]. \qquad (3.14)$$

On this basis, we define the entropy-generation number N_S as the ratio of the lost exergy divided by the total exergy invested during the time interval $0 - t$:

$$N_S(\theta, \tau, N_{tu}) = \frac{T_0}{E_x}\int_0^t \dot{S}_{gen} dt = 1 - \frac{\tau\eta_I - \ln(1 + \tau\eta_I)}{\theta[\tau - \ln(1 + \tau)]}. \qquad (3.15)$$

The entropy-generation number takes values in the range 0 to 1, the $N_S = 0$ limit representing the theoretical case of reversible operation. Note the relation

$N_S = 1 - \eta_{\mathrm{II}}$, where η_{II} is the second-law efficiency of the installation during the charging process.

Charts of the $N_S(\theta, \tau, N_{\mathrm{tu}})$ surface show that N_S decreases steadily as the heat-exchanger size (N_{tu}) increases (Bejan, 1982). This effect is expected. Less expected is the fact that N_S goes through a minimum as the dimensionless time θ increases. For example, the optimal time for minimum N_S can be calculated analytically in the limit $\tau \ll 1$, where Equation (3.15) becomes

$$N_S = 1 - [1 - \exp(-y\theta)]^2/\theta \qquad (3.16)$$

and the optimal time is

$$\theta_{\mathrm{opt}} = 1.256[1 - \exp(-N_{\mathrm{tu}})]^{-1}. \qquad (3.17)$$

In conclusion, for the common range of N_{tu} values (1 to 10), the optimal dimensionless charging time is consistently a number of order 1. This conclusion continues to hold as τ takes values greater than 1. When the changing time differs from the optimal value (i.e., when $\theta \to 0$, or $\theta \to \infty$), N_S approaches 1. In the short-time limit ($\theta \ll \theta_{\mathrm{opt}}$), the entire exergy content of the hot stream is destroyed by heat transfer to the liquid bath, which was initially at environmental temperature T_0. In the long-time limit ($\theta \gg \theta_{\mathrm{opt}}$), the external irreversibility takes over: the used stream exits the heat exchanger as hot as it enters ($T_{\mathrm{out}} = T_\infty$), and its exergy content is destroyed entirely by the heat transfer (or mixing) with the T_0 atmosphere. The traditional (first-law) rule of thumb of increasing the time of communication between heat source and storage material is counterproductive from the point of view of avoiding the destruction of exergy.

3.5 Aquifer Thermal Energy Storage

An aquifer is a groundwater reservoir. The word aquifer derives from the Latin *aqua* (water) and *ferre* (to carry). The material in an aquifer is highly permeable to water, and the surrounding layer consists of more impermeable materials such as clay or rock. Water from precipitation seeps continuously into the aquifer, and flows slowly through it until finally reaching the lake or the sea. Aquifers are often used as sources of fresh water.

Aquifer systems have received worldwide attention because of their potential for large-scale and long-term TES. In its most common form, the technique involves storing excess heat in the aquifer, and recovering it later during periods of heat demand. With growing concerns about global warming, the concept is receiving renewed attention as a viable means of conserving energy and reducing fossil fuel use. In developing aquifer thermal energy storage (ATES) systems, the physical processes governing the behavior of thermal energy transport in groundwater must be well understood.

Aquifers have large volumes, often exceeding millions of cubic meters. They consist of about 25% water, and have a high TES capacity. When heat extraction and charging performance is good, high heating and cooling power can be achieved. The amount of energy that can be stored in an aquifer depends on local conditions such as allowable temperature change, thermal conductivity, and natural groundwater flows.

Figure 3.10 illustrates the operation principle of simple ATES systems. Both heating and cooling cycles for a building are considered. Wells have been drilled to transport water to underground aquifers, which are underground spaces filled with water-bearing rock, sand, or gravel. Well spacing, depth, and size are the features that define the aquifer.

An aquifer storage system can be used for storage periods ranging from long to short, including daily, weekly, seasonal, or mixed cycles. To avoid undesired permanent changes of the temperature level in the aquifer, the input and output of heat must be of the same magnitude at least after a number of cycles. The system should be designed to be adjustable in case long-term energy flows do not balance. The TES capacity should be sized to match the heating and cooling loads. Extensive investigations and test runs are usually needed to predict the performance of an ATES before the design of the energy system. Such preparatory work can be costly. Aquifers with groundwater, or the ground itself (soil, rock), can be used as a storage medium in ATES systems. Aquifer stores are most suited for high-capacity systems. The existing capacities range in size from less than 50 to over 10,000 kW.

Possible heat sources for many promising ATES systems can be divided into two main groups.

Renewable energy: Solar heat can be used with ATES to supply heat to district heating networks, along with backup auxiliary heating systems. Solar heat can also be used with heat pumps, thus avoiding the need for auxiliary heating. Another ATES use is in geothermal heating, which allows storage of

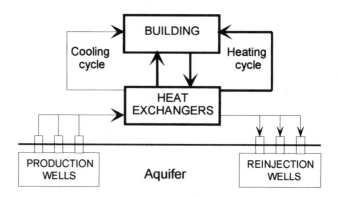

Fig. 3.10. Schematic of an aquifer thermal energy storage system.

excess production in summertime and to cover peak demand in winter, or for storing waste heat from geothermal power plants.

Waste or excess heat: Storage of waste heat from cogeneration or industrial processes may be needed during seasonal or other cycles. ATES can also be applied as backup for processes that use industrial waste heat, to cover the demand for heat during periods when the industrial process is interrupted (for production breaks, repairs, cleaning, etc.). Similarly, ATES can be used for load leveling in district heating systems, where the store is always charged at times of low heat demand, and is unloaded during peak heating periods.

3.6 Latent Heat Storage

Most practical systems with phase-change energy storage use solutions of salts in water. Several problems are associated with such designs. For example, supercooling of the phase-change material may take place instead of crystallization with heat release. This problem can be partially avoided by adding small crystals as nucleating agents. In addition, it is difficult to build a heat exchanger to handle the agglomeration of crystals of varying sizes, which float in the liquid. Finally, the system operation cannot be reversed completely.

A latent heat TES system must possess at least three components (Abhat, 1983): a heat storage substance that undergoes a phase transition within the desired operating temperature range, and where most of the added heat is stored as latent heat; a containment for storing the substance; and a heat-exchange surface for transferring heat from the heat source to the storage substance, and from the latter to the heat sink (e.g., from a solar collector to the latent-heat TES substance, to the load loop).

Latent heat TES is a promising storage technique because it provides a high-energy storage density, which is second only to chemical energy storage, and can store and release heat at a constant temperature, corresponding to the phase-transition temperature of the heat-storage medium. An important class of materials capable of storing energy through phase change is paraffin waxes. These have the advantage of very high stability over repeated cycles of latent TES operation without degradation.

Important design criteria to be met by the phase-change storage material are: high transition enthalpy per unit mass, ability to fully reverse the transition, adequate transition temperature, chemical stability and compatibility with the container (if present), limited volume change with the transition, nontoxicity, and low cost in relation to the application.

Common phase-change materials (PCMs) are water/ice, salt hydrates, and certain polymers. Because energy densities for latent heat TES exceed those for sensible heat TES, the latent heat devices are smaller and lighter, and have lower storage losses. In addition to ice and water mixtures, eutectic salts have been used as storage media for many decades. Perhaps the oldest application of a PCM for TES was the use of seat warmers for British railroad cars in the late 1800s. During cold winter days, PCM sodium thiosulfate pentahydrate (which melts and freezes at 44.4°C) filled into metal or rubber bags was used. Other early applications of PCMs included "eutectic plates" used for cold storage in trucking and railroad transportation applications. Another application was the thermal control of electronic packages for aerospace technology.

A large number of organic compounds suitable to be storage media for heating, ventilating and airconditioning (HVAC) applications have been investigated in recent years (Table 3.4). The most promising candidates are the normal paraffins. Their solid–liquid transitions (fusions) meet several important criteria satisfactorily: the heat of fusion has a mean value of 35 to 40 kcal/kg, there are no problems with reversing the phase change, and the phase-change temperatures vary considerably with the number of carbon atoms in the chains. Also, the normal paraffins are chemically inert, nontoxic, and available at reasonably low cost. The change of volume during transition, which is on the order of 10%, could pose problems in the design.

Zeolites are naturally occurring minerals. Their high heat of adsorption and ability to hydrate and dehydrate while maintaining structural stability have been found to be useful in various thermal storage and solar refrigeration systems. This hygroscopic property, coupled with the rapid exothermic reaction that occurs when zeolites are taken from a dehydrated to a hydrated form (when the heat of adsorption is released), makes natural zeolites effective storage materials for solar and waste heat energy.

Low energy density and time of availability have been key problems in the use of solar energy and waste heat. Zeolite systems are capable of utilizing solar heat, industrial waste heat, and heat from other sources, thereby converting underutilized resources into useful energy. The capacity of natural zeolites to store thermal energy and adsorb the water vapor used in that energy interaction comes from their honeycomb structure and resultant high internal surface area.

Application of PCMs in the temperature range 0 to 120°C is of interest for a variety of low-temperature applications, such as space heating, domestic hot water production, heat-pump assisted space heating, greenhouse heating, and solar cooling. The development of dependable TES systems for these and other applications requires a good understanding of heat-of-fusion storage materials and heat exchangers.

Commercial paraffins are characterized by two phase transitions (solid–liquid and solid–solid) which occur over a large temperature range depending on the paraffin type. For example, n-paraffins are usually preferred next to their iso-counterparts, as the desired solid-to-liquid phase transition is generally restricted to a narrow temperature range. Fatty acids are organic materials with excellent melting and freezing characteristics, and may have a good future if their costs can be reduced. Inorganic salt hydrates, on the other hand,

Table 3.4. Thermophysical data of phase-change materials

Compound	Melting temp. (°C)	Heat of fusion (kJ/kg)	Thermal conductivity (W/mK)	Density (kg/m³)
Inorganics				
$MgCl_2 \cdot 6H_2O$	117	168.6	0.570 (liquid, 120°C) 0.695 (solid, 90°C)	1450 (liquid, 120°C) 1569 (solid, 20°C)
$Mg(NO_3)_2 \cdot 6H_2O$	89	162.8	0.490 (liquid, 95°C) 0.611 (solid, 37°C)	1550 (liquid, 94°C) 1636 (solid, 25°C)
$Ba(OH)_2 \cdot 8H_2O$	78	265.7	0.653 (liquid, 87.7°C) 1.255 (solid, 23°C)	1937 (liquid, 84°C) 2070 (solid, 24°C)
$Zn(NO_3)_2 \cdot 6H_2O$	36	146.9	0.464 (liquid, 39.9°C) —	1828 (liquid, 36°C) 1937 (solid, 24°C)
$CaBr_2 \cdot 6H_2O$	34	115.5	— —	1956 (liquid, 35°C) 2194 (solid, 24°C)
$CaCl_2 \cdot 6H_2O$	29	190.8	0.540 (liquid, 38.7°C) 1.088 (solid, 23°C)	1562 (liquid, 32°C) 1802 (solid, 24°C)
Organics				
Paraffin wax	64	173.6	0.167 (liquid, 63.5°C) 0.346 (solid, 33.6°C)	790 (liquid, 65°C) 916 (solid, 24°C)
Polyglycol E400	8	99.6	0.187 (liquid, 38.6°C) —	1125 (liquid, 25°C) 1228 (solid, 3°C)
Polyglycol E600	22	127.2	0.189 (liquid, 38.6°C) —	1126 (liquid, 25°C) 1232 (solid, 4°C)
Polyglycol E6000	66	190.0	— —	1085 (liquid, 70°C) 1212 (solid, 25°C)

Table 3.4. *Continued*

Compound	Melting temp. (°C)	Heat of fusion (kJ/kg)	Thermal conductivity (W/mK)	Density (kg/m³)
Fatty acids				
Stearic acid	69	202.5	—	848 (liquid, 70°C)
				965 (solid, 24°C)
Palmitic acid	64	185.4	0.162 (liquid, 68.4°C)	860 (liquid, 65°C)
			—	989 (solid, 24°C)
Capric acid	32	152.7	0.153 (liquid, 38.5°C)	878 (liquid, 45°C)
			—	1004 (solid, 24°C)
Caprylic acid	16	148.5	0.149 (liquid, 38.6°C)	901 (liquid, 30°C)
			—	981 (solid, 13°C)
Aromatics				
Biphenyl	71	119.2	—	991 (liquid, 73°C)
			—	991 (liquid, 73°C)
Naphthalene	80	147.7	0.132 (liquid, 83.8°C)	976 (liquid, 84°C)
			0.341 (solid, 49.9°C)	1145 (solid, 20°C)

Source: Adapted from Lane (1988).

must be carefully examined (with the aid of phase diagrams) for congruent, "semi-congruent," or incongruent melting substances. Incongruent melting in a salt hydrate may be "modified" to overcome decomposition by adding suspension media, or extra water, or other substances that shift the peritectic point. The use of salt hydrates in hermetically sealed containers is normally recommended. Also, the employment of metallic surfaces to promote heterogeneous nucleation in a salt hydrate is seen to reduce the supercooling of most salt hydrates to a considerable extent. Thermal cycling and corrosion behavior are also of importance while choosing materials, because of their influence on the life of a latent heat store.

Numerous organic and inorganic PCMs melt with a high heat of fusion in the temperature range 0 to 120°C. For employment as heat storage materials in TES systems, however, PCMs must possess additional desirable features (Abhat, 1983; Dincer *et al.*, 1997b).

Thermodynamic criteria: The melting point is at the desired operating temperature, there is high latent heat of fusion per unit mass, so that a smaller amount of material stores a given amount of energy, there is high density, so that less volume is occupied by the material, there is high specific heat, so that significant sensible TES can also occur, there is high thermal conductivity, so that small temperature differences are needed for charging and discharging the storage, congruent melting, (i.e., the material should melt completely), so that the liquid and solid phases are homogeneous (this avoids the difference in densities between solid and liquid that causes segregation, resulting in changes in the chemical composition of the material), and there are small volume changes during phase transition, so that a simple containment and heat exchanger can be used.

Kinetic criteria: There is little or no supercooling during freezing (i.e., the melt should crystallize at its freezing point). This can be achieved through a high rate of nucleation and growth rate of the crystals. Supercooling may be suppressed by introducing a nucleating agent or a cold trigger in the storage material.

Chemical criteria: These include chemical stability, no susceptibility to chemical decomposition, so that a long operation life is possible, noncorrosive behavior to construction materials, and nontoxic, nonflammable, and nonexplosive characteristics.

Technical and economic criteria: These include simplicity, applicability, compatibility, reliability, commercial availability, and low cost.

To improve the performance of latent heat TES with PCMs, nucleating agents and thickeners have been used to prevent subcooling and phase separation. Extended heat transfer surfaces can be used to enhance the heat transfer from the PCM to the heat transfer tubes. Although many studies on the latent TES systems have been performed at relatively low temperatures (below 100°C) for TES in home heating and cooling units, few studies have been undertaken for higher temperature heat (above 200°C), as is the case in some solar energy systems and intermediate-temperature latent heat

TES. Magnesium chloride hexahydrate ($MgCl_2 \cdot 6H_2O$), with a melting temperature 116.7°C, is an attractive high-temperature PCM in terms of cost, material compatibility, and thermophysical properties (Choi and Kim, 1995). Sharma *et al.* (1992) showed that the acetamide–sodium bromide eutectic is a promising latent heat TES material which could find uses in commercial and laundry water heating, process heating, domestic water and air heating, crop drying, and food warming.

Sodium acetate trihydrate deserves special attention for its large latent heat of fusion-crystallization (264 to 289 kJ/kg) and its melting temperature of 58 to 58.4°C. However, this substance exhibits significant subcooling, preventing most practical large-scale applications, even though attempts have been made to find ways of suppressing this phenomenon.

Simulations of a PCM wall as a TES in a passive direct gain solar house suggest that the PCM melting temperature should be adjusted from the climate-specific optimum temperature to achieve maximum performance of the storage. A nonoptimal melting temperature significantly reduces the latent heat storage capacity. For example, a 3°C departure from the optimal temperature causes a 50% loss (Dincer and Dost, 1996).

The type of energy storage material can be chosen based on thermodynamic optimization (Lim *et al.*, 1992a). In the model of Figure 3.11 the hot stream of initial temperature T_∞ comes in contact with a single phase-change material through a finite thermal conductance UA, assumed known, where A is the heat transfer area between the melting material and the stream, and U is the overall heat transfer coefficient based on A. The phase-change material (solid or liquid) is at the melting point T_m. The stream is well mixed at the temperature T_{out}, which is also the temperature of the stream exhausted into the atmosphere (T_0).

The steady operation of the installation modeled in Figure 3.11 accounts for the unsteady (cyclic) operation in which every infinitesimally short storage (melting) stroke is followed by a short energy retrieval stroke: $\dot{m}$ is stopped, and the recently melted phase-change material is solidified to its original state by the cooling effect provided by the heat engine positioned between T_m and T_0. In this way, the steady-state equivalent model of Figure 3.11 represents the complete cycle, that is, storage followed by retrieval.

The steady cooling effect of the power plant can be expressed in two ways:

$$\dot{Q}_m = UA(T_{\text{out}} - T_m), \tag{3.18}$$

$$\dot{Q}_m = \dot{m}c_P(T_\infty - T_{\text{out}}). \tag{3.19}$$

By eliminating T_{out} between these two equations we obtain

$$\dot{Q}_m = \dot{m}c_P \frac{N_{\text{tu}}}{1 + N_{\text{tu}}}(T_\infty - T_m), \tag{3.20}$$

in which N_{tu} is the number of heat transfer units of the heat exchanger surface $N_{\text{tu}} = UA/(\dot{m}c_P)$.

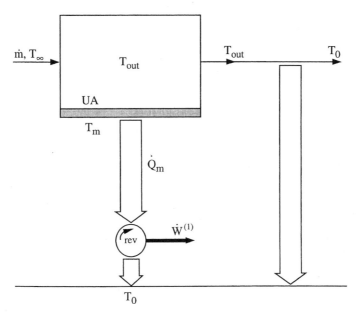

Fig. 3.11. The generation of power using a phase-change material and a hot stream that is ultimately discharged into the ambient (Lim *et al.*, 1992a).

The objective is the minimization of entropy generation, or the calculation of the maximum rate of exergy (useful work, $\dot{W}$ in Figure 3.11) that can be extracted from the phase-change material. For this, we model as reversible the cycle executed by the working fluid between T_m and T_0,

$$\dot{W} = \dot{Q}_m \left(1 - \frac{T_0}{T_m} \right) \tag{3.21}$$

and, after combining with Equation (3.20), we obtain

$$\dot{W} = \dot{m}c_p \frac{N_{\text{tu}}}{1 + N_{\text{tu}}} (T_\infty - T_m) \left(1 - \frac{T_0}{T_m} \right). \tag{3.22}$$

By maximizing $\dot{W}$ with respect to T_m, that is, with respect to the type of phase-change material, we obtain the optimal melting and solidification temperature:

$$T_{m,\text{opt}} = (T_\infty T_0)^{1/2}. \tag{3.23}$$

The maximum power output that corresponds to this optimal choice of phase-change material is

$$\dot{W}_{\text{max}} = \dot{m}c_P T_\infty \frac{N_{\text{tu}}}{1 + N_{\text{tu}}} \left[1 - \left(\frac{T_0}{T_\infty} \right)^{1/2} \right]^2. \tag{3.24}$$

The same results, Equations (3.23) and (3.24), could have been obtained by minimizing the total rate of entropy generation. One way to improve the power output of the single-element installation of Figure 3.11 is by placing the exhaust in contact with one or more phase-change elements of lower temperatures (Lim *et al.*, 1992a).

The thought that unites the examples of Figures 3.9 and 3.11 with many more applications of thermodynamic optimization is that the physical result of global optimization of thermodynamic performance is *structure* (configuration, topology, geometry, architecture, pattern). This structure-generating principle deserves to be pursued further, in increasingly more complex system configurations, as we show in Chapter 5. The generation of structure in engineering has been named *constructal design*; the idea that the same principle accounts for the generation of shape and structure in natural flow systems is *constructal theory* (Bejan, 2000).

3.7 Cold Thermal Energy Storage

Cooling capacity can be stored either by chilling or freezing water (or such other materials as glycol and eutectic salts). Water is the storage material of choice for a variety of practical and thermodynamic reasons, including its ready availability, relative harmlessness, and its compatibility with a wide availability of equipment for its storage and handling.

Cold thermal energy storage (CTES) is an innovative way of storing nighttime off-peak energy for daytime peak use. In many locations the demand for electrical power peaks during summer. Air-conditioning is the main reason, in some areas accounting for as much as half of the power demand during the hot midday hours when electricity is most expensive. Because at night utilities have spare power-generating capacity, electricity generated during this off-peak period is much cheaper. In essence, one can produce air conditioning during the day using electricity produced at night. CTES has become one of the primary means of addressing the electrical power imbalance between high daytime demand and high nighttime abundance.

CTES has many advantages. Lower nighttime temperatures allow the refrigeration equipment to operate more efficiently than during the day, reducing energy consumption. Lower chiller capacity is required, which leads to lower equipment costs. Furthermore, by using off-peak electricity to store energy for use during peak demand hours, daytime peaks of power demand are reduced, sometimes deferring the need to build new power plants. CTES systems are most likely to be cost effective when the maximum cooling load is much greater than the average load; the utility rate structure has high demand charges, ratchet charges, or a high differential between on- and off-peak energy rates; an existing cooling system is being expanded; an existing tank is available; limited onsite electric power is available; and/or backup cooling capacity and cold-air distribution are desirable.

For example, an ice-ball system uses chillers to build ice at night. The ice balls float in a glycol solution that runs through chillers in the evening. The chillers are set at -7.5 to $-6.5°C$. The ice balls are frozen in the storage tanks, and the glycol circulates around the ice balls. Chilled glycol is pumped into the bottom of the tank to freeze the ice balls, and warms as it rises and extracts heat from the ice balls. Later, the cycle is reversed and the glycol is pumped into the top of the tank and past the ice balls. The cold glycol solution then passes through heat exchangers and is delivered to the chilled water system of the building.

Two strategies are used for charging and discharging a storage to meet cooling demand during peak hours: full storage and partial storage.

The full-storage strategy consists of shifting the entire peak cooling load to off-peak hours (Figure 3.12). The system is typically designed to operate on the hottest anticipated days at full capacity during all nonpeak hours. This strategy is most attractive when peak demand charges are high or the peak period is short. Full storage (load-shifting) designs are those that use storage to decouple completely the operation of the heating or cooling generating equipment from the peak heating or cooling load. The peak heating or cooling load is met through the use (i.e., the discharging) of storage, while the heating or cooling generating equipment is idle. Full storage systems are likely to be economically advantageous only under the following conditions: spikes in the peak load curve are of short duration, time-of-use energy rates are based on short-duration peak periods, there are short overlaps between peak loads and peak energy periods, large cash incentives are offered for using TES, and high peak demand charges apply.

Partial storage means that the chiller operates to meet only a part of the peak-period cooling load, and the rest of the load is met by drawing from storage. The chiller is sized at a smaller capacity than the design load. Partial storage systems may operate as load-leveling or demand-limiting operations. In a load-leveling system (Figure 3.12b), the chiller is sized to run at its full

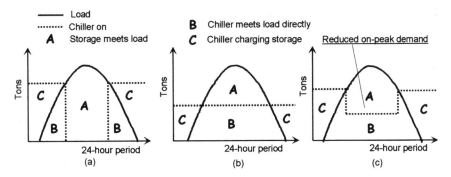

Fig. 3.12. Operating strategies: (a) full storage; (b) partial storage load leveling; and (c) partial storage demand-limiting.

capacity for 24 hours on the hottest days. The strategy is most effective where the peak cooling load is much higher than the average load. In a demand-limiting system, the chiller runs at reduced capacity during peak hours, and is often controlled to limit the peak demand charge of the facility (Figure 3.12c). Demand savings and equipment costs are higher than they would be for a load-leveling system, and lower than for a full storage system. Partial storage is often the most economical option, and consequently it is used in the majority of thermal storage installations. Although partial storage does not shift as much load (on a design day) as a full storage system, partial storage systems can have lower initial costs, particularly if the design incorporates smaller equipment by using low-temperature water and cold-air distribution systems.

The storage medium determines the size of the storage tank and the size and configuration of the HVAC system and components. The main options include chilled water, ice, and eutectic salts. Ice systems offer the densest storage capacity but have the most complex charge and discharge equipment. Water systems offer the lowest storage density, and are the least complex. Eutectic salts have in-between characteristics. Additional details can be found in Dincer and Rosen (2002).

To increase compactness, CTES systems have been developed so that they utilize latent heat, usually solid–liquid transitions. The most commonly used today are systems based on water and ice. A major drawback of ice thermal storage is that the production of ice requires that the chiller evaporator temperature must be lower than that for a water CTES, and, as a consequence, the chiller capacity and coefficient of performace (COP) are decreased. For water CTES the evaporation temperature of an ordinary chiller is in the vicinity of $0°C$. For ice CTES, however, evaporator temperatures are generally below $-10°C$, so that the capacity and COP are reduced to about 56% and, respectively, 71%, of those of a water CTES system. A capacity drop leads to an increase in the size and cost of the chiller, whereas a COP drop leads to an increase in energy cost. To control both costs, it is desirable to store the minimum amount of ice required by the available space.

Ice CTES systems can be economically advantageous and require less space than water CTES systems. An ice CTES system with heat pump is composed of a heat pump, an ice-making system, a storage tank, and an air conditioning system, which can be a conventional central system. Two basic types of commercial ice storage systems exist: static (ice building) and dynamic (ice shucking). In the static system, ice is formed on the cooling coils within the storage tank itself (ice-on-coil). Such systems are normally manufactured in packaged units, and are connected to the chilled water system of the building. Dynamic systems, which are becoming popular, make ice in chunks or crushed form (ice harvesting) and deliver it for storage in large pits similar to those used in chilled water systems. Dynamic systems may also include the formation of an ice glycol slurry, which can be stored in a tank. Static ice storage systems are designed to form ice on the surface of evaporator tubes, and to store it until chilled water is needed for cooling. The ice is melted by the warm return water, thereby recooling the water before it is pumped back to the coils in the building. Other static systems use a brine that circulates through tubes in an ice block or around containers filled with frozen water. These systems have the advantage that they are not open to the atmosphere.

PCMs are currently receiving increasing attention for use in CTES. Suitable for CTES are eutectic salts that undergo liquid–solid phase changes at temperatures as high as 8.3°C, and absorb and release large amounts of energy during phase change. Stored in hermetically sealed plastic containers, PCMs change to solids as they release heat to water or another fluid that flows around them. At such temperatures, chillers can operate more efficiently than at the low temperatures required by ice CTES systems. PCMs also have about three times the storage capacity of a typical chilled water CTES system. The choice of ice, chilled water, or PCMs depends on need. If low temperatures are needed, ice is the likely choice. If more conventional temperatures are needed and space is available, chilled water CTES can be installed. If space is limited, low temperatures are not required, and an easy-to-maintain system is needed, then PCM storage may be the right choice.

Two commercially available materials that enhance the ice-water phase-change process are eutectic salts and gas hydrates. Eutectic salts are mixtures of inorganic salts, water, and additives. Gas hydrates are produced by mixing gas with water. Both materials work by raising the temperature at which water freezes. They have the advantage of a freezing point of 8.3°C or 8.8°C, which reduces refrigerator power requirements.

PCMs also provide most of the storage space advantages associated with ice storage systems. By freezing and melting in the 8.3 to 8.8°C range, the PCMs can be easily used in conventional chilled water systems with centrifugal or reciprocating chillers. The storage tank can be placed above or below the level of the chiller system. In addition, chiller power requirements are reduced when phase-change materials are used for TES, because evaporative temperatures remain fairly constant.

Gas hydrates, still in the development stage for large commercial installations, have some advantages over eutectic salts. Gas hydrates have high latent heat values, which lead to size and weight advantages. Gas hydrates require only one-half to one-third of the space, and are approximately one-half the weight of an equivalent eutectic salt system. It is worth adding that methane hydrates are a naturally occurring sediment, which is recognized as a future fuel source; see Section 8.3.

The hydrated salt most commonly used for CTES applications changes phase at 8.3°C, and is often encapsulated in plastic containers. The material is a mixture of inorganic salts, water, and nucleating and stabilizing agents. It has a latent heat of fusion of 95.36 kJ/kg and a density of 1490 kg/m^3. A CTES using this PCM latent heat of fusion requires a capacity of about 0.155 m^3/ton-hour for the entire tank assembly, including piping headers.

The selection of a CTES system should account for the load, that is, the discharge rate, the energy discharged at all times during the discharge cycle, and the required storage inlet and outlet temperatures. Based on this information, a performance data schedule should be established to describe the system needs (Dincer and Rosen, 2002). Rating a CTES system on only energy storage capacity does not accurately account for the purpose of the system. The CTES system must be capable of the same performance as the conventional chiller that it supplements or replaces. In particular, CTES must provide stored energy when the building requires it. Performance specifications that recognize this complexity can help in achieving successful CTES projects. The selection among available system options depends upon site-specific factors. Some of the factors that enter into this decision are (Wylie, 1990): space availability, efficiency, chilled water temperatures, refrigeration compressor size, and maintenance costs.

3.8 Porous Medium Model of a Storage System with Phase-Change Material

A good understanding of the heat transfer processes that are involved is essential for predicting the behavior and performance of a TES, and for avoiding costly system overdesign. Modeling the thermal behavior of latent heat storage systems is much more complex than the modeling of sensible heat storage systems. Complexities are due to the nonlinear motion of the solid–liquid interface, the possible presence of buoyancy-driven flows in the melt, the conjugate heat transfer between the encapsulated PCM and the heat transfer fluid (HTF) in the flow channels, and the volume expansion of the PCM upon melting or solidification. For a general discussion of modeling and numerical techniques for the solution of solid–liquid phase-change, the reader is referred to the books of Crank (1975) and Alexiades and Solomon (1993), and to the recent review by Voller (1997).

In this section we outline the modeling of a PCM system as a porous medium with flow through it. The main component of a packed-bed thermal energy storage unit is the insulated vessel containing a large number of PCM capsules (Figure 3.13). This unit may be used, for example, in solar thermal and waste recovery systems. A heat transfer fluid flows around the capsules for heat storage and heat recovery. Detailed modeling of the heat transfer and fluid flow processes that take place in such a complex arrangement of PCM capsules is impracticable. The porous medium approach is more effective (Ismail and Stuginsky, 1999; Adebiyi et al., 1996; Saborio-Aceves et al., 1994; Goncalves and Probert, 1993; Beasley et al., 1989; Kondepudi et al., 1988; Torab and Beasley, 1987).

The basic assumption on which porous medium models rest is that the PCM capsules and the fluid can be modeled (averaged) as a continuous medium, and not as a medium comprised of independent components. As a result, coupled heat transfer equations may readily be formulated for both the PCM and the HTF. As an illustrative example, in the system depicted

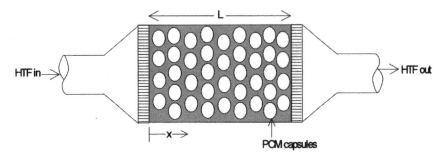

Fig. 3.13. Thermal energy storage system modeled as a porous medium with unidirectional flow.

in Figure 3.13, it is assumed that the thermophysical properties of the PCM and HTF are temperature-independent. Heat generation and chemical reactions are not present in the bed. The fluid flow through the void spaces is regarded as plug flow and conditions are uniform in the transverse direction. Consequently, transverse dispersion is neglected and the energy conservation equations become one-dimensional in space. Radiation heat transfer is negligible and the thermal gradients within the PCM capsules are ignored. The energy equations for the HTF and PCM can be written as

$$\phi(\rho c)_{\text{HTF}} \left(\frac{\partial T_{\text{HTF}}}{\partial t} + v_{\text{HTF}} \frac{\partial T_{\text{HTF}}}{\partial x} \right)$$

$$= \frac{\partial}{\partial x} \left(k_{\text{HTF}} \frac{\partial T_{\text{HTF}}}{\partial x} \right) + U_{\text{HTF}} A_{\text{bed}} (T_{\text{PCM}} - T_{\text{HTF}}), \qquad (3.25)$$

$$(1 - \phi)(\rho c)_{\text{PCM}} \frac{\partial T_{\text{PCM}}}{\partial t}$$

$$= \frac{\partial}{\partial x} \left(k_{\text{PCM}} \frac{\partial T_{\text{PCM}}}{\partial x} \right) + U_{\text{HTF}} A_{\text{bed}} (T_{\text{HTF}} - T_{\text{PCM}}), \qquad (3.26)$$

where U_{HTF} is the heat transfer coefficient between HTF and PCM, which is assumed constant, A_{bed} is the heat transfer area of particles (per unit bed volume), v_{HTF} is the mean HTF flow velocity, and ϕ is the void fraction (porosity) of the bed. The absorption of release of latent heat during phase change is taken into account via the relation between the heat capacity of the PCM and temperature (Goncalves and Probert, 1993). Equations (3.25) and (3.26) represent the two-temperature porous medium model, according to which every point inside the system is represented by two temperatures, T_{HTF} and T_{PCM}. The HTF and PCM are not in local thermal equilibrium. The initial and boundary conditions are specified by (Lacroix, 2002):

$$T_{\text{HTF}}(x, 0) = T_{\text{PCM}}(x, 0) = T_{\text{ini}}; \qquad T_{\text{HTF}}(0, t) = T_{\text{inlet}};$$

$$\frac{\partial T_{\text{HTF}}(L, t)}{\partial x} = 0; \qquad \frac{\partial T_{\text{PCM}}(0, t)}{\partial x} = \frac{\partial T_{\text{PCM}}(L, t)}{\partial x} = 0.$$

The finite-difference equations are obtained upon integrating the governing Equations (3.25) and (3.26) over the gridpoint cluster shown in Figure 3.14. We are assuming v_{HTF} constant and uniform throughout the system. We are using central differencing in space and forward differencing in the time domain. The resulting form of the finite-difference scheme for the HTF becomes

$$a_P(T_{\mathrm{HTF}})_P = a_W(T_{\mathrm{HTF}})_W + a_E(T_{\mathrm{HTF}})_E + b, \qquad (3.27)$$

where

$$a_W = \frac{k_{\mathrm{HTF}}}{\delta x_w} + \varepsilon(\rho C)_{\mathrm{HTF}} v_{\mathrm{HTF}}; \qquad a_E = \frac{k_{\mathrm{HTF}}}{\delta x_e};$$

$$a_P = a_W + a_E + a_P^0 + U_{\mathrm{HTF}} A_{\mathrm{bed}} \Delta x;$$

$$a_P^0 = \frac{\varepsilon(\rho C)_{\mathrm{HTF}} \Delta x}{\Delta t}; \qquad b = a_P^0 T_P^0 + U_{\mathrm{HTF}} A_{\mathrm{bed}} \Delta x (T_{\mathrm{PCM}})_P.$$

The corresponding form for the PCM is

$$a_P(T_{\mathrm{PCM}})_P = a_W(T_{\mathrm{PCM}})_W + a_E(T_{\mathrm{PCM}})_E + b, \qquad (3.28)$$

where

$$a_W = \frac{k_{\mathrm{PCM}}}{\delta x_w}; \qquad a_E = \frac{k_{\mathrm{PCM}}}{\delta x_e}; \qquad a_P = a_W + a_E + a_P^0 + U_{\mathrm{HTF}} A_{\mathrm{bed}} \Delta x;$$

$$a_P^0 = \frac{(1-\varepsilon)(\rho C)_{\mathrm{PCM}} \Delta x}{\Delta t};$$

$$b = a_P^0 T_P^0 + U_{\mathrm{HTF}} A_{\mathrm{bed}} \Delta x (T_{\mathrm{HTF}})_P. \qquad (3.29)$$

The old (known) values of the HTF and PCM temperatures are denoted by T_{HTF}^0 and T_{HTF}^0, respectively, and the new values are denoted by T_{HTF} and T_{PCM}. Because the above equations are coupled via the source term b, both are solved iteratively at a given time step, with a tri-diagonal matrix solver.

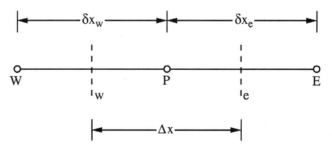

Fig. 3.14. Gridpoint cluster in one dimension.

This model has been used to predict the thermal behavior of latent heat TES systems containing many PCM capsules (Adebiyi et $al.$, 1996; Saborio-Aceves et $al.$, 1994). These studies also documented system-performance effect due to the temperature and mass flow rate of the flue gas, the porosity of the bed, and the dimensions of the vessel. Other factors (e.g., the thermal mass of the containment vessel wall) have been studied by coupling the model with the heat diffusion equation for the wall. Ismail and Stuginsky (1999) have also extended the above model to account for thermal diffusion in the transverse direction of the bed.

Here we study the effect of key parameters (e.g., T_{HTF}, T_{PCM}, ϕ, x, A_{bed}, v) on a PCM capsule with a heat transfer fluid flowing through it. The fluid temperature decreases along the PCM capsule, and the PCM temperature increases. However, the increase in the PCM temperature is not as significant as a decrease in the fluid temperature.

Figure 3.15 shows the variations of T_{HTF} and T_{PCM} versus the length of the porous PCM storage system. The void fraction was varied in the range $0.3 \leq \phi \leq 0.7$. The T_{HTF} and T_{PCM} curves show that the ϕ effect is imperceptible, and that T_{PCM} is nearly constant in the longitudinal direction. We have not studied the effect of ϕ on v_{HTF}.

Figure 3.16 shows the temperature distributions at different fluid velocities. The HTF temperature approaches the PCM temperature as v decreases,

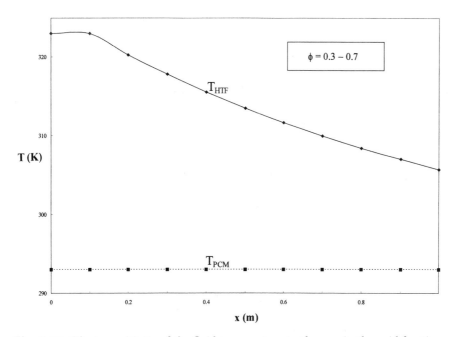

Fig. 3.15. The insensitivity of the fluid temperature to changes in the void fraction, and the nearly constant temperature of the phase-change material.

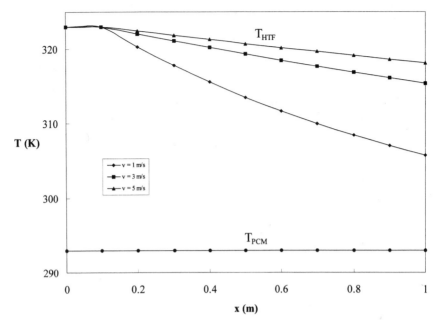

Fig. 3.16. The effect of the fluid velocity on the fluid temperature, and the nearly constant temperature of the phase-change material.

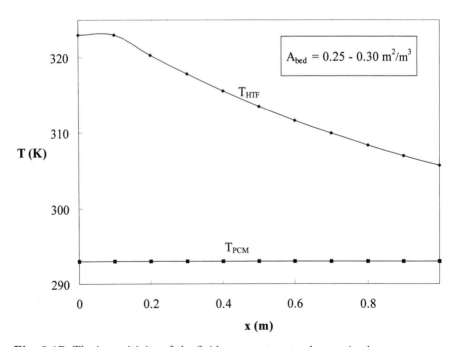

Fig. 3.17. The insensitivity of the fluid temperature to changes in the contact area per unit volume, and the nearly constant temperature of the phase-change material.

and T_{PCM} is nearly constant along x. Figure 3.17 shows the distribution of T_{HTF} and T_{PCM} versus the length of the porous PCM storage system at two different A_{bed} values, where A_{bed} is the coolant area of the particles per unit volume. The effect of A_{bed} is negligible.

The mathematical complexity of the heat transfer models increases when emphasis is placed on the physical phenomena occurring inside the PCM. In the porous medium approach, models focus on the overall thermal behavior of the system. The details of the heat transfer and phase-change phenomena inside individual PCM capsules are lumped into empirical coefficients, and the resulting mathematical models are much simpler. At the other end of the spectrum, models on convection-dominated phase change attempt to predict the heat transfer and the buoyancy-driven flows inside the PCM capsules. The resulting mathematical models are, in this case, considerably more elaborate (Lacroix, 2002). The accurate prediction of the transient heat transfer in the presence of phase change is of paramount importance, because it is at the heart of calculating the system capacity (Sahin and Dincer, 2000).

3.9 Fuel Cell Principles and Operation

Hydrogen is one of the most promising energy carriers for the future. The use of hydrogen as a fuel is inherently very clean. Hydrogen consumed by either combustion or a fuel cell produces only water as an end product. The high temperatures of combustion may stimulate the production of NOx from nitrogen and oxygen in the air, but this problem is common in fuels, and can be controlled. Unlike other fuels, hydrogen contains no pollutant-producing elements, and cannot produce SO_2, CO, CO_2, volatile organic chemicals, and the like. Fuel cells, which employ hydrogen to produce electricity, can be used to power a wide variety of applications for both stationary and mobile power generation. The interest in fuel cells has been motivated by their high efficiency even in small-scale installations, and by their low waste emissions.

The power generation principle of the fuel cell was discovered over 160 years ago by a Welsh judge, Sir William Grove. Until recently, fuel cell use was confined to the laboratory and to space applications, where fuel cells provide electricity, heat, and water. They have done so since the 1960s, when they were chosen over riskier, less reliable options. At that time, the technology was immature and far too expensive for terrestrial applications. Fuel cell technology is considered clean, quiet, and flexible and is already beginning to serve humanity in a variety of useful ways. Recently, interest in fuel cells has increased sharply, and progress toward commercialization has accelerated. As costs fall to competitive levels, practical fuel cell systems are becoming available and are expected to take a growing share of the markets for automotive power and generation equipment.

Fuel cells are efficient power producers, and create electricity in one simple step, with no moving parts, and at a very low temperature (at least in the case of proton exchange membrane fuel cells, or PEMFC). Because fuel cells do not burn fossil fuels, they emit none of the acid rain or smog-producing pollutants that are the inevitable by-product of burning coal, oil, or natural gas.

Here we outline the fundamentals of fuel cells. For additional details readers are directed to Hirschenhofer *et al.* (1998). A fuel cell operates like a battery, however, unlike the battery, it does not run down or require recharging. The fuel cell produces energy in the form of electricity and heat as long as a fuel is supplied. Fuel cells convert the chemical energy of a reaction directly into electrical energy without combustion where the oxygen comes from the air. Normally, in a fuel cell gaseous fuels are fed continuously to the anode (i.e., the negative electrode) and an oxidant (i.e., oxygen from air) is fed continuously to the cathode (i.e., the positive electrode). As a result, electrochemical reactions occur at the electrodes to produce electric current. The only by-products are water and heat (Figure 3.18a). No pollutants are produced if pure hydrogen is used. However, very low levels of nitrogen oxides are emitted, but usually in the undetectable range. The CO_2 emissions from electrochemical conversion are relatively low because of the high efficiency, and are in concentrated form. The basic physical structure or building block of a fuel cell consists of an electrolyte layer in contact with a porous anode and cathode on either side. A schematic representation of a fuel cell with the reactant/product gases and the ion conduction flow directions through the cell is shown in Figure 3.18b.

The most common classification of fuel cells is by the type of electrolyte used in the cells and includes (Dincer, 2002a): phosphoric acid fuel cell (PAFC), molten carbonate fuel cell (MCFC), solid oxide fuel cell (SOFC), proton exchange membrane (polymer) electrolyte fuel cell (PEMFC or PEFC), alkaline fuel cell (AFC), and direct methanol fuel cell (DMFC). In summary, there are a number of different types of fuel cells that are being developed. The characteristics of each type differ with respect to operating temperature, available heat, tolerance to thermal cycling, power density, tolerance to fuel impurities, and so on. These differences make each technology suitable for particular applications. They are also at very different stages of development. Some have not yet fully emerged from the laboratory. Table 3.5 shows the most common fuel cells and their operating temperature ranges and applications.

Figure 3.19 shows an expanded view of a PAFC in a fuel cell stack with its essential components. Individual fuel cells are combined and interconnected to provide the necessary voltage level. Here, the interconnect becomes a separator plate to provide an electrical series connection between adjacent cells, and a gas barrier that separates the fuel and oxidant of adjacent cells. The rest of the fuel cell consists of the structure for distributing the reactant gases across the electrode surface, serving as a mechanical support, electrolyte reservoirs for liquid electrolyte cells, and cells.

The impact of variables such as temperature, pressure, and gas constituents on fuel cell performance needs to be assessed in order to predict how the cell will interact with the power plant system supporting it. In the

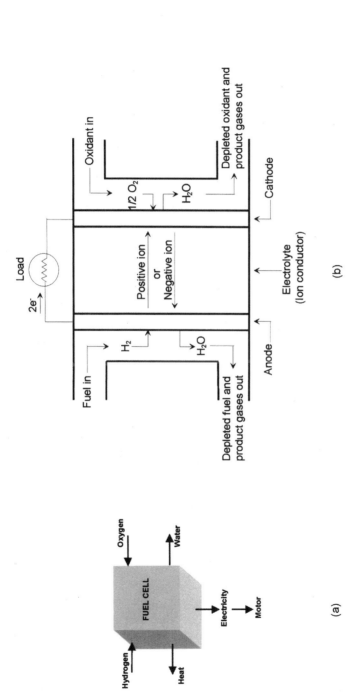

Fig. 3.18. (a) Operation of a fuel cell, converting hydrogen and oxygen (from the air) into electricity, water, and heat (Dincer, 2002a); (b) schematic of the individual fuel cell (after Hirschenhofer *et al.*, 1998).

Table 3.5. Common fuel cells and their characteristic parameters

Type	Operation temperature range (°C)	Application
Solid oxide (SOFC)	500–1000	All sizes of CHP[a]
Direct methanol (DMFC)	50–100	Buses, cars, appliances, small CHP
Polymer electrolyte (PEFC)	50–100	Buses, cars
Phosphoric acid (PAFC)	200	Medium CHP
Molten carbonate (MCFC)	600	Large CHP
Alkaline (AFC)	50–250	Space vehicles

[a]CHP: combined heat and power.
Source: Adapted from Wiens (2000).

fuel cells, the performance of the electrodes is greatly affected by their porosity, which appears to be the key parameter in performance improvement. This is due to the fact that the current densities obtained from smooth electrodes are generally in the single-digit range of $mA\,cm^{-2}$, or even smaller because of rate-limiting issues such as the available area of the reaction sites.

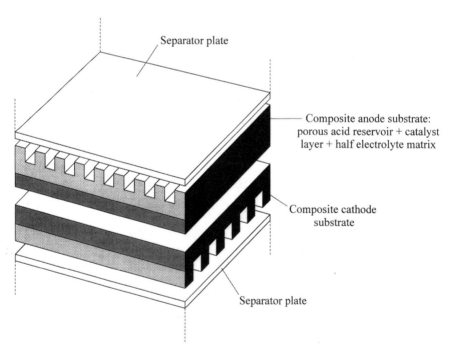

Fig. 3.19. Expanded view of a basic fuel cell repeated unit in a fuel cell stack (after Hirschenhofer *et al.*, 1998).

The porous electrodes used in fuel cells achieve much higher current densities as a result of the high surface area of electrode relative to the geometric plate (as projected) area. Note that the large contact surface increases the number of reaction zones significantly, and the optimized electrode structure results in favorable mass transport properties. In an idealized porous gas fuel cell electrode, high current densities at reasonable polarization are obtained when the liquid (electrolyte) layer on the electrode surface is sufficiently thin so that it does not significantly slow the transport of reactants to the electroactive zones, and a stable three-phase interface is then established as the gas/electrolyte/electrode surface. When an excessive amount of electrolyte exists in the porous electrode structure, the electrode is considered flooded, and this increases the concentration polarization to a larger value.

The porous electrodes in fuel cells are used for three purposes (Hirschenhofer et al., 1998): to provide a surface site where gas/liquid ionization or de-ionization reactions can take place, to conduct ions away from or into the three-phase interface once they are formed (therefore an electrode must be made of materials with large electrical conductivity), and to provide a physical barrier that separates the bulk gas phase from the electrolyte. The catalytic function of electrodes is more important in low-temperature fuel cells, and less so in high-temperature fuel cells because ionization reaction rates increase with temperature.

The porous electrodes, particularly in a low-temperature fuel cell, normally consist of a composite structure that contains (i) platinum electrocatalyst on carbon black with a large surface area, and (ii) a polytetrafluoroethylene binder. Here, polytetrafluoroethylene is a hydrophobic agent which serves as the gas-permeable phase, and carbon black is an electron conductor which provides a high surface area to support the electrocatalyst. The electrocatalyst is platinum, which promotes the rate of electrochemical reaction (oxidation/reduction) for a given surface area. Note that the carbon black also has a certain degree of hydrophobicity, depending upon the surface properties of the material.

3.10 Fuel Cell Structure and Performance

For optimum design and analysis of a fuel cell, both ideal and actual performances should be determined. In the actual case, the losses (irreversibilities) are calculated and deducted from the ideal performance. The actual performance of the fuel cell changes with its type and operating conditions. Here, we follow the treatment presented by Hirschenhofer et al. (1998).

The electrochemical reactions that take place in a fuel cell directly affect its ideal performance which is particularly defined by the Nernst potential (so called: cell voltage). Table 3.6 lists the overall cell reactions corresponding to the individual electrode reactions given in Table 3.7, along with the corresponding form of the Nernst equation. The Nernst equation represents

Table 3.6. Electrochemical reactions in fuel cells[a]

Fuel cell type	Anode reaction	Cathode reaction
PEMFC	$H_2 \longrightarrow 2H^+ + 2e^-$	$1/2O_2 + 2H^- + 2e^- \longrightarrow H_2O$
AFC	$H_2 + 2(OH)^- \longrightarrow 2H_2O + 2e^-$	$1/2O_2 + H_2O + 2e^- \longrightarrow 2(OH)^-$
PAFC	$H_2 \longrightarrow 2H^+ + 2e^-$	$1/2O_2 + 2H^- + 2e^- \longrightarrow H_2O$
MCFC	$H_2 + CO_3^= \longrightarrow H_2O + CO_2 + 2e^-$ $CO + CO_3^= \longrightarrow 2CO_2 + 2e^-$	$1/2O_2 + CO_2 + 2e^- \longrightarrow CO_3^=$
SOFC	$H_2 + O^= \longrightarrow H_2O + 2e^-$ $CO + O^= \longrightarrow CO_2 + 2e^-$ $CH_4 + 4O^= \longrightarrow 2H_2O + CO_2 + 8e^-$	$1/2O_2 + 2e^- \longrightarrow O^=$

[a]CO: carbon monoxide, H_2: hydrogen, CO_2: carbon dioxide, H_2O: water, $CO_3^=$: carbonate ion, O_2: oxygen, e^-: electron, OH^-: hydroxyl ion, H^+: hydrogen ion. Source: Adapted from Hirschenhofer et al. (1998).

the relationship between the ideal standard potential (E°) for the cell reaction and the ideal equilibrium potential (E) at other temperatures and partial pressures of reactants and products. For any temperature and pressure, the ideal voltage can be calculated using these equations based on the standard ideal potential as given in Table 3.8. It is clear from the Nernst equation that at a given temperature the ideal cell potential increases with an increase in reactant pressure. In fact, higher pressure will result in better fuel cell performance.

Table 3.7. Fuel cell reactions and the corresponding Nernst equations[a]

Cell reactions	Nernst equation
$H_2 + 1/2O_2 \longrightarrow H_2O$	$E = E^\circ + (RT/2F)\ln[P_{H_2}/P_{H_2O}]$ $+ (RT/2F)\ln[P_{O_2}^{1/2}]$
$H_2 + 1/2O_2 + CO_2^c \longrightarrow H_2O + CO_2^a$	$E = E^\circ + (RT/2F)\ln[P_{H2}/P_{H_2O}(P_{CO_2})^a]$ $+ (RT/2F)\ln[P_{O_2}^{1/2}(P_{CO_2})^c]$
$CO + 1/2O_2 \longrightarrow CO_2$	$E = E^\circ + (RT/2F)\ln[P_{CO}/P_{CO_2}]$ $+ (RT/2F)\ln[P_{O_2}^{1/2}]$
$CH_4 + 2O_2 \longrightarrow 2H_2O + CO_2$	$E = E^\circ + (RT/8F)\ln[P_{CH_4}/P_{H_2O}^2 P_{CO_2}]$ $+ (RT/8F)\ln[P_{O_2}^2]$

[a]a: anode, c: cathode, E: equilibrium potential, E°: standard potential, P: gas pressure, R: universal gas constant, T: temperature, F: Faraday's constant. Standard conditions are one atmosphere and $25°C$.
[b]The standard Nernst potential (E°) is the ideal cell voltage at standard conditions. It does not include losses that are found in an operating fuel cell. Thus it can be thought of as the open circuit voltage.
Source: Adapted from Hirschenhofer et al. (1998).

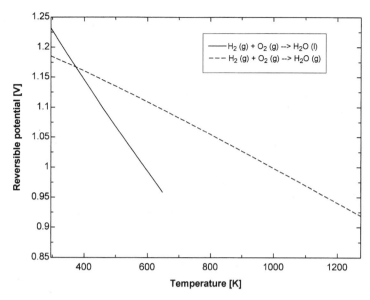

Fig. 3.20. Variation of fuel cell reversible (ideal) potential with temperature.

Figure 3.20 illustrates the linear change of reversible potential (ideal standard potential) with cell temperature. Here, it is clear that at high temperature cells, E° corresponds to a reaction where the water product is in a gaseous state, and E° is less than 1.229 volts at standard conditions when gaseous water is the product. Table 3.8 shows the impact of temperature on E for the oxidation of H_2. The first case in this table is for open circuit voltage in the absence of a fuel cell, that is, in the absence of the losses associated with operating a fuel cell.

The thermal efficiency of a fuel cell is defined as the ratio of useful energy production to the change in stored chemical energy (considered thermal energy) as a result of the reaction of the fuel with an oxidant. Note that a fuel cell converts chemical energy directly into electrical energy, and the maximum theoretical efficiency is not bound by the Carnot cycle. It therefore becomes

$$\eta_{\text{actual}} = \text{Useful Energy}/\Delta H. \qquad (3.30)$$

Table 3.8. Ideal voltage as a function of cell temperature

Cell type		PEFC	PAFC	MCFC	SOFC
Temperature (°C)	25	80	205	650	1100
Ideal voltage	1.18	1.17	1.14	1.03	0.91

Source: Adapted from Hirschenhofer *et al.* (1998).

Here, the efficiency values range from 60 to 90%, respectively. As an example, a hydrogen fuel cell with water vapor as the product has a maximum possible operating efficiency of 80% at an operating temperature of 100°C, and 60% at 1000°C (Mench et al., 2001).

In the ideal case of fuel cell, the ideal efficiency is defined as the ratio of the change in Gibbs free energy of the reaction (ΔG), i.e., useful electric energy, to the change in stored energy:

$$\eta_{\text{ideal}} = \Delta G / \Delta H, \tag{3.31}$$

where $\Delta G = \Delta H - T\Delta S$, leading to the conclusion that the difference between ΔG and ΔH is proportional to the change in entropy. The most widely used efficiency of a fuel cell is based on the change in the standard free energy of the cell reaction,

$$H_2 + 1/2O_2 \longrightarrow H_2O_{(l)} \tag{3.32}$$

where the product water is in liquid form. At standard conditions of 25°C (298 K) and 1 atm, the chemical energy ($\Delta H = \Delta H_0$) in the hydrogen/oxygen reaction is 285.83 kJ/mole, and the free energy available for useful work is 237.14 kJ/mole. Thus the thermal efficiency of a fuel cell operating reversibly on pure hydrogen and oxygen at standard conditions becomes (Hirschenhofer et al., 1998): $\eta_{\text{ideal}} = (237.14/285.83) = 0.83$.

The actual efficiency of the fuel cell is defined as the ratio of the operating cell voltage to the ideal cell voltage, due to the fact that the actual cell voltage is less than the ideal cell voltage because of the losses associated with cell polarizations and the ohmic loss. The thermal efficiency of the fuel cell is then written as:

$$\eta_{\text{ideal}} = \frac{\text{Useful Energy}}{\Delta H} = \frac{\text{Useful Power}}{\Delta G / \eta_{\text{ideal}}} = \frac{\text{Volts}_{\text{actual}} \times \text{Current}}{\text{Volts}_{\text{ideal}} \times \text{Current}/0.83}$$

$$= \frac{0.83 \times V_{\text{actual}}}{V_{\text{ideal}}}. \tag{3.33}$$

After substituting $V_{\text{ideal}} = 1.229$ V, Equation (3.33) becomes

$$\eta_{\text{ideal}} = \frac{0.83 \times V_{\text{actual}}}{V_{\text{ideal}}} = \frac{0.83 \times V_{\text{actual}}}{1.229} = 0.675 V_{\text{actual}} = 0.675 V_{\text{cell}} \tag{3.34}$$

A fuel cell can be operated at different current densities, expressed as mA/cm^2, or A/cm^2 or A/m^2. The corresponding cell voltage then determines the fuel cell efficiency. Decreasing the current density increases the cell voltage, thereby increasing the fuel cell efficiency as shown with the polarization curve in Figure 3.21. This is the standard measure of performance for fuel cell systems. Due to losses resulting from undesired species crossover from one electrode through the electrolyte and internal currents, the actual open circuit voltage is below the theoretical value. Also, there are three major

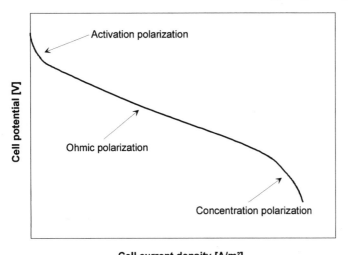

Fig. 3.21. Generalized polarization curve for a fuel cell.

classifications of losses that result in a drop from the open circuit voltage: activation polarization, ohmic polarization, and concentration polarization. The operating voltage of a fuel cell can be represented as the departure from ideal voltage caused by these polarizations,

$$V_{\text{cell}} = E - \eta_{a,a} - \eta_{a,c} - \eta_r - \eta_{m,a} - \eta_{m,c} = E - \eta_a - \eta_r - \eta_m, \quad (3.35)$$

where E is the open circuit potential of the cell, and η_a, η_r, and η_m represent activation, ohmic (resistive), and mass concentration polarization. Activation and concentration polarization occur at both anode and cathode locations, whereas the resistive polarization represents ohmic losses throughout the fuel cell.

Activation polarization is a measure of the catalyst effectiveness at a given temperature, and is represented by the Tafel equation at each electrode (Bard and Faulkner, 1980);

$$\eta_a = \eta_{a,a} + \eta_{a,c} = \frac{RT}{nF\alpha} \ln\left(\frac{i}{i_0}\right)_a + \frac{RT}{nF\alpha} \ln\left(\frac{i}{i_0}\right)_c. \quad (3.36)$$

Here α is the charge transfer coefficient, which can be different between anode and cathode, and which represents the portion of the applied electrical energy that is used to change the rate of electrochemical reaction. The number n is the number of exchange electrons per mole of reactant, and F is Faraday's constant. The exchange current density (i_0) represents the activity of the electrode for a particular reaction at equilibrium.

At increased current densities, a linear region is evident on the polarization curve, and in this zone the reduction in voltage is dominated by internal ohmic losses as

$$\eta_r = i(\Sigma r_k). \tag{3.37}$$

Each r_k value is the area-specific resistance of individual cell components, including the ionic resistance of the electrolyte, and the electric resistance of bipolar plates, cell interconnects, contact resistance between mating parts, and any other cell components through which electrons flow.

In Figure 3.21 the concentration polarization region is completely a mass-transport-related phenomenon, and the mass concentration polarization (η_m) based on the Tafel expression then becomes

$$\eta_m = -\frac{RT}{n\mathrm{F}} \ln\left(1 - \frac{i}{i_l}\right), \tag{3.38}$$

where i_l is the limiting current density (i.e., the maximum current at zero surface concentration of reactant).

As an illustrative example, we study how the different polarizations in a SOFC operating at 1073 K and 1 atm, respectively change with current density. The variations of such polarizations with current density are then shown in Figure 3.22, which is based on the data of Table 3.9. The results show that ohmic and cathode activation polarizations contribute to the major irreversible losses in the fuel cell over the normal operating conditions. The cathode activation polarization is higher than anode activation due to its lower exchange current density. Because the exchange current density directly affects

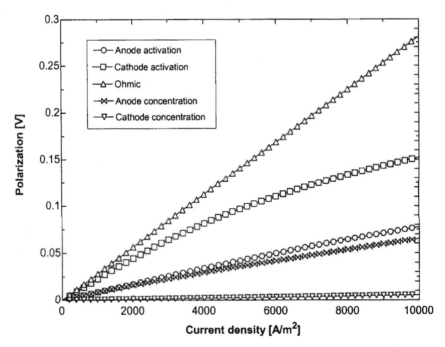

Fig. 3.22. Variation of different polarizations with current density in a SOFC at 1073 K and 1 atm.

Table 3.9. Fuel cell parameters used in Figure 3.22.

Parameter	Value
Fuel cell operating temperature (K)	1073
Fuel cell operating pressure (atm)	1
Hydrogen temperature (K)	298
Air temperature (K)	298
Anode exchange current density (A/m^2)	5300[*]
Cathode exchange current density (A/m^2)	2000[*]
Fuel flow rate (kg/s)	1
Air flow rate (%)	600 theoretical air[*]
Fuel utilization rate (%)	75
Overall heat transfer coefficient (kW/m^{2}K)	0.05[*]

[*]Data from Chan *et al.* (2001).

the rate of electrochemical reaction at the anode and cathode, high exchange current density at the anode means high electrochemical reaction rate and low activation polarization. The cathode concentration polarization is negligible in comparison with anode concentration polarization, and anode concentration polarization becomes significant at large current densities, which is due to the slow rate of mass transfer resulting in the depletion of reactants in the vicinity of the reaction site, and thus are unable to meet the high demand required by the load.

For a fuel cell, the molar flow rate of species required for electrochemical reaction is

$$\dot{n}_{\text{reactant}} = \frac{iA}{n\text{F}}, \tag{3.39}$$

where i and A represent the current density and total electrode area, respectively.

The stoichiometric ratio for an electrode reaction is defined as the ratio of reactant provided to that needed for the electrochemical reaction of interest. Based on the above equations, there are some useful equations developed for the required mass flow rates as a function of fuel cell electrical power and operating conditions (Mench *et al.*, 2001). Here is an example for the DMFC cathode and anode:

$$\dot{m}_{\text{air}} = 3.56 \times 10^{-4} \frac{P_{\text{stack}}}{V_{\text{cell}}} \xi_c \quad \text{and}$$

$$\tag{3.40}$$

$$\dot{m}_{\text{fuel-solution}} = 1.71 \times 10^{-3} \frac{P_{\text{stack}}}{V_{\text{cell}}\text{Mol}} \xi_a,$$

where P_{stack} is the fuel cell stack electrical power output, V_{cell} is the operating cell voltage, Mol is the fuel solution molarity, and ξ_a and ξ_c are the anode and cathode stoichiometric ratios.

Note that in practice it is always desirable to reduce electrolyte thickness to reduce internal ohmic losses. However, the electrolyte must also be relatively impermeable to fuel and oxidizer to minimize reactant crossover, be stable in oxidizing and reducing environments over time, and maintain structural integrity at operating conditions.

Permeability of the electrolyte to the reactants results in mixed potentials at the electrodes, reducing performance, and possibly degrading the catalyst. The requirements for the electrode include low activation losses, long-term stability, and acceptable ionic/electronic conductivity. In low-temperature PEMFC systems, the catalyst material most frequently used is platinum, or platinum/ruthenium, whereas high-temperature SOFC systems (with greatly reduced activation polarization) utilize much cheaper catalyst materials such as nickel.

In fuel cells, thermal management is also a major issue. The energy that is released by electrochemical reaction is split between electrical and thermal components. The waste heat rate for fuel cell (e.g., H_2 PEMFC) systems is very high ($\sim$60% or more of the output) and can be written as follows:

$$P_{\text{waste}} = \dot{m}c_p\Delta T = niA(V_{0c} - V_{\text{cell}}),\tag{3.41}$$

where n is the number of cells in series in the stack, V_{cell} is the cell operating voltage and V_{0c} is the theoretical open circuit voltage and can be calculated from basic thermodynamic data, for example Newman (1999). Waste heat must be properly managed with cooling channels in the stack, which take up space and require parasitic pumping losses.

Water transport through the electrolyte occurs by diffusion, electro-osmotic drag, and hydraulic permeation resulting from a pressure difference across the anode and cathode. As a first approximation, diffusion through the electrolyte can be represented with Fick's law (Chapter 1). A more complete approach is presented in Section 4.8. Appropriate expressions relating to diffusion coefficients are available elsewhere, for example, Mench et al. (2001). Electro-osmotic drag of water through membranes has been studied recently by Ren and Gottesfeld (2001) and the drag coefficient was shown to be a nearly linearly increasing function of temperature from 20 to 120°C. Hydraulic permeation of water through the membrane is typically small for H_2 PEMFCs due to the low-pressure differences between the two sides. At the cathode surface, the oxygen reduction reaction shown in Figure 3.18b will result in water production proportional to the current density. Considering the cathodic water reduction reaction, and combining the different forms of water transport through the membrane, the molar water transport through the membrane, the molar water transport and creation at the cathode can be shown to be

$$j_{\text{H}_2\text{O}} = -D\frac{\Delta C_{c-a}}{\Delta x} + \frac{iA}{\text{F}}(\lambda_{\text{drag}} + 0.5) - \frac{aK}{\ell}\Delta P.\tag{3.42}$$

At a typical operating temperature of 70°C, λ_{drag} was determined by Ren and Gottesfeld (2001) to be around 3, or 86% of the current-dependent transport.

Membrane performance suffers without sufficient water, because PEM ionic conductivity is directly related to the degree of water content and temperature, as correlated by Springer *et al.* (2001). Alternatively, excessive water at the cathode can cause flooding, that is, liquid water accumulation at the cathode surface, which prevents oxygen access to the reaction sites. Flooding is most likely near the cathode exit under high current density, high humidification, low temperature, and low flow rate conditions. In most H_2 PEMFC systems, drying is more of a concern than flooding, and an external humidifier is needed. The maximum mass flow rate of water that can be removed from the fuel cell by gas flow can be shown by letting the exit relative humidity ϕ equal 1 in the following expression (Mench *et al.*, 2001),

$$\dot{n}_{H_2O,\text{removal}} = \dot{n}_{\text{others}} \left(\frac{P_{g,\text{sat}}\phi}{P - P_{g,\text{sat}}\phi} \right), \qquad (3.43)$$

where $\dot{n}_{\text{others}}$ is the molar flow rate of all other species in the flow besides water vapor, and P is the total pressure. Note that the consumption of oxygen at the cathode will result in a decrease in $\dot{n}_{\text{others}}$ along the flow path. This decrease can be nonnegligible, depending on stoichiometry and species.

The heat generation rate per cell in the stack, as posed by Thomas and Zalbowitz (1999), is based on the theoretical and experimental cell voltage and current. The equation is an energy rate balance between the cell products and reactants, the heat generation, and the measured electrical power.

$$P_{\text{total}} = P_{\text{heat}} + P_{\text{electrical}} \implies (V_{\text{ideal}} \cdot I_{\text{cell}}) = P_{\text{heat}} + (V_{\text{cell}} \cdot I_{\text{cell}})$$
$$\implies P_{\text{heat}} = (V_{\text{ideal}} - V_{\text{cell}}) \cdot I_{\text{cell}} \qquad (3.44)$$

This part deals with flow field aspects of fuel cells, particularly PEMFCs, where flow conditions are typically laminar, with Reynolds numbers on the order of 100 to 1000. The common length scale is on the order of 1 mm (the flow channel cross section), and the fluids are typically hydrogen and air. In general, the flow field should be designed to minimize pressure drop (reducing parasitic pump requirements), while providing adequate and evenly distributed mass transfer through the carbon diffusion layer to the catalyst surface for reaction. In H_2 PEMFCs the reaction is typically limited by the cathode kinetics. There is a nearly negligible anode activation polarization, unless the inlet flow of hydrogen is greatly diluted. As described by Ren and Gottesfeld (2001), at a current density of $1\,A/cm^2$, a 0.4 to 0.5 V voltage loss is required to overcome cathode activation polarization losses, whereas a loss of only 20 to 30 mV results at the hydrogen anode for the same current density. Because air is generally used as the oxidizer, significant cathode concentration polarization can also exist without adequate flow rates and properly designed flow fields. Three basic channel configurations of serpentine, parallel, and interdigitated flow are used in PEMFCs, although some small-scale fuel cells do not use a flow field and rely on diffusion processes from the environment.

In serpentine flow paths, the flow exhibits relatively high-pressure drop. In addition, reactant consumption can lead to additional pressure drop. If not designed properly, some flow channels with a high local pressure may be adjacent to channels with low pressure. In this case, resultant pressure mismatch can lead to reactant bypass, a phenomenon of undesired reactant penetration through the backing layer, underneath a current-collecting rib, and into another flow channel.

Here, we follow the treatment presented by Mench *et al.* (2001). Assuming laminar flow, the pressure drop through the channel situated a distance L from the bypass location to the bypass emergence point can be derived from laminar flow theory:

$$\Delta P_{\text{flow path}} = \frac{32 L V \mu}{d_h^2}, \tag{3.45}$$

where V is the channel bulk flow velocity. An estimate of the pressure drop through the porous backing layer can be estimated based on Darcy's law:

$$\Delta P_{\text{backing}} = \frac{U_{\text{avg}} \mu}{K} \Delta S, \tag{3.46}$$

where ρ is the density of the fluid, U_{avg} is the bulk average velocity through the porous medium, ΔS is the porous medium thickness in the flow direction (equivalent to the current collecting rib shoulder width), and K is the permeability of the backing layer. The backing layer permeability is a highly nonlinear function of porosity, and can be found from the Carman–Kozeny relation [cf. the paragraph preceding Equation (1.9)],

$$K = \frac{d_s^2 \phi^3}{180 (1 - \phi)^2}, \tag{3.47}$$

where ϕ is the porosity of the porous medium (backing layer) and d_s is the diameter of the theoretical spherical particles assumed in the porous medium model. A dimensionless parameter termed the bypass ratio (BPR) can be defined as the ratio of pressure drop through the channel to the backing layer:

$$\text{BPR} = \frac{32 L V K}{d_h^2 U_{\text{avg}} \Delta S}. \tag{3.48}$$

The channel bulk flow velocity and the average bypass velocity (U_{avg}) will be related, but the basic physics can be understood with this relationship. If BPR is much greater than unity, significant bypass can occur; if BPR is much less than unity, significant bypass can be avoided. However, this concept has yet to be verified experimentally. It is interesting to note that in Equation (3.48), channel pressure, flow viscosity, and temperature should have little or no effect, whereas the hydraulic diameter, channel length, current collecting rib thickness, and backing layer permeability are the key physical parameters. The hydraulic diameter d_h and the permeability K dominate as a result of their nonlinear relationship with BPR. Therefore, backing layer compression to reduce effective porosity by current collecting ribs is also a major factor in avoiding reactant bypass. A porous medium model of heat and fluid flow in a PEM fuel cell was described recently by Yuan *et al.* (2002).

3.11 The Concept of Exergy-Cost-Energy-Mass (EXCEM) Analysis

An analytical point of view that summarizes the thermodynamic design issues discussed in this chapter is the EXCEM analysis concept illustrated in Figure 3.23. Of the quantities represented by EXCEM, only mass and energy are subject to conservation laws. Cost increases or remains constant, while exergy decreases or remains constant. Balances can be written for each of the EXCEM quantities. The application of EXCEM analysis requires an understanding of the appropriate balance equations for each of the EXCEM quantities (Rosen and Dincer, 2003). Mass and energy, being subject to conservation laws (neglecting nuclear reactions), can be neither generated nor consumed. Consequently, the general balance equation for each of these quantities becomes [cf. Equation (3.1)]

$$\text{Mass input} - \text{Mass output} = \text{Mass accumulation}, \tag{3.49}$$

$$\text{Energy input} - \text{Energy output} = \text{Energy accumulation}. \tag{3.50}$$

Entropy is created during a process due to irreversibilities:

$$\text{Entropy input} + \text{Entropy generation} - \text{Entropy output}$$

$$= \text{Entropy accumulation}. \tag{3.51}$$

By combining the conservation law for energy and the nonconservation law for entropy, the following general balance equation for exergy can be obtained,

$$\text{Exergy input} - \text{Exergy output} - \text{Exergy consumption}$$

$$= \text{Exergy accumulation}. \tag{3.52}$$

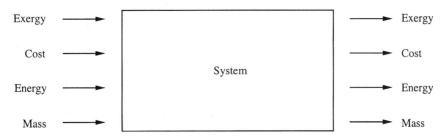

Fig. 3.23. Overview of the EXCEM analysis methodology.

Note that exergy is consumed due to irreversibilities, and that exergy consumption is proportional to entropy generation. Equations (3.50) and (3.52) demonstrate the main difference between energy and exergy: energy is conserved whereas exergy, a measure of work potential, can be consumed.

Cost is an increasing quantity. The general balance equation for cost is

$$\text{Cost input} + \text{Cost generation} - \text{Cost output} = \text{Cost accumulation}, \quad (3.53)$$

where cost input, output, and accumulation represent, respectively, the cost associated with all inputs, outputs, and accumulations for the system. Cost generation corresponds to the appropriate capital and other costs associated with the creation and maintenance of a system. The "cost generation rate" term in a differential cost balance represents the total cost generation spread out over the operating life of the system. The "amount of cost generated" term in an integral cost balance represents the fraction of the total cost generation accounted for in the time interval under consideration.

In summary, EXCEM analyses can aid in the design, improvement, and optimization of processes, and in the formulation of energy, economic, and environmental policies. Examples are given in Rosen and Dincer (2003).

3.12 Exergy, Environment, and Sustainable Development

We close with a discussion of the contemporary importance of the exergy-based engineering presented in this chapter. From the pure thermodynamics point of view, exergy represents the maximum amount of work that can be produced by a system or a flow of matter or energy as it comes to equilibrium with a reference environment. Exergy is a measure of the potential of the system or flow to cause change, as a consequence of not being completely in equilibrium with the reference environment. Unlike energy, exergy is not subject to a conservation law (except in the limit of ideal or reversible processes). Rather, exergy is consumed or destroyed due to irreversibilities in all real processes. The exergy consumption during a process is proportional to the entropy created due to irreversibilities associated with the process [cf. Equation (3.5)].

During the past decade exergy-related studies have received much attention in various disciplines, ranging from chemical to mechanical engineering, from environmental engineering to ecology, and so on. Some key features of exergy are identified to highlight its importance in a wide range of applications (Dincer, 2002b). Exergy is a powerful and effective tool for:

(1) designing and analyzing energy systems by combining the conservation of mass and energy principles with the second law of thermodynamics;
(2) furthering the goal of more efficient energy resource use by assessing meaningful efficiencies and enabling the locations, types, and true magnitudes of wastes and losses to be determined;

(3) revealing whether (and by how much) it is possible to design more efficient energy systems by reducing the inefficiencies in existing systems;

(4) addressing the impact on the environment of energy resource utilization; and

(5) helping to achieve sustainable development.

The link between exergy and the environment is particularly intriguing. Energy production, transformation, transport, and use have an impact on the earth's environment. The exergy of a quantity of energy or a substance can be viewed as a measure of its usefulness, quality, or potential to cause change. For the same reason, exergy promises to be an effective measure of the potential of a substance to have an impact on the environment. In practice, a thorough understanding of what exergy is, and how it provides insights into the efficiency and performance of energy systems, is required for engineers, scientists, and policy makers working in the area of energy systems and the environment. This need is particularly important because energy and environmental policies are likely to play an increasingly prominent role in the future in a broad range of local, regional, and global environmental concerns.

People have long been intrigued by the implications of the laws of thermodynamics on the environment. One myth speaks of Ouroboros, a serpent-like creature that survived and regenerated itself by eating only its own tail. By neither taking from nor adding to its environment, this creature was said to be completely environmentally benign and self-sufficient. It is useful to examine this creature in light of the thermodynamic principles recognized today. Assuming that Ouroboros was an isolated system (i.e., it received no energy from the sun or the environment, and emitted no energy during any process), Ouroboros' existence would have violated neither the conservation law for mass nor the first law of thermodynamics. However, unless it was a reversible creature, Ouroboros' existence would have violated the second law (which states that exergy is reduced for all real processes), because Ouroboros would have had to obtain exergy externally, to regenerate the tail it ate into an equally ordered part of its body (or it would ultimately have dissipated itself to an unordered lump of mass). Thus Ouroboros would have to have had an impact on its environment (Rosen and Dincer, 2001).

The most appropriate link between the second law and environmental impact appears to be exergy, in part because it is a measure of the departure of the state of a system from that of the environment. The magnitude of the exergy of a system depends on the states of both the system and the environment. This departure is zero only when the system is in equilibrium with its environment. The concept of the environment as it applies to exergy analysis is discussed in detail by Dincer (2002a).

To understand the relation between exergy and the environment, consider the following aspects.

Order destruction: The destruction of order, or the creation of chaos, is a form of environmental damage. Entropy is fundamentally a measure of chaos, and exergy of order. A system of high entropy is more chaotic or disordered than one of low entropy, and relative to the same environment, the exergy of an ordered system is greater than that of a chaotic one.

Resource degradation: The degradation of resources found in nature (e.g., fossil fuels) is a form of environmental damage. It is known that a material resource (found in nature or created artificially) that is in a state of disequilibrium with the environment has exergy as a consequence of this disequilibrium. Two principal approaches exist to reduce the environmental impact associated with resource degradation.

Increased efficiency: Increased efficiency preserves exergy by reducing the exergy necessary for a process, and therefore reduces environmental damage. Increased efficiency also reduces exergy emissions and hence plays a role in environmental damage.

Using external exergy resources (e.g., solar energy): The earth is an open system subject to a net influx of exergy from the sun. It is the exergy delivered with solar radiation that is valued; all the energy received from the sun is ultimately radiated out to the universe. Environmental damage can be reduced by taking advantage of the openness of the earth and utilizing solar radiation (instead of degrading resources found in nature to supply exergy demands).

Waste exergy emissions: The exergy associated with waste emissions can be viewed as a potential for environmental damage in that the exergy of the wastes, as a consequence of not being in stable equilibrium with the environment, represents a potential to cause change. The emitted exergy may cause changes that are damaging to the environment, for example, the death of fish and plants due to the release of certain substances in stack gases, as these substances react and come to equilibrium with the environment, even though in some cases the changes may be perceived as beneficial (e.g., the increased growth rate of fish and plants near the cooling-water outlets of thermal power plants).

Because of the relation between exergy and both energy and environment, it is clear that exergy is a central concept in sustainable development. Furthermore, exergy methods can be used to enhance sustainability. Contemporary researchers recognize that one important element in obtaining sustainable development is the use of exergy analysis. By noting that energy can never be "lost," because it is conserved according to the first law of thermodynamics, whereas exergy can be lost because of internal irreversibilities, current research suggests that exergy losses, particularly those due to the use of non-renewable energy forms, should be minimized in order to obtain sustainable development. Furthermore, environmental effects associated with emissions and resource depletion can be expressed in terms of one exergy-based indicator, which is based on physical principles.

Sustainable development also includes economic viability. Thus the methods relating exergy and economics also reinforce the link between exergy and sustainable development. The objectives of most analyses integrating exergy

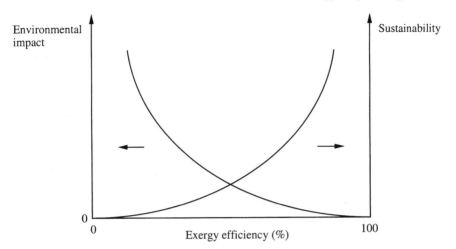

Fig. 3.24. Qualitative illustration of the relation among the environmental impact and sustainability of a process, and its exergy efficiency.

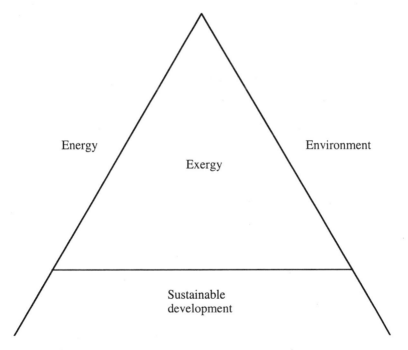

Fig. 3.25. The interdisciplinary nature of exergy.

and economics include the determination of the appropriate allocation of economic resources so as to optimize the design and operation of a system, and the economic feasibility and profitability of a system. Exergy-based economic analysis methods are referred to as thermoeconomics, second-law costing, cost accounting, and exergoeconomics. The links among sustainable development, technology, and resource utilization are explored in Dewulf *et al.* (2000) and Lems *et al.* (2002).

Figure 3.24 shows qualitatively the relation among exergy, sustainability, and environmental impact. Sustainability increases and environmental impact decreases as the exergy efficiency of a process increases. The two limiting cases are significant. As exergy efficiency approaches 100%, environmental impact approaches zero, because exergy is only converted from one form to another without losses (either through internal consumption or through waste emission). Sustainability approaches infinity because the process approaches reversibility. In the second extreme, exergy efficiency approaches 0%, and sustainability approaches zero because exergy-containing resources are used but nothing is accomplished. Environmental impact approaches infinity because, to provide a fixed service, an ever-increasing quantity of resources must be used and a correspondingly increasing amount of exergy-containing wastes are emitted.

Based on these ideas and the interdisciplinary nature of exergy, Rosen and Dincer (2001) proposed that exergy is the confluence of energy, environment, and sustainable development, as illustrated in Figure 3.25. The basis for this treatment is the interdisciplinary character of exergy and its relation to each of these disciplines.

4

Environmental and Civil Engineering

4.1 The Energy–Environment Interface

Realistic models of energy systems demand the treatment of installations and their flowing surroundings together, more so when the installations are large and their spheres of impact greater. The interface between energy systems and the environment is formed by flows, environmental flows. In the opening sections of this chapter we review the fundamentals of some of the most important types of flows that govern the behavior of environmental fluids (air, water) and fluid-saturated porous media.

On the fluid-flow side, the most important class is the "free stream" turbulent flows—shear layers, jets, plumes—flows that are free from wall effects because they are situated sufficiently far from solid boundaries. These are reviewed in Bejan (1995a). Most of the flows that effect the interaction between our installations and their environments are highly turbulent. They are high Reynolds number flows. The smoke plume swept by the wind in the wake of an industrial area and the water jets discharged by a city into the river are examples of the impact of our presence on what surrounds us. Free-stream flows rely on turbulent mixing to diffuse away our refuse and, in this way, to minimize the effect that high concentrations of such refuse (heat, species) might have on the biosphere. Turbulent flow is a more effective mixing mechanism than laminar flow; consequently, the occurrence of turbulence in any flow is anticipated on the theoretical basis provided by the constructal principle (Bejan, 2000).

The fundamentals of the modeling of environmental flows through fluid-saturated porous media are outlined in Sections 4.2 through 4.4. This field experienced significant growth during the past two decades, and now is one of the most active in thermal sciences (Nield and Bejan, 1999). Its development is comparable with that of classical convection (transport by the flow of a pure fluid): governing principles are in place, experimental data continue to stimulate improvements in the governing principles, and there is an abundance of practical applications. From an environmental standpoint, the fundamental aspects that are covered in this book are relevant to understanding the spreading of contaminated fluids through the ground, the leakage of heat through the walls of buildings, and the flow of geothermal fluids through the earth's porous crust.

4.2 Wakes: Concentrated Heat Sources in Forced Convection

Free stream flows are "free" from interaction with solid boundaries. We review the main results that are available for calculating the rate of spreading or mixing through wakes, plumes, and penetrative convection. The flow description is based on volume-averaging the relevant properties (velocity, temperature) of the porous medium with fluid in its pores. The fluid is single phase. It is assumed that the smallest (infinitesimal-like) volume for which the following description is valid meets the condition that it is a representative elementary volume (Bear, 1972; Greenkorn, 1983; Nield and Bejan, 1999), which means that its length scale is at least one order of magnitude greater than the pore length scale. The representative elementary volume assumption is particularly well suited for the description of environmental flows such as water seepage through soil or sand. It is further assumed that the fluid-saturated porous medium is homogeneous and isotropic.

Figure 4.1 shows two key configurations in a saturated porous medium, the thermal wake behind a point heat source, $q[W]$, and the wake behind a line source, $q'[W/m]$. The convection and diffusion of heat in the wake are governed by Equation (2.5) or its equivalent in cylindrical coordinates. The transversal length scale of the wake is the δ_T scale given by Equation (2.10).

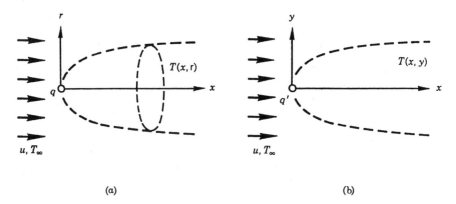

(a)   (b)

Fig. 4.1. Thermal wakes in a porous medium with forced convection: (a) point heat source; (b) line heat source.

There are two key questions in the description of a thermal wake; the rate of lateral spreading, which is answered by Equation (2.10), and the temperature decay in the downstream direction. The latter is represented by the longitudinal variation of the wake temperature averaged over the constant-x plane. It results from the conservation of energy in a control volume of transversal dimension δ_T. The energy conservation statement for Figure 4.1a, that is, for a cylinder of radius δ_T and length x, is given by

$$q \sim \rho c_p U_\infty \delta_T^2 (T_c - T_\infty),   \tag{4.1}$$

where $T_c(x)$ is the wake centerline temperature, and $(T_c - T_\infty)$ is the order of magnitude of the wake excess temperature relative to the free stream temperature. Combining Equations (4.1) and (2.10) we obtain the longitudinal distribution of the temperature in the wake,

$$T_c(x) - T_\infty \sim \frac{q}{kx}.   \tag{4.2}$$

This result shows that the wake excess temperature decreases as x^{-1} in the downstream direction. The corresponding result for the wake behind the line source (Figure 4.1b) is given by

$$T_c(x) - T_\infty \sim \frac{q'}{(\rho c_p)_f (U_\infty \alpha x)^{1/2}}.   \tag{4.3}$$

The scaling laws (4.2) and (4.3) agree in an order of magnitude sense with the formulas derived from the similarity solutions to the same problems (e.g., Bejan, 1995a).

4.3 Plumes: Concentrated Heat Sources in Natural Convection

When the fluid motion is driven by buoyancy, the flow field and the temperature field are coupled, and are both driven by the concentrated source. Environmental applications include geothermal convection, flows in the vicinity of buried nuclear waste and electric cables, and the effect of underground nuclear explosions. When the heated region above the point heat source is slender enough, it is permissible to study its development based on boundary layer theory (Bejan, 1995a; Rees and Hossain, 2001). As shown in Figure 4.2, we attach a cylindrical system of coordinates (r, z) and (v_r, v_z) to the plume so that the z-axis passes through the heat source and points against gravity.

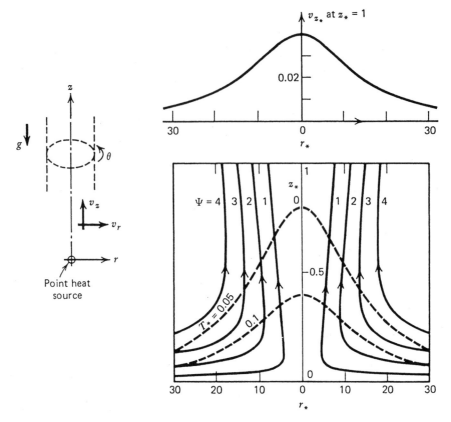

Fig. 4.2. Slender plume above a point heat source in a fluid-saturated porous medium (Bejan, 1995a; from Bejan, *Convection Heat Transfer*, 2nd ed., Copyright © 1995 Wiley. This material is used by permission of John Wiley & Sons, Inc.).

The governing equations for this θ-symmetric convection problem are

$$\frac{\partial v_r}{\partial r} + \frac{v_r}{r} + \frac{\partial v_z}{\partial z} = 0, \tag{4.4}$$

$$v_r = -\frac{K}{\mu}\frac{\partial P}{\partial r}, \qquad v_z = -\frac{K}{\mu}\left(\frac{\partial P}{\partial z} + \rho g\right), \tag{4.5}$$

$$v_r\frac{\partial T}{\partial r} + v_z\frac{\partial T}{\partial z} = \alpha\left[\frac{1}{r}\frac{\partial}{\partial r}\left(r\frac{\partial T}{\partial r}\right) + \frac{\partial^2 T}{\partial z^2}\right]. \tag{4.6}$$

Inside the slender flow region we can write $r \sim \delta_T$ and $z \sim H \gg \delta_T$, where δ_T is the length scale of the plume radius; therefore, after eliminating the pressure terms, Equations (4.5) to (4.6) reduce to

$$\frac{\partial v_z}{\partial r} = \frac{Kg\beta}{\nu}\frac{\partial T}{\partial r}, \tag{4.7}$$

$$v_r\frac{\partial T}{\partial r} + v_z\frac{\partial T}{\partial z} = \frac{\alpha}{r}\frac{\partial}{\partial r}\left(r\frac{\partial T}{\partial r}\right). \tag{4.8}$$

The scale analysis of these two equations dictates

$$v_z \sim \frac{Kg\beta}{\nu}\Delta T \quad\text{and}\quad v_z \sim \frac{\alpha H}{\delta_T^2}, \tag{4.9}$$

where ΔT is the plume-ambient temperature difference $T - T_\infty = $ function (z). A third scaling relation follows from the fact that energy released by the point source $q[W]$ is convected upward through the plume flow,

$$q \sim \rho c_p v_z \delta_T^2 \Delta T. \tag{4.10}$$

Combining relations (4.9) and (4.10) yields the wanted plume scales

$$v_z \sim \frac{\alpha}{H}\,\text{Ra}, \quad \delta_T \sim H\text{Ra}^{-1/2}, \quad \Delta T \sim \frac{q}{kH}, \tag{4.11}$$

where Ra is the Rayleigh number based on source power, $\text{Ra} = Kg\beta q/(\alpha\nu k)$. In conclusion, the scale analysis suggests the following dimensionless variables.

$$z_* = \frac{z}{H}; \quad r_* = \frac{r}{H}\text{Ra}^{1/2}; \quad T_* = \frac{T - T_\infty}{(q/k)H};$$

$$v_{z*} = \frac{v_z}{(\alpha/H)\text{Ra}}; \quad v_{r*} = \frac{v_r}{(\alpha/H)\text{Ra}^{1/2}}. \tag{4.12}$$

The corresponding governing equations and boundary conditions are

$$\frac{\partial v_{r*}}{\partial r_*} + \frac{v_{r*}}{r_*} + \frac{\partial v_{z*}}{\partial z_*} = 0; \tag{4.13}$$

$$\frac{\partial v_{z*}}{\partial r_*} = \frac{\partial T_*}{\partial r_*}; \tag{4.14}$$

$$v_{r*}\frac{\partial T_*}{\partial r_*} + v_{z*}\frac{\partial T_*}{\partial z_*} = \frac{\partial^2 T_*}{\partial z_*^2}; \tag{4.15}$$

$$v_{r*} = 0, \quad \frac{\partial T_*}{\partial r_*} = 0, \quad\text{at } r_* = 0; \tag{4.16}$$

$$v_{z*} \longrightarrow 0, \quad T_* \longrightarrow 0, \quad\text{as } r_* \longrightarrow \infty. \tag{4.17}$$

Integrating Equation (4.13) subject to the $r_* \to \infty$ boundary conditions, we find that $v_{z*} = T_*$. Replacing T_* by v_{z*} in Equations (4.15) and (4.17), we obtain a problem identical to the boundary layer treatment of a laminar round jet discharging into a constant-pressure reservoir (Bejan, 1995a). Applied to the present problem, the method consists of introducing the similarity variable η and streamfunction profile $F(\eta)$ such that

$$\eta = \frac{r_*}{z_*}; \qquad \psi = z_* F(\eta); \tag{4.18}$$

$$v_{r*} = -\frac{1}{r_*}\frac{\partial \psi}{\partial z_*}; \qquad v_{z*} = \frac{1}{r_*}\frac{\partial \psi}{\partial r_*}. \tag{4.19}$$

The similarity solution is (Bejan, 1995a)

$$F = \frac{(C\eta)^2}{1 + (C\eta/2)^2}, \tag{4.20}$$

where constant $C = 0.141$ is determined from the conservation of energy in every constant-z cut across the plume

$$q = \int_0^{2\pi} \int_0^\infty \rho c_p v_z (T - T_\infty) r \, dr \, d\theta \text{ (constant)}. \tag{4.21}$$

In conclusion, the solution has the form

$$T_* = v_{z*} = \frac{2C^2}{z_*}\frac{1}{1 + (C\eta/2)^2};$$

$$v_{r*} = \frac{C}{z_*}\frac{C\eta - 1/4(C\eta)^3}{[1 + (C\eta/2)^2]^2} \qquad \psi = 4z_* \ln\left[1 + \left(\frac{C\eta}{2}\right)^2\right]. \tag{4.22}$$

Figure 4.2 shows the traces of the ψ, $T_* = $ constant surfaces left in any $\theta = $ constant cut through the point source. Note that the $T_* = $ constant trace has the same shape as the vertical velocity profile v_z*, in accordance with the first of Equation (4.22). The flow and temperature field is presented in dimensionless form (r_*, z_*): whether the actual flow field is *slender* is determined by the slenderness condition $\delta_T/H < 1$ which, based on Equation (4.11), is the same as the order of magnitude requirement $\mathrm{Ra}^{1/2} > 1$.

4.4 Penetrative Convection

A distinct class of buoyancy-driven flows is associated with the interaction between finite-size isothermal porous layers and localized sources of heat. The basic configuration can be modeled as a two-dimensional layer of size $H \times L$, with three sides at one temperature. The fourth side is permeable and in communication with a reservoir (fluid or porous medium) of different temperature. We consider two possible orientations of this configuration; a shallow layer with lateral heating (Figure 4.3a), and a tall layer with heating from below or cooling from above (Figure 4.3b). In both cases, natural convection penetrates the porous medium over a length dictated by the strength of the buoyancy effect (the Rayleigh number) and not by the geometric ratio H/L.

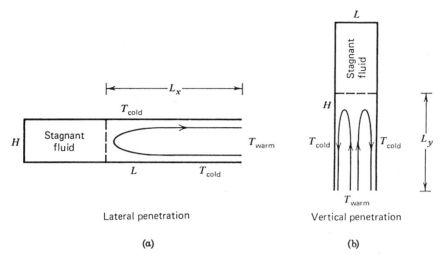

Fig. 4.3. Horizontal and vertical penetration of natural convection into an isothermal porous space heated from one end (Bejan, 1981).

The lateral penetration length L_x (Figure 4.3a), can be derived from the equations for the conservation of mass, momentum, and energy, which are represented in order-of-magnitude terms by

$$\frac{u}{L_x} \sim \frac{v}{H}, \tag{4.23}$$

$$\frac{u}{H}, \frac{v}{L_x} \sim \frac{Kg\beta}{\nu} \frac{\Delta T}{L_x}, \tag{4.24}$$

$$u\frac{\Delta T}{L_x} \sim \alpha\frac{\Delta T}{L_x^2}, \quad \alpha\frac{\Delta T}{H^2}. \tag{4.25}$$

These three equations determine the unknown scales u, v, and L_x. Assuming that the flow penetrates the layer such that $L_x > H$, it is easy to show that

$$L_x \sim H \, \mathrm{Ra}_H^{1/2}. \tag{4.26}$$

The convective heat transport between the isothermal porous layer and the heat reservoir positioned laterally scales as

$$Q \sim (\rho c_P)_f Hu\Delta T \sim k\Delta T \, \mathrm{Ra}_H^{1/2}, \tag{4.27}$$

where $\mathrm{Ra}_H = Kg\beta H\Delta T/(\alpha\nu)$. This heat transfer result demonstrates that the actual length of the porous layer (L) does not influence the heat transfer rate; Q and L_x are set by the Rayleigh number Ra_H. The actual flow and temperature patterns associated with the lateral penetration phenomenon can be determined analytically as a similarity solution (Bejan, 1981). The penetration length and heat transfer rate predicted by the similarity solution are

$L_x = 0.158H \, \mathrm{Ra}_H^{1/2}$ and $Q/(k\Delta T) = 0.319 \, \mathrm{Ra}_H^{1/2}$. The effect of anisotropy in the medium, and the effect of temperature variation along the horizontal walls of the porous layer are also documented in Bejan (1981).

In the vertical layer of Figure 4.3b the bottom wall is permeable and in communication with a different reservoir. Fluid motion sets in as soon as a ΔT is imposed between the bottom surface and vertical walls. Fluid motion is present because no matter how small the ΔT, the porous medium experiences a finite-temperature gradient of order $\Delta T/L$ in the horizontal direction near the heated wall.

Let L_y be the distance of vertical penetration. The balances for mass, momentum, and energy are

$$\frac{u}{L} \sim \frac{v}{L_y};\tag{4.28}$$

$$\frac{u}{L_y}, \frac{v}{L} \sim \frac{Kg\beta}{\nu} \frac{\Delta T}{L};\tag{4.29}$$

$$u\frac{\Delta T}{L} \sim \alpha\frac{\Delta T}{L^2}, \quad \alpha\frac{\Delta T}{L_y^2}.\tag{4.30}$$

Assuming vertical penetration over a distance L_y greater than L, we conclude that

$$L_y \sim L\,\mathrm{Ra}_L,\tag{4.31}$$

where Ra_L is the Rayleigh number based on L, namely, $\mathrm{Ra}_L = Kg\beta L\Delta T/(\alpha\nu)$. The net heat transfer rate through the bottom wall of the system scales as

$$Q_y \sim (\rho c_P)_f Lv\Delta T \sim k\Delta T \, \mathrm{Ra}_L.\tag{4.32}$$

In conclusion, both L_y and Q_y are proportional to Ra_L, unlike the corresponding quantities in the case of lateral penetration, which are proportional to $\mathrm{Ra}_H^{1/2}$. Once again, the imposed temperature difference (ΔT) and the transversal dimension of the layer (L) determine the longitudinal extent (L_y) of the penetrative flow. The physical height of the porous layer (H) does not influence the phenomenon as long as it is greater than L_y.

The phenomenon of partial vertical penetration was studied in the cylindrical geometry as a model of geothermal flows (Bejan, 1980a). The vertical penetration length and net heat transfer rate are $L_y/r_0 = 0.0847 \, \mathrm{Ra}_{r0}$ and $Q_y/(r_0 k\Delta T) = 0.255 \, \mathrm{Ra}_{r0}$, where $\mathrm{Ra}_{r0} = Kg\beta r_0\Delta T/(\alpha\nu)$ is the Rayleigh number based on the dimension normal to the penetrative flow (the cylindrical well radius r_0).

Environmental flows can be modeled in considerably more complex and diverse configurations than the free flows illustrated above. A distinct class is the flows that depend greatly on the solid walls that confine them, in the extreme, the flows that occur in completely enclosed spaces (Sections 2.9 and 2.10). The most complex configurations are in between. The study of

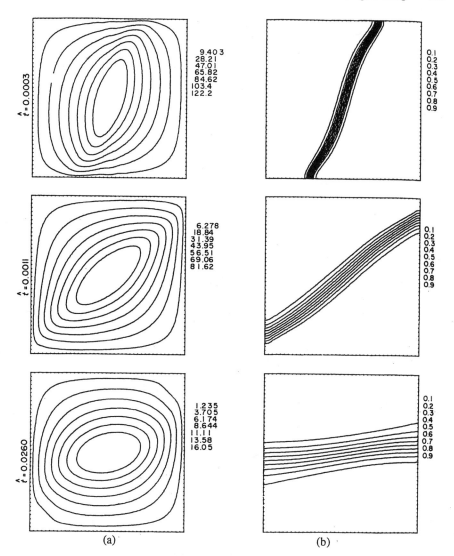

Fig. 4.4. The horizontal spreading and layering of thermal and chemical deposits in a porous medium where natural convection is driven solely by temperature gradients (Ra = 1000, $H/L = 1$, $\phi/\sigma = 1$): (a) streamlines; (b) isotherms, or isosolutal lines for Le = 1 (Zhang and Bejan, 1987).

flows that are partially or temporarily free while confined by walls benefits from a good understanding of both extremes, free flows and enclosed flows. In closing, let us consider one such an example, namely, the time-dependent Darcy flow due to the presence in a porous medium of two distinct regions (different temperatures and species concentrations) separated initially by a vertical interface (Figure 4.4). In time, the two regions share a counterflow

that brings the entire space to thermal and chemical equilibrium. The space
is two-dimensional with height H and horizontal dimension L.

As an example of how two dissimilar adjacent regions come to equilibrium
by convection, Figure 4.4 shows the evolution of the flow, temperature, and
concentration fields of a relatively high Rayleigh number flow driven by a
buoyancy effect that is due entirely to density changes caused by temperature changes. This is also known as heat transfer-driven flow. As the time
increases, the warm fluid (initially on the left-hand side) migrates into the
upper half of the system. The thermal barrier between the two thermal regions
is smoothed gradually by thermal diffusion. Furthermore, as the Lewis number decreases, the sharpness of the concentration dividing line disappears,
as the phenomenon of mass diffusion becomes more pronounced (Zhang and
Bejan, 1987).

In the case of heat transfer-driven flows, the time scale associated with the
end of convective mass transfer in the horizontal direction is

$$\hat{t} \approx \left(\frac{\phi}{\sigma}\right)\left(\frac{L}{H}\right)^2 \text{Ra}^{-1}, \quad \text{Le Ra} > \frac{\phi}{\sigma}\left(\frac{L}{H}\right)^2, \tag{4.33}$$

$$\hat{t} \approx \left(\frac{\phi}{\sigma}\right)\left(\frac{L}{H}\right)^2 \text{Le}, \quad \text{Ra} < \frac{\phi}{\sigma}\left(\frac{L}{H}\right)^2. \tag{4.34}$$

The dimensionless time $\hat{t}$ is defined as $\hat{t} = (\alpha t)/(\sigma H^2)$. Values of $\hat{t}$ are
listed on the side of each frame of Figure 4.4. The Rayleigh number is
$\text{Ra} = Kg\beta H\Delta T/(\alpha\nu)$, where ΔT is the initial temperature difference between
the two regions. The Lewis number definition is $\text{Le} = \alpha/D$, where D is the
mass diffusivity of the chemical species. The time criteria (4.33) and (4.34)
have been tested numerically along with the corresponding time scales for
approach to thermal equilibrium.

4.5 Aerosol Transport and Collection in Filters

Air is composed of gases, variable amounts of water vapor, and microscopic
and submicroscopic particles. These particles may be solid or liquid. They contain organic or inorganic matter, and substances that are harmless or toxic.
A gaseous medium in which particles are suspended is known as an aerosol
(Figure 4.5a).

Aerosol particles may originate naturally (e.g., dust, salt, pollen, microbes,
viruses, etc.), or as a result of industrial activity, incineration, and combustion
processes. Particles with different origins and types of chemical composition
mix together in the air (Figure 4.6). They have been classified by size into three
conventionally defined modes (Whitby and Sverdrup, 1980): nuclei, accumulation, and coarse-particle modes (Figure 4.7). The nuclei and accumulation
modes together account for fine particles, or fine particulate matter.

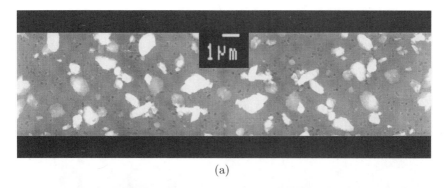

(a)

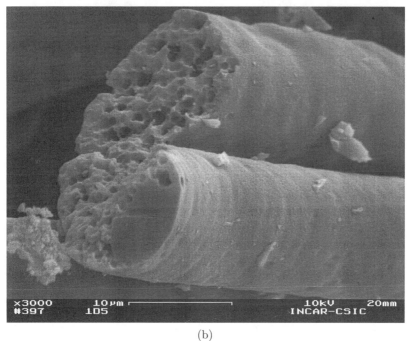

(b)

Fig. 4.5. (a) Aerosol particles with different sizes; (b) activated carbon fiber (courtesy of J. M. V. Nabais).

The nuclei mode consists primarily of combustion particles emitted directly into the atmosphere from motor vehicles, especially diesel-powered vehicles, as well as particles formed in the atmosphere by gas-to-particle conversion. Significant concentrations of particles in nuclei mode are not always present, but are usually found near freeways. These very fine particles rapidly attach to particles in the accumulation mode or to each other.

The accumulation mode includes combustion and photochemical smog particles and attached nuclei mode particles. Removal mechanisms, such as

Fig. 4.6. Approaching sand and dust storm in Tunisia (courtesy of N. A. Ferreira).

settling, deposition on surfaces, or attachment to rain droplets, are ineffective in this size range. Accumulation mode particles are about the same size as the wavelength of visible light, and consequently these particles scatter and absorb visible light. They account for most of the visibility effects of urban particulate pollution, and may affect the earth's climate by reflecting solar radiation. The particle size range 0.01 to 0.1 µm contains the majority of atmospheric particles (Hinds, 1999).

Coarse particles include windblown dust, salt particles from sea spray, and those that are mechanically generated, such as particles from construction sites.

Atmospheric aerosols have not only a wide range of particle sizes (ranging from tenths of a nanometer to several micrometers), but also a very large

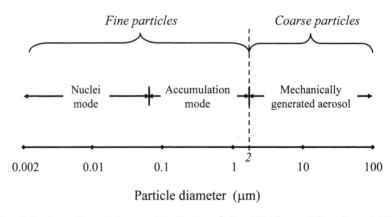

Fig. 4.7. Aerosol particle size distribution (after Whitby and Sverdrup, 1980).

range of particle concentrations, from fewer than 100 particles per cm^3 in clean background conditions, to more than 10^5 particles per cm^3 in urban areas with high levels of pollution (Heintzenberg, 1989; Heintzenberg and Covert, 1990).

A change in aerosol particle concentration affects air quality, impairs visibility (Garror, 1999), and may affect the health of local residents (Friedlander, 1977; Wilson and Spengler, 1996). When inhaled, micrometric particles ($>1\,\mu m$) are captured in the upper respiratory tract, or are trapped in the airways of the lungs (cf. Section 6.1.3): they are subsequently cleared by means of coughing or swallowing. Submicrometric particles that may be drawn into the lungs and reach the gas-exchange surfaces of the alveolar region are regarded as posing a significant health risk. Results from epidemiological studies show a significant positive association between submicrometric particle concentration and mortality rates (Ozkaynak and Thurston, 1987; Schwartz and Dockery, 1992). Biogenic particles, such as bacteria, fungi, and pollen, can also be responsible for various infectious and allergic diseases (Malmberg, 1990; Lacey and Dutkiewicz, 1994; Brandão and Lopes, 1991).

The presence of aerosol particles in buildings is generally regarded as posing significant problems. Of particular concern is their potential for causing damage to paintings and other works of art (Nazaroff and Cass, 1991), as well as leading to the failure of semiconductors and other electronic components (Liu and Ahn, 1987).

The removal of particles from air streams is a serious problem that scientists are making great efforts to resolve (Davies, 1973; Dorman, 1973). The most common method used for the removal of aerosol particles is filtration, which is generally a simple, highly efficient, and economical process. Both micrometric and submicrometric aerosol particles can be effectively collected by means of a porous bed (Gutfinger and Tardos, 1979; Tien, 1989) and fibrous filters (Fuchs, 1964; Davies, 1973). The use of activated carbon granules or fibers (Figure 4.5b) in the manufacturing of filters is of particular interest because such fibers allow the removal of particles and undesirable gases and odors (Miguel, 2003b). This is due to the high adsorption characteristics of activated carbon materials (cf. Section 1.1).

The filtration of aerosol particles by means of porous filters occurs when an air stream containing particles passes through a filter, depositing particles on the surfaces of collecting elements (e.g., granules or fibers). Two different deposition modes have been identified: surface mode and depth mode (within the thickness of the filter). The latter forms the subject of this section, because it is the main factor leading to the accumulation of micrometric and submicrometric aerosol particles in filters. In filtration operations, there are certain major concerns:

(1) The filter should demonstrate a high ability to capture particles that pass through the filter medium (i.e., the filter should be characterized by low penetration); and

(2) The removal and retention of suspended particles from the gas stream causes the filter medium to experience a decrease in its ability to allow air to flow through it (i.e., deterioration of filter permeability). As a result, the pressure drop required for maintaining a constant airflow through the filter increases as the deposition process continues. The life of the filter depends greatly on this phenomenon.

There are five mechanisms that govern the deposition of aerosol particles in filters (Figure 4.8). When the aerosol particles that follow fluid streamlines touch the surface of filter collecting elements (granules, fibers) and become attached to it, the mechanism is known as *interception*. This applies to particles in streamlines that approach the collector at a distance that is less than the average radius of particles. Another possibility is when particles deviate from fluid streamlines: this greatly enhances the probability of their hitting collector elements and remaining there. The deviation from streamlines is caused by a combination of the following mechanisms:

(1) *Inertial impaction*: This occurs when aerosol particles with large inertia are unable to adjust to the abruptly changing streamlines in the neighborhood of a fiber or bed.

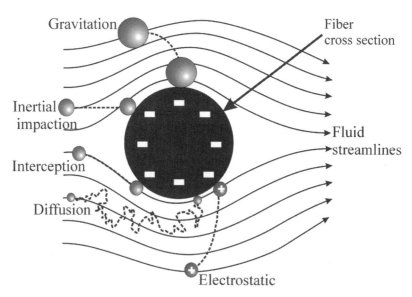

Fig. 4.8. Schematic representation of particle deposition on a collector (Miguel, 2003b).

(2) *Gravitation*: This settling of aerosol particles is due to gravity, and is especially important where there are heavy particles and slow air flows.
(3) *Diffusion*: Fine aerosol particles tend to move in random directions (Brownian motion) instead of following air streamlines, and thereby collide with and stick to collector elements.
(4) *Electrostatic*: Aerosol particles do not follow air streamlines when Coulomb forces act between particles and filter collector elements.

The inertial and gravitational mechanisms are characteristic of micrometric particles, as they are more effective for particle sizes above $2\,\mu m$. Diffusion is the dominant mechanism for submicrometric particles, especially for particles below $0.2\,\mu m$. When present, Coulomb forces are effective with respect to the deposition of micrometric and submicrometric particles.

4.6 Filter Efficiency and Filtration Theories

The efficiency (φ) with which aerosol particles are collected in a filter can be calculated from the concentrations of particles upstream and downstream of the filter, c_u and c_d, respectively,

$$\varphi = 1 - \frac{c_d}{c_u}. \tag{4.35}$$

Alternatively, some authors prefer to use the penetration of aerosol particles, which is defined as

$$\Pi = 1 - \varphi = \frac{c_d}{c_u}. \tag{4.36}$$

Parameters Π and φ depend, among other factors, on filter thickness, filter microstructure, particle characteristics, and the mass of particles existing within the filter. It has been demonstrated by Leers (1957) that in most practical circumstances the penetration by aerosol particles declines as the filter thickness (H) increases,

$$\Pi = \exp\left(-\frac{H}{h}\right). \tag{4.37}$$

The characteristic filtration length h does not depend on H (Shapiro *et al.*, 1991), but depends on several other factors, most notably on the microstructure of the filter and on particle size (Davies, 1973).

Filtration theories have been developed since the middle of the 20th century. The classical "single-collector element" filtration model (Pich, 1966; Davies, 1973) considers aerosol transport only around a single fiber or a single bed. The collection efficiency of particles is defined by the superposition of mechanisms responsible for the deposition of aerosol particles, as described in the preceding paragraph. The collection efficiency of the filter is obtained by combining the single-collector efficiency with macroscopic parameters for the entire filtration process.

According to classical filtration theory, the characteristic filtration length (h) can be expressed as

$$h = \frac{2r_c}{s\alpha\eta},$$ (4.38)

where r_c is the particle-loaded collector radius, η is the collector-element (single-element) efficiency, α is the filter solidity $(1 - \phi)$, and s is a coefficient that can be taken as equal to $4/[\pi(1 - \alpha)]$ for fibrous materials (Hinds, 1999; Flagan and Seinfeld, 1988), and equal to 1.5 for granular beds (D'Ottavio and Goren, 1983) or 1.88 (Schmidt, 1978).

The collector-element efficiency η depends on the partial efficiencies of the distinct aerosol capture mechanisms, and can be computed (compounded) according to

$$\eta = 1 - (1 - \eta_r)(1 - \eta_I)(1 - \eta_D)(1 - \eta_G)(1 - \eta_{el}).$$ (4.39)

Table 4.1 lists the partial efficiencies for the deposition of spherical aerosol particles due to the mechanisms of interception (η_r), inertial impaction (η_I), diffusion (η_D), gravitation (η_G), and electrostatics (η_{el}).

Filters collect not only spherical or quasispherical particles, but also particles that are elongated (e.g., fibrous particles). The filtration of fibrous aerosol particles (e.g., asbestos or anthropogenic mineral fibers) is of great importance because such particles pose significant health risks (Spurny, 1997). Fibrous particles are very slender and undergo rotational motion, which is generally coupled with translational motion. As a consequence of their elongated shapes, their spatial orientation in the airflow cannot be disregarded (note that spatial orientation is not a factor in the case of spherical particles). The trajectory and orientation of fibrous particles in air flow can be obtained based on the methods developed by Gallily et al. (1986), Gradon et al. (1988), Foss et al. (1989), Bernstein and Shapiro (1994), and Asgharian and Cheng (2002).

Filters also collect biogenic particles, such as bacteria, fungi, or pollen. Biogenic particles differ from mineral particles in several respects, including density and surface characteristics. The evaluation of their partial deposition efficiencies has been limited to empirical studies. In general, the results suggest that the differences between the partial efficiencies of mineral and biogenic particles with the same shape and size are not significant (Jankowaska et al., 2000). Differences in efficiencies may be significant only when the collection of particles is due to the presence of electrical forces (Baumgartner, 1987). These differences may be due to the fact that most biogenic particles contain a high percentage of water, and have dielectric constants larger than mineral particles of the same size.

During filtering, aerosol particles are collected in the filter and cause continuous changes in porosity, structure, and surface characteristics. Consequently, the penetration Π of the filter also undergoes continuous changes (Leers, 1957; Billard et al., 1963; Payet et al., 1992). In order to estimate the effect of deposition of solid aerosol particles in filters, Zhao et al. (1991)

Table 4.1. Collection mechanisms and efficiencies of a clean (not loaded) single collector[a]

Mechanism	Filter	Efficiency	Range of validity	Source
Diffusion	Fibrous	$\eta_D = 1.6[(1-\alpha)/\text{Ku}]^{1/3}\text{Pe}^{-2/3}$	Pe < 100	Lee and Liu (1982)
	Fibrous	$\eta_D = 1.6[(1-\alpha)/\text{Ku}]^{1/3}\{1+0.388\text{Kn}\,[(1-\alpha)\text{Pe}/\text{Ku}]^{1/3}\text{Pe}^{-2/3}\}$	Accounts for the slip flow effect	Liu and Rubow (1990)
	Fibrous	$\eta_D = 4.89[(1-\alpha)/\text{Ku}]^{0.54}\text{Pe}^{-0.92}$	Pe < 50	Rao and Faghri (1988)
	Fibrous	$\eta_D = 1.8[(1-\alpha)/\text{Ku}]^{1/3}\text{Pe}^{-2/3}$	100 < Pe < 300	Rao and Faghri (1988)
	Fibrous	$\eta_D = [2-\ln(\text{Re}_c)]^{1/3}\text{Pe}^{-2/3}$	$\text{Re}_c > 1$	Loeffler (1971)
	Granular bed	$\eta_D = 4\{[2+1.5\alpha+1.5(8\alpha-3\alpha^2)^{1/2}]/[(1-\alpha)(2-3\alpha)]\}^{1/3}\text{Pe}^{-2/3}$	$\text{Re}_c < 1$; $\alpha < 0.67$; Pe ≥ 1000	Sinkar (1975)
	Granular bed	$\eta_D = 4[1.1/(1-\alpha)]\text{Pe}^{-2/3}$	$\text{Re}_c < 1$; $0.3 < \alpha < 0.65$	Tan et al. (1975)
Electrical	Fibrous	$\eta_{el} = \xi Z_p \psi(1-\alpha)/(800\,\mu\alpha r_p u)$	Charged particles and filter under electric field	Zhao et al. (1991)
	Granular bed	$\eta_{el} = 4[\xi Z_c\psi/(6\pi r_p\mu u)]/[1+\xi Z_c\psi/(6\pi r_p\mu u)]$	Charged particles and filter under electric field	Tien (1989)
Gravity	Fibrous	$\eta_G = \text{Ga\,St}$		Tardos and Pfeffer (1980)
	Granular bed	$\eta_G = \text{Ga\,St}$		Tardos and Pfeffer (1980)
Inertia	Fibrous	$\eta_I = 0.075\,\text{St}^{6/5}$		Friedlander (1977)
	Fibrous	$\eta_I = 1/\{1+\text{St}^{-1}[1.53-0.23\ln(\text{Re}_c)+0.0167(\ln(\text{Re}_c))^2]\}+2I_r/(3\text{St})$	$0.8 < \text{St} < 2$; $\text{Re}_c < 1$; $I_r < 0.2$	Suneja and Lee (1974)
	Fibrous	$\eta_I = (\text{St}^4 + 1.62\times10^{-4})/\{\text{St}[1.03\,\text{St}^3 + (1.14 + 0.0404\ln(\text{Re}_c)\text{St}^2 + 0.0148\ln(\text{Re}_c) + 0.201]\}$	$30 \leq \text{Re}_c \leq 40$; $0.07 \leq \text{St} \leq 5$	Ilias and Douglas (1989)
	Granular bed	$\eta_I = (1+0.04\text{Re}_c)\text{St}$		Pendse and Tien (1982)
	Granular bed	$\eta_I = 2\{\text{St}[1+1.75\text{Re}_c/(150\alpha)]\}^{3.9}/\{4.3\times10^{-6}+\{\text{St}[1+1.75\text{Re}_c/(150\alpha)]\}^{3.9}\}$	$15\alpha/(1.75\text{Re}_c + 150\alpha) < \text{St} < 30\alpha/(1.75\text{Re}_c + 150\alpha)$	Gal et al. (1985)
Interception	Fibrous	$\eta_r = 0.6[(1-\alpha)/\text{Ku}]^{1/3}[I_r/(1+I_r)]^2$		Lee and Liu (1982)
	Fibrous	$\eta_r = 0.6[(1-\alpha)/\text{Ku}]^{1/3}[I_r/(1+I_r)^{1/2}]^2(1+2\text{Kn}/I_r)$	Accounts for the slip flow effect	Liu and Rubow (1990)
	Granular bed	$\eta_r = 1.5[2+1.5\alpha+1.5(8\alpha-3\alpha^2)^{1/2}]/[(1-\alpha)(2-3\alpha)]I_r^2$	$\text{Re}_c < 1$; $\alpha < 0.67$; Pe ≥ 1000	Sinkar (1975)
	Granular bed	$\eta_r = 1.5[1.1/(1-\alpha)]^3 I_r^2$	$\text{Re}_c < 1$; $0.3 < \alpha < 0.65$	Tan et al. (1975)

[a] In this table the dimensionless numbers are defined as follows: Ga $= r_c g/u^2$; Kn $= \lambda/r_p$; Ku $= -0.5\ln\alpha - 0.75 + \alpha - 0.25\alpha^2$; Pe $= 12\pi\mu r_c r_p u/(\xi K_{Bt} T)$; Re$_c = 2\rho r_c u/\mu$; and St $= 2\xi\rho_p u r_p^2/(9\mu r_c)$; $\xi = 1 + 1.32\times10^{-3}[6.32 + 2.01\exp(-166.44\,pr_p)]/pr_p$.

developed a methodology for the circumstances in which particles deposited in the filter do not dramatically alter the internal structure of the filter, as shown in Figure 4.9. In other words, the deposited particles only increase the collector size, and particles collected in different collectors do not touch. This model may be especially adequate for the deposition of small particles during the first stages of filtration. Denoting the mass of particles collected per filter area as M_p, the particle-loaded collector radius and filter solidity are given by

$$r_c = r_{co}\left(1 + \frac{M_p}{\rho_p \alpha_o H}\right)^{1/2}, \tag{4.40}$$

$$\alpha = \alpha_o\left(1 + \frac{M_p}{\rho_p \alpha_o H}\right), \tag{4.41}$$

where r_{co} and α_o are the collector radius and filter solidity when the filter is free of deposited particles. Consequently, the penetration Π can be obtained from the relations for clean filters [Equations (4.37) and (4.38)] if the collector radius and the filter solidity in these relations are replaced by Equations (4.40) and (4.41).

Particle deposition also occurs on particles that have previously been captured. These agglomerates of particles radically alter the internal structure of the filter (Figure 4.10). In order to account for these circumstances, Jung and Tien (1991) suggested that for granular filters the efficiency behaves as

$$\eta = \eta_o\left[1 + a_{\eta 1}\left(\frac{M_p}{\rho_p H}\right)^{a_{\eta 2}}\right]. \tag{4.42}$$

(a)   (b)

Fig. 4.9. (a) Clean filter and (b) filter loaded during the first stages of filtration.

Fig. 4.10. Filter having agglomerates of particles.

For experiments performed at $0.0017 \leq \text{St} \leq 0.038$ and $0.00172 \leq I_r \leq 0.008$, it was found that the coefficients are correlated by

$$a_{\eta 1} = 0.095\text{St}^{-1.48}I_r^{0.432}10^{3a_{\eta 2}}, \qquad a_{\eta 2} = 0.442\text{St}^{-0.347}I_r^{0.24}, \qquad (4.43)$$

where I_r is the interception parameter, which is the ratio between the particle radius and the collector radius. An empirical equation relating the collector-element efficiency η to mass loaded (accumulated) in fibrous filters was developed by Emi *et al.* (1982),

$$\eta = \eta_o \left[1 + \left(\frac{M_p}{0.316b_{\eta 1}\eta_o u^{0.25}} \right)^{b_{\eta 2}} \right], \qquad (4.44)$$

where $b_{\eta 1}$ and $b_{\eta 2}$ are, respectively, 0.0027 and 1.15 (filter mesh 200), or 0.0015 and 1.23 (filter mesh 325), and 0.0011 and 1.34 (filter mesh 500). The mesh size indicates the number of fibers per inch in one of the directions (warp or weft). The penetration Π for the particle-loaded filter is obtained from Equations (4.37) and (4.38) using η from Equations (4.42) and (4.44).

Experiments performed by Payet *et al.* (1992) with submicrometer oil particles showed an increase in particle penetration during the filtering operation, in contrast to penetration by solid aerosols. Payet *et al.* (1992) stressed that this increase in penetration is not attributable to a decrease in collector-element efficiency η, but rather to the combination of a decrease in the number of collecting elements available for filtration (fibers or granular bed), and an increase in air velocity within the filter. In fact, if a number of collecting elements cease to participate in the collection of particles, this represents a decrease in filter solidity. Assuming that the volume of liquid retained in the filter (V_{liq}) is distributed uniformly, the volume fraction of collecting elements that is rendered ineffective is

$$\frac{V_{\text{liq}}}{V_f} = \frac{m_{\text{liq}}}{\rho_{\text{liq}}V_f},$$

(4.45)

and filter solidity is given by

$$\alpha = \alpha_o \left(1 - \frac{m_{\text{liq}}}{\rho_{\text{liq}}V_f}\right),$$

(4.46)

where α_o is filter solidity before loading. In sum, the penetration of particles can be obtained from the preceding relationships if the filter solidity is replaced by Equation (4.46).

An advanced class of "integrated porous filter models" appeared in the 1990s, and represents an important movement in the theoretical approach to the filtration process (Shapiro and Brenner, 1990; Shapiro et al., 1991; Quintard and Whitaker, 1995). These models deal with the process of filtration in the global structure of a porous filter, and use equations to describe the processes of particle transport and particle capture, permitting the calculation of the filter efficiency.

Shapiro and Brenner (1990) suggested that the filtration of submicrometer aerosol particles by means of the use of porous filters can be studied as a convective dispersion phenomenon. The macroscale or coarse-scale aerosol transport equation is

$$\frac{\partial \overline{c_w}}{\partial t} + U^* \cdot \nabla \overline{c_w} - D^* : \nabla \nabla \overline{c_w} + \omega^* \overline{c_w} = 0,$$

(4.47)

where $\overline{c_w}$ is the average aerosol concentration, and the coarse-scale coefficients U^* (aerosol velocity), D^* (aerosol dispersivity), and ω^* (deposition rate coefficient) depend on Reynolds number, Peclet number, interception parameter, and filter solidity. Assuming that aerosol filtration is steady and one-dimensional in the x-direction, Equation (4.47) reduces to

$$U_x^* \frac{\partial \overline{c_w}}{\partial \bar{x}} - D_x^* \frac{\partial^2 \overline{c_w}}{\partial \bar{x}^2} + \omega_x^* \overline{c_w} = 0.$$

(4.48)

This formulation requires conditions that are applicable at the boundaries (Shapiro and Brenner, 1990),

$$uc_u = U_x^* \overline{c_w} - D_x^* \frac{\partial \overline{c_w}}{\partial \bar{x}} \quad \text{at } x = 0 \text{ (upstream of the filter)}, \quad (4.49)$$

$$\frac{\partial \overline{c_w}}{\partial \bar{x}} = 0 \quad \text{at } x = H \text{(downstream of the filter)}, \quad (4.50)$$

which means that the aerosol flux upstream of the filter is transformed into a convective and a dispersive flux, Equation (4.49), and the aerosol concentration downstream of the filter is constant, Equation (4.50). Equation (4.48) and boundary conditions (4.49) and (4.50) lead to the solution

$$\overline{c_w} = \frac{c_u u}{U_x^*} \frac{a_1 \exp\left[a_2\left(\overline{x_w} - \overline{H_o}\right)\right] - a_2 \exp\left[a_1\left(\overline{x_w} - \overline{H_o}\right)\right]}{a_1^2 \exp(-a_2\overline{H_o}) + a_2^2 \exp(-a_1\overline{H_o})} \tag{4.51}$$

with

$$a_1 = 0.5 \left[1 + \left(1 + 4\frac{\omega_x^* D_x^*}{U_x^{*2}}\right)^{1/2}\right],$$

$$a_2 = 0.5 \left[1 - \left(1 + 4\frac{\omega_x^* D_x^*}{U_x^{*2}}\right)^{1/2}\right], \tag{4.52}$$

$$\overline{x_w} = x\frac{U_x^*}{D_x^*} \qquad \overline{H_o} = H\frac{U_x^*}{D_x^*}. \tag{4.53}$$

Combining Equations (4.51) to (4.53) with Equation (4.36) allows us to express the penetration of aerosol particles as

$$\Pi = \frac{a_1 - a_2}{a_1^2 \exp(-a_2\overline{H_o}) + a_2^2 \exp(-a_1\overline{H_o})}. \tag{4.54}$$

The following limiting cases are worth highlighting.

(1) When $\omega_x^* D_x^* \gg U_x^{*2}$, the penetration can be expressed as

$$\Pi = \frac{2U_x^*}{(\omega_x^* D_x^*)^{1/2}} \exp\left[-\left(\frac{\omega_x^*}{D_x^*}\right)^{1/2} H\right]. \tag{4.55}$$

This limit corresponds to aerosol filtration performed at very low Peclet numbers. It is still not known (proven) if this limit represents a practically important particle filtration regime.

(2) When $\omega_x^* D_x^* \ll U_x^{*2}$, aerosol penetration is described by

$$\Pi = \exp\left(-\frac{\omega_x^*}{U_x^*} H\right). \tag{4.56}$$

In this limit, the penetration is independent of the dispersivity. The similarity between Equations (4.56) and (4.37) leads to the conclusion that both equations are valid over the same range. In addition, the characteristic filtration length h may be expressed in terms of ω^* and U^*,

$$h = \frac{U_x^*}{\omega_x^*}. \tag{4.57}$$

The characteristic filtration length was calculated by Shapiro et al. (1991) for different Peclet numbers, interception parameters, and filter solidities, and is shown in Figure 4.11. The length h decreases when the interception parameter and filter solidity increase, but increases when the Peclet number increases. The influence of the interception parameter is stronger as the Peclet number increases. This means that the deposition of aerosol particles becomes less efficient with increasing particle size and decreasing filter solidity.

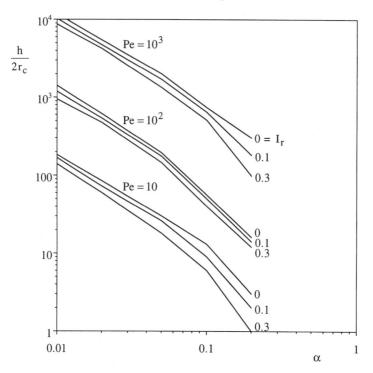

Fig. 4.11. The characteristic filtration length h versus the filter solidity obtained for a porous medium consisting of circular cylinders arranged in periodic square arrays (adapted from Shapiro *et al.*, 1991).

4.7 Pressure Drop, Permeability, and Filter Performance

For a fluid moving with a slow steady velocity, the pressure drop Δp through a filter with the permeability K and the volume-averaged velocity u are related in accordance with Darcy's law, $\Delta p = (-\mu/K)Hu$ (cf. Section 1.3). At sufficiently high pore Reynolds numbers, Darcy's law breaks down and, instead, the Dupuit–Forchheimer relationship is applicable (Miguel, 1998a; Nield and Bejan, 1999; see also Section 1.3)

$$\Delta p = -\frac{\mu}{K} H \left(1 + \frac{\rho c_F K^{1/2} u}{\mu}\right) u. \qquad (4.58)$$

The dimensionless factor is $c_F = 0.043/(1 - \alpha)^{2.13}$ for a fibrous filter (Miguel, 1998a), and $0.1429/(1 - \alpha)^{3/2}$ for a granular bed (Ergun, 1952), where α is the solidity of the filter. The permeability of the filter can be determined based on empirical and semiempirical analyses described in published research papers. Table 4.2 provides a summary of useful expressions for calculating the permeability of clean filters.

Table 4.2. Permeability of clean (not loaded) filters

Filter	Equation	Range of validity	Source
Fibrous	$r_c^2[16\alpha^{3/2}(1 + 56\alpha^3)]$	$0.004 < \alpha < 0.3$	Davies (1973)
Fibrous	$r_c^2/(16\alpha^{3/2})$	$\alpha < 0.006$	Davies (1973)
Fibrous	$(-0.738 - 0.5\ln\alpha + \alpha - 0.887\alpha^2 + 2.038\alpha^2)r_c^2/(4\alpha)$	$\alpha \ll 1$; square arrangement	Sangani and Acrivos (1982)
Fibrous	$(-0.745 - 0.5\ln\alpha + \alpha - 0.25\alpha^2)r_c^2/4\alpha$	$\alpha \ll 1$; staggered arrangement	Sangani and Acrivos (1982)
Fibrous	$91(1 - \alpha)^3 r_c^2/(3125\alpha^2)$	$0.42 < \alpha < 0.9$	Miguel (1998b)
Fibrous	$5.56 r_c^2/[e^{10.1\alpha} - 1]$	$0.05 \le \alpha \le 0.4$	Koponen et al. (1998)
Granular bed	$2.47 \times 10^{-4} r_c^2$	$25\,\mu\text{m} \le r_c \le 500\,\mu\text{m}$	Nutting (1930)
Granular bed	$(1 - \alpha)^3 r_c^2/(45\alpha^2)$	$\alpha > 0.5$	Carman (1937)
Granular bed	$4(1 - \alpha)^3 r_c^2/(150\alpha^2)$	$\alpha > 0.5$	Ergun (1952)
Granular bed	$(1 - \alpha)^3 r_c^2/(9a\alpha^2), \quad a = 5.35 \pm 0.23$	—	Fand et al. (1987)

Models have also been proposed for estimating the effect of the deposition of solid aerosol particles on the permeability of the filter. For granular filters, the permeability can be estimated from Jung and Tien (1991),

$$K = \frac{K_o}{1 + a_{\kappa 1}\alpha_p^{a_{\kappa 2}}}, \tag{4.59}$$

with

$$a_{\kappa 1} = 0.348\text{St}^{-1.2}I_r^{0.86}10^{3a_{\kappa 2}}, \tag{4.60}$$

$$a_{\kappa 2} = 3.51\text{St}^{-0.092}I_r^{0.275}. \tag{4.61}$$

Parameters $a_{\kappa 1}$ and $a_{\kappa 2}$ correspond to experimental data obtained, respectively, in the range $0.0017 \leq \text{St} \leq 0.038$ and $0.00172 \leq I_r \leq 0.008$. Parameter α_p is the packing fraction of particles, which is defined as the volume of deposited particles per unit volume of filter $M_p/\rho_{cp}H$, where ρ_{cp} represents the density of cake of particles in the filter (usually fixed to a value of particle density). The permeability of fibrous filters can be expressed as (Bergman et al., 1978)

$$K = \frac{1}{16\left(\dfrac{\alpha}{r_c} + \dfrac{\alpha_p}{r_p}\right)\left(\dfrac{\alpha}{r_c^2} + \dfrac{\alpha_p}{r_p^2}\right)^{1/2}}. \tag{4.62}$$

This approach is based on the empirical model developed by Davies (1973) for clean fibrous filters. The idea behind Equation (4.62) is that the collected particles form dendrites that behave like new fibers. This equation applies to those fibrous filters with solidity lower than 0.006. For filters with high solidity $(0.006 < \alpha < 0.3)$, the permeability can be determined from

$$K = \frac{1}{16\left(\dfrac{\alpha}{r_c} + \dfrac{\alpha_p}{r_p}\right)\left(\dfrac{\alpha}{r_c^2} + \dfrac{\alpha_p}{r_p^2}\right)^{1/2}[1 + 56(\alpha + \alpha_p)^3]}. \tag{4.63}$$

The filtration of solid aerosol particles can occur under varying humidity conditions. A number of researchers (Gupta et al., 1993; Miguel and Silva, 2001; Miguel, 2003a) have carried out extensive laboratory tests to quantify this effect on permeability as a function of particle hygroscopicity and size. Results have shown an increase in permeability with increasing humidity for the particle sizes tested, both for nonhygroscopic particles and for hygroscopic particles below the deliquescent point (the point of conversion of a solid substance into a liquid, as the result of the absorption of water vapor from the air). Above the deliquescent point, the permeability is greatly reduced because of the change of phase experienced by particles.

The increase in permeability with air humidity seems to cause a high compact arrangement of particles within the filter. In order to account for the effect of humidity on the density of the cake particles, Miguel (2003a) proposed that

$$\rho_{cp} = \frac{\rho_p M_p}{a_M M_p + (a_\Theta \Theta + a_{M\Theta})\rho_p H}, \tag{4.64}$$

where Θ is the humidity, and a_M, a_Θ, and $a_{M\Theta}$ are coefficients determined empirically (Table 4.3).

Unlike the research on the filtration of solid aerosols, the research on the permeability of filters loaded with liquid aerosols or liquid–solid mixed aerosols is rather limited. Experiments performed by Novick et $al.$ (1993) with diethylene glycol and dioctyl phthalate particles reveal that the variation of permeability with mass loading is practically not affected by the size of liquid particles. They also describe the results of tests involving a solid-and-liquid combined aerosol in high-efficiency particulate air filters. These results indicate that for a given mass loading of particles, a solid-and-liquid combined aerosol can affect the permeability of the filter more significantly than pure-solid or pure-liquid aerosol particles.

A high-quality air filter is characterized by high efficiency in collecting particles, coupled with high permeability. The importance of defining a performance factor for filters is twofold: it allows the comparison of different filter media, and it provides information regarding the need for filter replacement, which could be a crucial safety aspect.

In research studies, filter performance (or filter quality) is often defined as the ratio of the negative logarithm of penetration to the pressure drop across the filter (Miguel, 2003a,b). The figure of merit resulting from this definition has the advantage that it can be calculated directly from parameters that can be measured simply. Its drawback is that it is not a dimensionless factor (it has the units Pa^{-1}). Its magnitude depends on the system used, and filters must be compared for specified filtration velocity, particle diameter, and particle loading. To overcome this problem, filter performance is defined as the dimensionless ratio of collector-element efficiency (η) to the drag coefficient (C_D),

$$\Omega_{\eta C} = \frac{\eta}{C_D}. \tag{4.65}$$

The drag coefficient is defined as

$$C_D = \frac{2\Delta p}{\rho u^2}. \tag{4.66}$$

By substituting Equations (4.37) and (4.38) into Equation (4.65) we obtain

$$\Omega_{\eta C} = -\frac{r_c \rho u^2 \ln \Pi}{s\alpha H \Delta p}. \tag{4.67}$$

Based on dimensional analysis and the Buckingham pi theorem, Miguel (2003a) has suggested an alternative relationship for evaluating filter performance as the product of the number of particles caught per unit of filter area (N) and filter permeability (K),

$$\Omega_{NK} = NK, \tag{4.68}$$

where

$$N = \frac{M_p}{\rho_p V_p}. \tag{4.69}$$

Table 4.3. The coefficients a_M, a_Θ, and $a_{M\Theta}$ for Equation (4.64)

Aerosol particle	Mean particle diameter (μm)	a_M	a_Θ	$a_{M\Theta}$	Range of validity
Sodium chloride	0.55	0.893	−0.00272	0.00118	$0.32 \leq \Theta \leq 0.61$
Sodium chloride	1.30	0.917	−0.00136	0.00072	$0.32 \leq \Theta \leq 0.61$
Alumina	0.82	0.979	−0.00103	0.00073	$0.53 \leq \Theta \leq 0.90$
Alumina	1.42	1.010	−0.00094	0.000309	$0.32 \leq \Theta \leq 0.90$
Alumina	6.02	0.958	−0.00129	0.00099	$0.32 \leq \Theta \leq 0.90$

For granular filters, the use of Equations (4.69) and (4.60) in conjunction with Equation (4.68) leads to

$$\Omega_{NK} = \frac{M_p K_o}{\rho_p V_p (1 + a_{\kappa 1} \alpha_p^{a_{\kappa 2}})}. \tag{4.70}$$

For fibrous filters, substituting Equations (4.69), (4.62), and (4.63) into Equation (4.68) yields

$$\Omega_{NK} = \frac{M_p}{16 \rho_p V_p \left(\dfrac{\alpha}{r_c} + \dfrac{\alpha_p}{r_p} \right) \left(\dfrac{\alpha}{r_c^2} + \dfrac{\alpha_p}{r_p^2} \right)^{1/2}} \qquad (\alpha < 0.006), \tag{4.71}$$

$$\Omega_{NK} = \frac{M_p}{16 \rho_p V_p \left(\dfrac{\alpha}{r_c} + \dfrac{\alpha_p}{r_p} \right) \left(\dfrac{\alpha}{r_c^2} + \dfrac{\alpha_p}{r_p^2} \right)^{1/2} \left[1 + 56(\alpha + \alpha_p)^3 \right]}$$

$$(0.006 < \alpha < 0.3). \tag{4.72}$$

Figure 4.12 shows the ratio $\Omega_{NK}/\Omega_{NK,\text{max}}$ calculated for deposition of aerosol particles with several diameters (d) in fibrous filters. When d increases, the maximum value of $\Omega_{NK}/\Omega_{NK,\text{max}}$ drifts toward higher values of mass of particles collected per filter area (M_p).

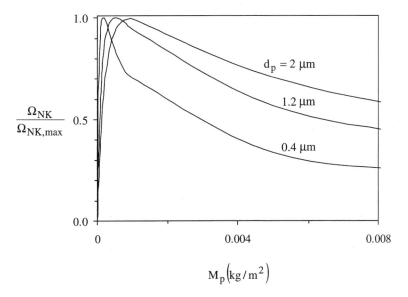

Fig. 4.12. Ratio between the performance Ω_{NK} and the maximum performance $\Omega_{NK,\text{max}}$ versus the mass of particles collected per filter area.

The filter performance Ω_{NK} can also be expressed in terms of filter efficiency (η) and loading time (t). If the concentration of the aerosol particles upstream of the filter and the filtration velocity are constant, then

$$M_p = \eta c_u \rho_p V_p u t. \tag{4.73}$$

Assuming that the effect of humidity on the permeability is neglected, the substitution of Equation (4.73) into Equations (4.70) and (4.71) yields

$$\Omega_{NK} = \frac{\eta c_u u t K_o}{1 + a_{\kappa 1} \left(\dfrac{\eta \rho_p c_u V_p u t}{\rho_{cp} H} \right)^{a_{\kappa 2}}}, \tag{4.74}$$

$$\Omega_{NK} = \frac{\eta c_u u t}{16 \left(\dfrac{\alpha}{r_c} + \dfrac{\eta \rho_p c_u V_p u t}{\rho_{cp} H r_p} \right) \left(\dfrac{\alpha}{r_c^2} + \dfrac{\eta \rho_p c_u V_p u t}{\rho_{cp} H r_p^2} \right)^{1/2}}. \tag{4.75}$$

Note that Equations (4.74) and (4.75) are valid only for the loading time, before the clogging of the filter.

4.8 Ionic Transport

In this and the next two sections we focus on the penetration (ingress) of ionic species through a porous medium. This phenomenon is of interest in many fields. For example, researchers in geophysics study the diffusion of salts or pollutants in soils (Shackelford, 1991). Advances have also been made by biologists working on electrodiffusion processes in charged membranes (Helfferich, 1962; Brumleve and Buck, 1978; Horno and Castilla, 1994), as described in Chapter 6. Civil engineers study the diffusion of ions (chloride, sulfate) through reinforced concrete, and their effect on the durability of materials.

We begin with the transport of species through a porous medium, which was presented in general terms in Sections 1.5 to 1.7. Brownian motion, or the random movement of particles in a solution, is the reason why when a concentration difference is maintained between two regions of a solution, global transport of species (diffusion) proceeds from high concentrations toward low concentrations. The driving force for the movement of a species is the gradient of the chemical potential μ (Bockris and Reddy, 2001),

$$\mu = \mu_0 + RT \ln a, \tag{4.76}$$

where μ_0 is the chemical potential in a reference state, R the universal ideal gas constant, T the absolute temperature, and a the activity of the species. The latter is linked to the species concentration by (Helfferich, 1962)

$$a = \gamma c, \tag{4.77}$$

where γ is the activity coefficient of the solution. A gradient of chemical potential induces a flux of matter in order to drive the system to equilibrium,

$$\mathbf{j} = -l \nabla \mu. \tag{4.78}$$

In this equation l is a phenomenological coefficient proportional to the species concentration via the absolute mobility v,

$$l = vc. \tag{4.79}$$

At this point we introduce the Einstein relation, which links the diffusion coefficient of a species in an infinite solution to the absolute mobility (Atkins, 1998)

$$D = vRT. \tag{4.80}$$

Combining Equations (4.76) to (4.80), and assuming that the solution is infinitely diluted ($\gamma = 1$), we obtain Fick's first law (Fick, 1855),

$$\mathbf{j} = -D\nabla c. \tag{4.81}$$

This equation can be written for every species that is present in the solution.

The above description holds for a certain species in the solution. The diffusion coefficient D is a characteristic of that species. If we want to describe the transfer through a porous medium, we have to take into account the medium itself. Indeed, the species penetrates the solid through its pores, which have highly diverse geometries. Consequently, the size of the pores, and the tortuosity of the paths influence (slow down) the diffusion of the species through the porous medium. Among the definitions available in the literature, we use (Van Brakel and Heertjes, 1974)

$$D_e = \phi \frac{\tau}{T^*} D, \tag{4.82}$$

where ϕ, τ, and T^* are, respectively, the porosity, the constrictivity, and the tortuosity of the solid material, and D_e is the effective diffusion coefficient of the species. Based on Equation (4.82), Fick's first law for the porous medium becomes

$$\mathbf{j}_e = -D_e \nabla c, \tag{4.83}$$

where $\mathbf{j}_e$ is called the effective flux vector.

So far we have described the diffusion of molecules. We now turn our attention to the transport of ions in solution. Consider an electrolyte solution. Because ions are electrically charged particles, the electrostatic stresses that are present are due not only to interactions between the ions, but also to ion–solvent interactions. A local electrical field is created in order to maintain the electroneutral solution. This electrical field is also known as the liquid junction potential.

Let us take as an example sodium and chloride ions diffusing in water. The diffusion coefficient of chloride is higher than that of sodium. Therefore chloride will diffuse faster. Furthermore, because the chloride and sodium ions are electrically charged, an electrical field exists between them. This diffusion potential accelerates the diffusion of sodium and tends to slow down the diffusion of chloride (Figure 4.13). Note that we are not dealing any more with a chemical potential difference that drives the diffusion (which leads to

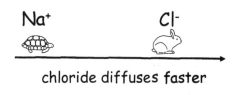

Fig. 4.13. The effect of the diffusion potential on the diffusion of sodium and chloride in water.

a Fick's law approach), but with an electrochemical potential difference. Consequently, Equation (4.76) must be replaced by a relation that accounts for both chemical and electrical potentials (Guggenheim, 1933),

$$\mu = \mu_0 + RT \ln a + z\mathrm{F}\psi, \qquad (4.84)$$

where z is the ion valence, F is Faraday's constant, and ψ is the electrical potential.

Equation (4.84) can be combined with Equations (4.77) to (4.80) to obtain a relation between the flux of matter, the ionic concentration, and the liquid junction potential,

$$\mathbf{j} = -D \left[\nabla c + \frac{z\mathrm{F}}{RT} c \nabla \psi \right]. \qquad (4.85)$$

Equation (4.85) is the Nernst–Planck equation, which can be written for every ion present in solution. In a way that parallels Equation (4.81), we introduce the effective diffusion coefficient D_{ei}, in order to describe the transfer of ions through the porous structure,

$$\mathbf{j}_{ei} = -D_{ei} \left[\nabla c_i + \frac{z_i \mathrm{F}}{RT} c_i \nabla \psi \right]. \qquad (4.86)$$

Subscript i indicates the ion type. The electrical potential created between the ions in the solution contained in the pores is called the membrane potential (Sen, 1989).

Consider next the mass conservation equation, (1.40). Here we assume that there is no species generation. Therefore, omitting the rate of species generation term, we write

$$\phi \frac{\partial c}{\partial t} + \mathbf{v} \cdot \nabla c = \nabla \cdot (-\mathbf{j}_e). \tag{4.87}$$

We saw that the diffusivity (D) and the effective diffusion coefficient (D_e) depend on position if the porous medium is not homogeneous. In addition, when the porous medium is a cementitious material, the diffusion coefficients are also functions of the degree of hydration (or age) of the material. When the cement hydration stops [it lasts approximately half a year, according to Tang (1996)], the diffusion coefficients become constant. Nilsson et al. (1996) proposed relations between the diffusion coefficients and the age of the cement-based material. When it is possible to neglect convection phenomena through the porous medium (saturated porous medium in the absence of a pressure gradient), Equation (4.87) becomes

$$\frac{\partial c}{\partial t} = -\frac{1}{\phi} \nabla \cdot \mathbf{j}_e \tag{4.88}$$

When the effective flux $\mathbf{j}_e$ is replaced by Equation (4.83), Equation (4.88) is referred to as Fick's second law of diffusion. An analytical solution of (4.88) based on error functions can be determined after invoking appropriate boundary and initial conditions. For example, assume that the representative elementary volume is made of a bundle of tubes (the pores) aligned with the direction x, so that the problem is one-dimensional, and Equation (4.88) becomes

$$\phi \frac{\partial c}{\partial t} = D_e \frac{\partial^2 c}{\partial x^2}. \tag{4.89}$$

The objective is to determine the penetration profile of a species through a porous medium saturated with a specified pore solution. Assume that the concentration of the species of interest is zero in the beginning $c(x, t = 0) = 0$. The semi-infinite porous medium is then exposed on its left boundary to a solution containing the species of interest, with the boundary condition $c(x = 0, t) = c_s$ (Figure 4.14). Based on these initial and boundary conditions, the solution of Equation (4.89) is [e.g., Crank (1975)]:

$$\frac{c(x, t)}{c_s} = \mathrm{erfc}\left(\frac{x}{2(t D_e / \phi)^{1/2}} \right). \tag{4.90}$$

The solution shown in Figure 4.14 is for the cases $t D_e / \phi = 0.00005$, 0.0005, and 0.005 m^2.

Recall that Equation (1.40) was obtained for the cases in which the flux of species follows Fick's first law of diffusion. It holds for particles, not ionic

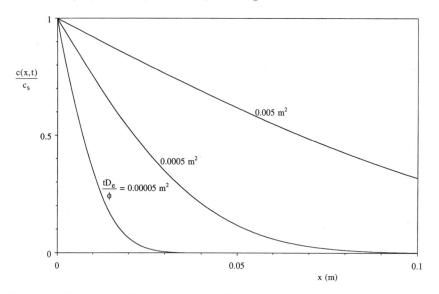

Fig. 4.14. Solution to Fick's second law [Equations (4.88) and (4.89)] when the porous medium is semi-infinite (one-dimensional problem).

species. For the latter, Fick's first law is replaced by the Nernst–Planck equation (4.86),

$$\phi \frac{\partial c_i}{\partial t} = \nabla \left[D_{ei} \left(\nabla c_i + \frac{z_i F}{RT} c_i \nabla \psi \right) \right]. \tag{4.91}$$

The right-hand side of this equation represents the flux of matter obtained from the Nernst–Planck equation. An analytical solution of Equation (4.91) does not exist. We show later how to tackle this problem numerically (cf. the end of Section 4.10).

4.9 Reactive Porous Media

In Chapter 1 we described flows through porous media, and mass transfer through porous media. The porous medium was modeled as the solid structure containing void spaces through which the fluid flows, and the molecular and ionic species diffuse. In this section we show how the "irregular porous media" found in nature become extremely complex when the solid structure is no longer an inert solid, but a reactive medium. What if the solid structure reacts with the pore solution? What kind of "reactions" do we expect? How do we account for them? We consider these questions from an applied environmental and civil engineering perspective and, because of this, our discussion is not exhaustive or general.

In the case of reinforced concrete, the durability of a structure depends on the concrete that protects the steel bars from corrosion. The degradation of the concrete may be due, for example, to carbonation, which is a chemical reaction between the cement hydrates and the CO_2 originating from the outside air. In brief, because the pore solution contained in cement-based materials is highly basic (pH $= 13$), the reinforcement bars are protected from corrosion by a thin oxide layer, which is called passive film (Baron and Ollivier, 1992). When carbonation occurs, the pH of the pore solution tends to diminish, and so does the natural protection of the steel bars. It is a factor that favors the initiation of corrosion. When CO_2 reacts with cement hydrates, calcium carbonate is created by precipitation, which tends to modify the porosity of the medium. In addition, because the carbonation reaction in some cases liberates water molecules, the saturation degree of the porous medium also changes (Saetta et al., 1995). It is important to note that, when it comes to modeling, the present challenge is to deal with the kinetics of carbonation, which is slower in the core material than at its surface, and the impact of carbonation on transport properties (porosity, pore geometry, diffusivity, permeability). These aspects are discussed further in Saetta et al. (1995), Sellier et al. (2000), and Chaussadent (1999), for example.

Carbonation is not the only chemical reaction that concerns civil engineers. One example is the penetration of sulfate, which reacts chemically with the cement hydrates. Sulfate ions may come from the ground, from sea water, or from the aggregates used in the mixing of concrete. The product of the chemical reaction causes an increase in the solid-phase volume and, consequently, leads to the formation of cracks (Baron and Ollivier, 1992).

Another example is the penetration of chloride through the concrete. If chloride ions reach the reinforcement bars, corrosion is initiated, and the degradation of the entire structure takes place. This is typically the case when a reinforced concrete structure is in contact with sea water (e.g., bridges, jetties), or when de-icing salts are sprayed over traffic structures. Therein lies the challenge: reinforced structures are an attractive economic choice when it comes to mechanical resistance. When exposed to sea water, however, their service life is shorter than that of plain concrete structures (Nagataki, 1995). Furthermore, the rehabilitation of structures deteriorated by chloride ingress is a multibillion-dollar expense (Skalny, 1987). This challenge drives the efforts made by researchers to understand and predict the penetration of chloride ions through saturated concrete.

The interactions of chloride with the porous structure during chloride ingress deserve close scrutiny. In spite of efforts made during the last two decades by many researchers, the mechanisms of chloride binding are still not fully understood. Chloride binding is both chemical and physical, the latter being the dominant mode of binding (Ramachandran et al., 1984). Chloride reacts chemically with unhydrated alumina phases and forms for example, Friedel's salt (Nagataki et al., 1993; Nilsson et al., 1996; Pavlik, 2000). Physical absorption occurs into the lattice of calcium silicate hydrates (C-S-H) where the chloride ions are trapped (Beaudoin et al., 1990; Wowra and Setzer, 2000). In addition, in spite of a pore surface that is charged negatively (see the end of this section), there are still a limited number of sites where chloride binding is possible (Diamond, 1986). Note that chloride binding is linked to many parameters such as cement composition, the pH of the pore solution, and the cation associated with the chloride ion (NaCl or CaCl$_2$, for example).

Chloride binding is usually described by a binding isotherm, which shows that at constant temperature the amount of bound ions is a nonlinear function of the free chloride concentration (c_{free}) in solution. Note also that chloride binding is assumed to be instantaneous. An example of a binding isotherm is presented in Figure 4.15. This relationship is commonly described by a Freundlich isotherm: $c_{\text{bound}} = \alpha c_{\text{free}}^{\beta}$, where α and β are empirical coefficients determined from experiments. Therefore the total amount of chloride in concrete is the sum of free ions in the pore solution and the chloride ions bound to the pore walls:

$$c_{\text{total}} = c_{\text{free}} + c_{\text{bound}}, \tag{4.92}$$

$$\frac{\partial c_{\text{total}}}{\partial t} = \frac{\partial c_{\text{free}}}{\partial t} + \frac{\partial c_{\text{bound}}}{\partial t}. \tag{4.93}$$

The ions bound to the pore walls do not participate in the corrosion process. The quantity of bound chloride (C_{mB}) is usually measured per unit of mass of material. Consequently, we write $c_{\text{bound}} = (\phi^{-1} - 1)\rho_s C_{mB}$, where ϕ is the material porosity, and ρ_s is the density of the material containing no water. Chloride binding must be taken into account in the mass conservation law,

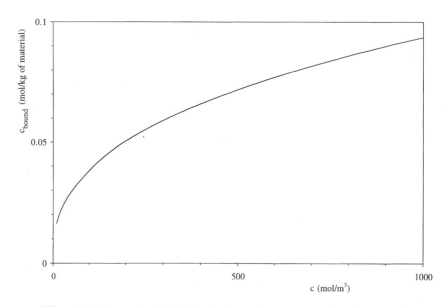

Fig. 4.15. Example of chloride isotherm for a cement-based material.

such that in a one-dimensional problem Equation (4.88) becomes

$$\frac{\partial c_{\text{free}}}{\partial t} = -\frac{1}{\phi}\frac{\partial j_e}{\partial x} - \frac{1-\phi}{\phi}\rho_s\frac{\partial C_{mB}}{\partial t}. \tag{4.94}$$

For the sake of simplicity, from now on we omit the subscript "free." We now introduce the binding capacity $\partial C_{mB}/\partial c_{\text{free}}$, which represents the ability of the material to bind chlorides when the chloride concentration changes in the pore solution. From Equation (4.94) we obtain

$$\frac{\partial c}{\partial t} = -\left(\phi + (1-\phi)\rho_s\frac{\partial C_{mB}}{\partial c}\right)^{-1}\frac{\partial j_e}{\partial x}. \tag{4.95}$$

So far, we did not consider whether the ionic interactions with the pore walls are reversible or irreversible. In their state-of-the-art book, Nilsson *et al.* (1996) make an allusion to a possible binding hysteresis at low chloride concentrations. Apart from this reference, there is a lack of data on chloride desorption and the possibility of hysteresis.

To learn more about this issue, which is important when the objective is to predict ionic penetration rates, it is useful to look at the field of nuclear energy. In the case of radio elements, binding occurs preferentially on calcium silicate hydrates (C-S-H). Consider the case of cesium ion. It binds within the interlayer space of C-S-H particles [cf. Richet (1992), Anderson *et al.* (1983), and Atkinson *et al.* (1984)]. These results concern cement pastes only. The authors claim that cesium binding is weak, reversible, and linear, meaning that the amount of bound cesium is proportional to the amount of free cesium in the pore solution. Some progress has been made recently: Frizon (2003) carried out experiments on cesium binding by using several components that constitute cement-based materials. He focused mainly on cement paste, sand and mortar. The mortar was prepared with the same sand, the same cement, and the same water-to-cement ratio. The study addressed both cesium sorption and cesium desorption. Results on binding are presented in Figure 4.16 (Lorente *et al.*, 2002b; Frizon *et al.*, 2003). This figure requires explanation. The activity of cesium 137 in the lixiviant solution is measured by gamma spectrometry. The activity $\mathcal{A}$ is commonly expressed in Becquerels (Bq), and represents the number of atoms disintegrated per second. The activity is related to the concentration via (Pannetier, 1982)

$$\mathcal{A} = N_A\frac{\ln 2}{T_{1/2}}Vc, \tag{4.96}$$

where N_A is the Avogadro constant $(6.023 \times 10^{23} \text{ mol}^{-1})$, V is the volume of solution, $T_{1/2}$ is the half-life period (30.15 years in the case of cesium 137, for example), and c is the concentration. Figure 4.16 shows a decrease in cesium activity in the pore solution during the first week. Steady state is reached after this time, meaning that all the sites available for cesium are saturated

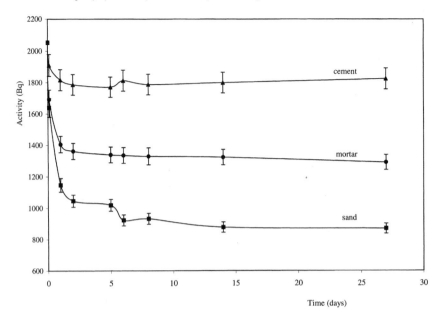

Fig. 4.16. Experimental results on cesium sorption.

by the cesium. The novelty of these results lies in the fact that cesium binding appears to be more important on sand, and hence on mortar, than on cement paste.

Plotted in Figure 4.17 are the results obtained during the desorption process that follows the sorption experiments. In the case of the cement paste, steady state is reached after a few days. The same cannot be said about sand and mortar. A comparison between Figures 4.16 and 4.17 shows that in the case of cement the amount of bound cesium (250 Bq) is released into the pore solution in the desorption test. This is not true for sand and for mortar, where an important amount of cesium remains bound during a desorption test. The experiments with mortar are still in progress. Nevertheless, it is now clear that the partial irreversibility of cesium binding must be taken into account.

In some materials the pore walls may acquire an excess of electrical charges through ionization reactions. The porous structure is not inert from an electrical standpoint (Hunter, 1981). The excess of charges is balanced by counterions attracted from the pore solution (Figure 4.18). This theory is known as the electrical double layer theory (EDL), which was developed independently by Gouy in 1910 and Chapman in 1913, and improved by Stern in 1924 and Parsons in 1954 [see Hunter (1981)]. The EDL theory states that close to the charged pore surface there exists a specific area the properties of which are different from the bulk properties of the pore solution. This area is divided into two regions: the compact layer and the diffuse layer.

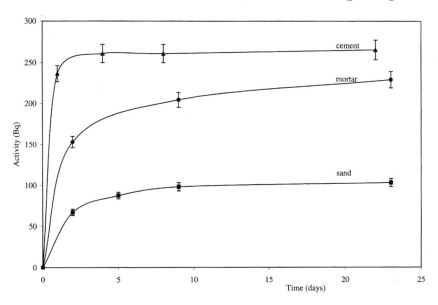

Fig. 4.17. Experimental results on cesium desorption.

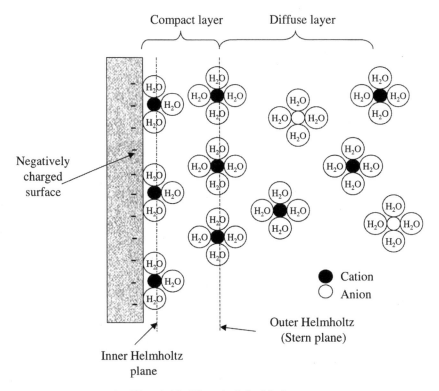

Fig. 4.18. Electrical double layer.

The compact layer itself is divided into the inner and outer Helmholtz planes. The inner Helmholtz plane (IHP) is the plane of the centers of the counter-ions adsorbed at the pore surface. In Figure 4.18 the counter-ions are partly unhydrated. The outer Helmholtz plane (OHP, or Stern plane) is the plane of the centers of counter-ions that are not specifically adsorbed on the charged surface, but are bound by electrostatic forces (Bockris and Reddy, 2001). Farther out from the wall, in the diffuse layer, co-ions and counter-ions diffuse at slower rates than in the bulk solution.

The electrical double layer will have an influence on the ionic transport. Its thickness—the Debye length—depends on the ionic strength of the solution, $1/2 \sum_i z_i^2 c_i$, where z_i and c_i are, respectively, the charge number and the concentration of each ion i. The more important the ionic strength, the thinner the Debye length and, consequently, the smaller the influence of the EDL on the ionic transport. More details regarding the theoretical description of the EDL can be found, for example, in Bockris and Reddy (2001).

When the fluid contained in the pores moves with respect to the solid surface (regardless of driving force), every particle is subjected to shear stresses. Hunter (1981) defined a surface of shear, which is very close to the solid surface, and inside of which the fluid is stationary. Actually, this surface of shear is meant to be identical to the outer Helmholtz plane. The average electrostatic potential in the surface of shear is the zeta-potential.

A consensus exists in the civil engineering community regarding the negative sign of the surface charge in the case of cement-based materials (see, e.g., Chatterji and Kawamura, 1992). Furthermore, it seems that the zeta-potential of cementitious materials depends on the nature of the pore solution, in intensity and sign (Chatterji, 1998).

4.10 Electrodiffusion

In the field of civil engineering, the notion of time must be understood on a very large scale. For example, in the case of a chloride diffusion test through a sample of concrete (Figure 4.19) the penetration depth of the chloride ions is approximately 17 mm after 10 months of exposure, and 22 mm after 23 months. These results must be interpreted in an order of magnitude sense, because they are influenced strongly by the composition of the concrete. Nevertheless, engineers would not wait 10 months to see the predicted chloride ingress. In the last two decades, a solution was proposed to shorten the duration of experiments (AASHTO, 1983): an external electrical current is created through the porous medium, accelerating ionic transport. This process is called electrokinetics. The principle of the experimental setup is shown in Figure 4.20. Thanks to the electromigration process, the testing time is reduced from several months to a few days.

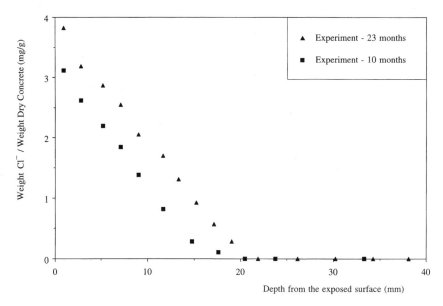

Fig. 4.19. Simulation of chloride penetration, showing the effect of the exposure time.

Electrokinetics is also used in decontamination processes, for example, in the remediation of concrete structures (Andrade *et al.*, 1995), where the corrosion of the reinforcement bars is due to chloride attack. An anode is attached to the surface of the concrete. This electrode is surrounded by liquid

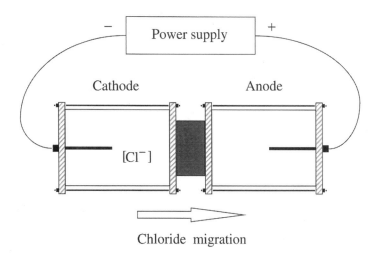

Fig. 4.20. Cross-section of the steady-state electromigration setup.

electrolyte. When a direct current is passed between the anode and the steel reinforcement, the latter acts as a cathode. The chloride ions move from the concrete to the anode. The method is also widely used in soil decontamination (Shackelford, 1991). Recently, it was applied to nuclear remediation (DePaoli *et al.*, 1997; Harris *et al.*, 1997; Frizon *et al.*, 2002). It involves three main mechanisms: electromigration, electro-osmosis, and electrophoresis (Masliyah, 1994). The last-named refers to the transport of charged colloidal particles in the stationary pore solution. Note that electrophoresis has a minor effect among the three mechanisms in the case of cement-based materials. Electro-osmosis is the global movement of the pore solution relative to a stationary surface. The ions in the pore solution move in response to the external electrical field. The liquid is dragged along with the ions. Due to the high ionic strength of the pore solution in concrete, electro-osmosis is not expected to be preponderant compared to electromigration (DePaoli *et al.*, 1997). Electromigration involves the movement of ions in the pore solution. It is the main mechanism in electrokinetics. Although the migration method was known for 30 years, its theoretical basis was established quite recently by Andrade (1993). The current density i is defined as

$$i = \frac{I}{A},$$
(4.97)

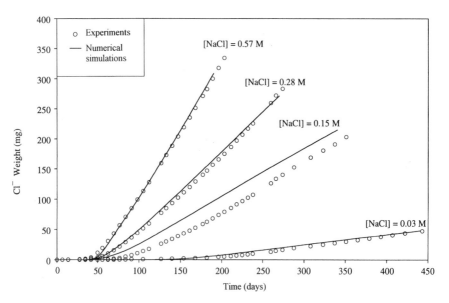

Fig. 4.21. Cumulative amount of Cl⁻ downstream, for various Cl⁻ concentrations upstream. Comparison between numerical simulations and experimental data from Bigas (1994).

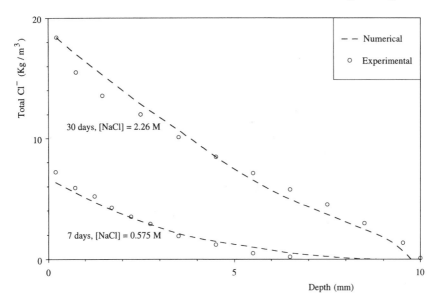

Fig. 4.22. Chloride concentration profiles. Comparison between numerical simulation and experimental data from Francy (1998).

where I is the imposed current, and A is the cross-sectional area of the sample through which the ions migrate. The current density equation is (Rubinstein, 1990):

$$i = \frac{I}{A} = F \sum_i z_i j_{ei}. \tag{4.98}$$

In summary, we saw how complex the problem becomes when the porous structure is a reactive medium. The problem of ionic transport through a saturated porous medium is described by the Nernst–Planck equation (4.86), the mass balance equation written for every ionic species (i),

$$\frac{\partial c_i}{\partial t} = - \left(\phi + (1 - \phi)\rho_s \frac{\partial C_{mBi}}{\partial c_i} \right)^{-1} \frac{\partial j_{ei}}{\partial x}, \tag{4.99}$$

and the current continuity equation (4.98). This way of describing the transport of ions in a porous medium is called the multispecies approach. It represents one of the newest developments as far as the civil engineering field is concerned. Indeed, in many cases ionic penetration profiles are determined using Fick's laws and the influence of the nature of the pore solutions (the other ions) is demonstrated (Tang and Nilsson, 1993). This approach was developed independently and simultaneously by three different research teams: Truc (2000) and Truc et al. (2000), Li and Page (1998, 2000), and Samson and Marchand (1999).

A sample of results obtained using the numerical model developed by Truc (2000), is given in Figures 4.21 and 4.22. Figure 4.21 shows a comparison between numerical results provided by the multispecies approach and experiments (Bigas, 1994). The ordinate shows the total amount of chloride ions measured in the downstream compartment of a diffusion test setup. Several chloride concentrations in the upstream cell were tested. As expected, the amount of chloride ions increases with the time of exposure. The numerical results follow the experimental trend very closely.

In Figure 4.22 we present an example of chloride concentration profiles. These profiles were determined experimentally for different upstream concentrations and different times of exposure (Francy, 1998). The experimental conditions became input data for the multispecies numerical program. Note that the calculated penetration depths are very close to experimental measurements.

4.11 Tree-Shaped Flow Networks

In this section we draw attention to another class of complex flow structures with multiple scales which is emerging in many sectors of engineering: tree-shaped, or dendritic distribution networks (Bejan, 2000). The purpose of a tree network is to make a flow connection between one point (source or sink) and an infinity of points (area or volume). In civil engineering applications such as the distribution of water and electricity, the area is approximated by a large number of evenly or unevenly distributed points [the consumers (Bejan and Lorente, 2001)]. Tree-shaped flows are the rule in nature, in animate and inanimate flow systems as well: lungs, vascularized tissues, nervous system, river basins and deltas, lightning, snowflakes, and so on. This is why their features have been studied first in physiology and geophysics [e.g., Murray (1926), Thompson (1942), Weibel (1963), and Schumm et al. (1987)].

The empirical view inherited from the natural sciences is that tree-shaped flows are examples of spontaneous self-organization and self-optimization. In contrast with this, the engineering view is theory: flow architectures such as the tree are the results of a process of optimization of global flow performance subject to constraints. They are deducible from the constructal principle of maximization of flow access through the generation of geometry (Bejan, 2000). Trees emerge in the same way as the flow systems with optimized internal structure (spacings), which are presented in Chapter 5. Tree-shaped flows persist in time, in nature, and in engineering because they are efficient. They are important fundamentally in engineering, because the relation among efficiency, complexity, and compactness is key to the progress toward design and integration of increasing numbers of smaller and smaller flow components. In this direction, the compact and complex flow system becomes a *designed porous medium*.

The optimization objective that generates a tree-shaped flow depends on the application. It is reasonable to argue that in live organs such as the lungs, the objective is minimal flow resistance, or minimal pumping power. The same thought is invoked in the numerical simulation of river basins (Rodriguez-Iturbe and Rinaldo, 1997). In engineering, tree networks are designed for a variety of purposes: minimal global resistance to fluid and heat flows in compact heat transfer devices (Bejan, 2000; Chen and Cheng, 2002), minimal losses in electric power distribution networks, minimal travel time in urban design, and minimal transportation costs in economics. These engineering developments, including the examples from the natural sciences, have been brought together under the constructal law of the maximization of flow access (Bejan, 2000). In the present treatment we focus on the main properties of the tree architecture, and the complexity and robustness of its design. These characteristics serve as the basis of the conceptual design of flow devices for the future.

One basic feature of tree-shaped flows is the existence of at least two dissimilar flow regimes, or flow paths: one with low resistance (e.g., channels), and the other with high resistance (e.g., diffusion). Maximum flow access is achieved when the many streams of the tree are organized such that the flows with high resistance inhabit the smallest scales of the flow structure, whereas the flows with low resistance are assigned to the larger scales. In the lung, for example, the high resistance flow is the diffusion of O_2 and CO_2 through the tissue that separates the smallest flow spaces of the structure (the alveoli). An entire sequence of flows with decreasing flow resistances (larger bronchial tubes) is distributed in a very special way over the larger scales: tubes become larger, resistances decrease, and longitudinal length scales increase. Flow resistances cannot be eliminated. They can be arranged or assembled—forced to coexist with each other in a finite volume—so that their global impact on performance is minimal. Flow diversity conspires with optimized organization, and the result is construction.

Figure 4.23 shows an outline of the fluid tree designed to flow between a volume and one point. The construction started from the smallest scale (the dark cube with D_0 in its center, Figure 4.23a), where a balance was reached between the flow along the tube of diameter D_0 and the diffusion through the solid in which the tube is embedded. The construction proceeded toward larger scales, by optimizing every geometric feature of each new (larger) construct. For example, the optimized tube diameters increase in a certain way, such as $D_{i+1}/D_i = 2^{1/3}$, if the flow is laminar, and if the construct is sufficiently large. The construct level is i, for example, $i = 3$ in Figure 4.23a, and $i = 6$ in Figure 4.23b. When the construct is sufficiently large, the optimal number of constituents (the constructs of the preceding size) is two. This is important: pairing, or bifurcation of tubes (dichotomy), is an optimized feature of the flow architecture. The relative dimensions of the tree flow of Figure 4.23 are summarized in Bejan (2000).

Optimized tree-shaped architectures provide a purely theoretical foundation for the existence of *allometric laws* in physiology. These are scaling relations that have been observed to exist between flow parameters and animal body size (Schmidt-Nielsen, 1984; Vogel, 1988). Global features that can be

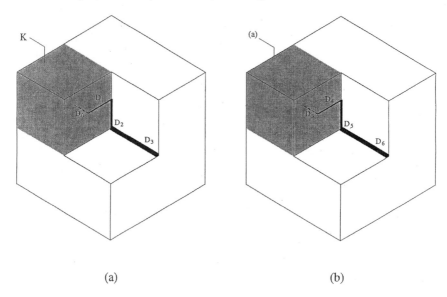

(a)   (b)

Fig. 4.23. Three-dimensional construct for flow between a volume and one point: the doubling of the outer dimension in going from the (a) optimized third construct to (b) the optimized sixth construct (Bejan, 2000; from Bejan, *Shape and Structure, from Engineering to Nature*, Copyright © Cambridge University Press. Reprinted with the permission of Cambridge University Press).

measured on designed bodies such as Figure 4.23 are the total volume, the total internal surface of all the tubes (A_i), the volume-averaged porosity, the total mass (M_i), and the cross-sectional area of the tree trunk (the thickest duct). Interesting relations emerge between these quantities when they are plotted against each other, as in Figure 4.24. The curves are drawn through the points represented by each level of assembly, or pairing (i). In this presentation, the level of the construct (i) can be hidden from view, in the same way that it is absent from the measurements reported by the biologist who examines complete organs from large and small animals. The level i was plotted on top of the curves for clarity, to show their origin. The ratio $\lambda = L_0/D_0$ is a number of order 1 (but greater than 1, such as 2 or 3), and accounts for the aspect ratio of the smallest duct in the fluid tree (e.g., alveolus).

In Figure 4.24 the internal area and total mass have been nondimensionalized based on L_0 as the length scale, $\hat{A}_i = A_i/L_0^2$ and $\hat{M}_i = M_i/(\rho L_0^3)$. In constructal theory the elemental (smallest) duct length is fixed, and does not depend on the macroscopic size of the system. Figure 4.24 shows that the total exposed surface of the tree of ducts is almost proportional to the mass of the construct. The bundle of curves is represented adequately by

$$\hat{A}_i \sim O(1)\hat{M}_i^{1.03}. \tag{4.100}$$

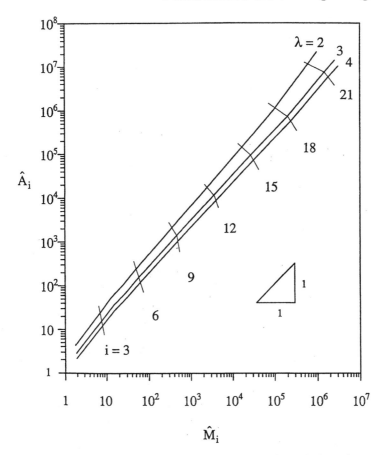

Fig. 4.24. The relation between the total internal surface of the tubes and the total mass of the tree flow system of Figure 4.23 (Bejan, 2000; from Bejan, *Shape and Structure, from Engineering to Nature*, © 2000 Cambridge University Press. Reprinted with the permission of Cambridge University Press).

The effect of the aspect ratio of the smallest tube ($\lambda = L_0/D_0$) is small. We may say that a macroscopic relation such as Equation (4.100) is robust with respect to the elemental architectural detail represented by λ.

More robustness with respect to λ is exhibited by the relations between root cross-sectional area and body mass, total volume and body mass, overall flow conductance and body mass, and total internal contact surface and total tube volume. These theoretical relations are reported in Bejan (2000). They are important in the background of fields such as physiology and zoology, where they are accepted empirically (first observed, then plotted, and later correlated). The relations anticipated based on theory agree with the known allometric laws not only qualitatively, but also quantitatively.

Geometry is the key to achieving maximum global performance under constraints, and to predicting along the way the performance of natural flow systems. This becomes more evident if we consider a simpler model of tree flow than that shown in Figure 4.23. To start with, the minimum-resistance geometry for fluid flow between two points is the straight duct with round cross-section. This shape can be derived based on variational calculus, and it holds for both laminar and turbulent flow. It is found throughout engineering and in many natural flow systems.

The round duct shape is a robust design feature. Shapes that are nearly round perform almost as well as the perfectly round shape. For example, in laminar flow through a cross-section shaped as a regular hexagon, the flow resistance exceeds the resistance through the round cross-section by only 3.7% (Bejan, 2000). Hexagonal ducts have the advantage that they can be packed to fill a volume completely.

In open channel flow (e.g., rivers) the optimal cross-section is a half-disc, such that the free surface is the horizontal diameter of a semicircle. Other cross-sectional shapes (e.g., rectangle, triangle) can be optimized for minimum resistance, and in all the cases the maximum river depth is a fraction of the river width. This scaling is supported by a large volume of river measurements (Leopold *et al.*, 1964). Robustness is also a characteristic of river cross-sections: other shapes (e.g., rectangle, triangle) that have been optimized for minimum resistance by choosing the depth/width ratios have nearly the same resistance as the half-disc shape deduced based on variational calculus.

As an aside, there is a large and much older body of information on optimal external shapes for least drag in liquid and gaseous flow. Everything that flies or swims, natural or manmade, has been optimized for minimum resistance. For a review of optimal aerodynamic and hydrodynamic shapes, their history and methodology (variational calculus), the reader is directed to Miele (1965).

Returning to the optimization of geometry for tree-shaped flow, Figure 4.23, the simplest tree is the T-shaped construct of round tubes shown in Figure 4.25 (Bejan *et al.*, 2000). The flow connects one point (source of sink) with two points. The constraints are: (i) the total duct volume $(D_1^2 L_1 + 2D_2^2 L_2 = \text{constant})$, and (ii) the total space allocated to the construct $(2L_2 L_1 = \text{constant})$. If the flow is laminar and fully developed, the minimization of the flow resistance subject to constraint (i) yields the ratio of tube diameters,

$$\frac{D_1}{D_2} = 2^{1/3}. \tag{4.101}$$

This ratio is an old result, which in physiology is known as Murray's law. The result is remarkable for its robustness: the optimal D_1/D_2 ratio is independent of the assumed tube lengths. It is also independent of the relative positions of the three tubes and is independent of geometry.

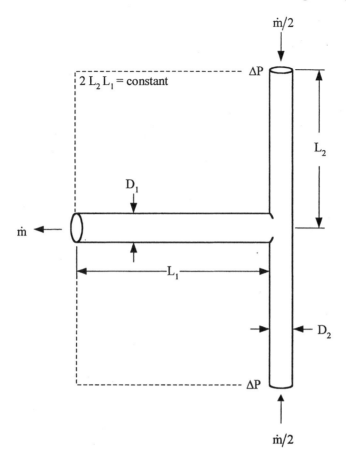

Fig. 4.25. T-shaped assembly of round tubes (Bejan *et al.*, 2000).

The second level of optimization, which consists of selecting the ratio L_1/L_2 subject to space constraint (ii) is new. The result

$$\frac{L_1}{L_2} = 2^{1/3} \tag{4.102}$$

shows that at the junction the tube lengths must change in the same proportion as the tube diameters. Equations (4.101) and (4.102) are a condensed summary of the geometric proportions found more laboriously in the optimization of three-dimensional flow constructs such as Figure 4.23, where the tube lengths increase by factors in the cyclical sequence $2, 1, 1, 2, 1, 1, \ldots$. The average of this factor for one step is $2^{1/3}$, which means that the optimization of the plane construct of Figure 4.25 is an abbreviated substitute for the optimization of the three-dimensional construct of Figure 4.23.

If the flow in the T-shaped construct is fully developed and turbulent, Equations (4.101) and (4.102) are replaced by (Bejan *et al.*, 2000)

$$\frac{D_1}{D_2} = 2^{3/7} \quad \text{and} \quad \frac{L_1}{L_2} = 2^{1/7}. \tag{4.103}$$

Unlike in the laminar case, in which the ratio D/L was preserved in going from each tube to its stem or branch, in turbulent flow the geometric ratio that is preserved is D/L^3: Note that Equations (4.103) yield $D_1/L_1^3 = D_2/L_2^3$.

The assembly of three tubes can be optimized further by giving the morphing geometry more degrees of freedom. One example is to allow the angle of confluence to vary. This alternative is outlined in Figure 4.26, where the total space constraint (ii) is a disc-shaped area with specified radius (r). It was found that when the flow is fully developed and laminar, the optimized flow architecture is represented by $D_1/D_2 = 2^{1/3}, \alpha = 0.654$ rad and $L_1 = L_2 = r$. In this configuration the tubes are connected in the center of the disc, and the angle between the two L_2 tubes is very close to $75°$.

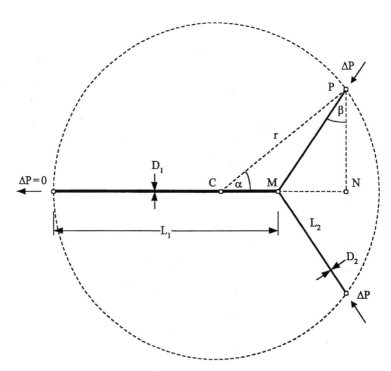

Fig. 4.26. Y-shaped assembly of round tubes (Bejan *et al.*, 2000).

Other simple flow constructs (e.g., open channels, gravity-driven flows), and other global constraints (e.g., tube wall material) lead to similar compact conclusions. Another way to see the robustness of the results is to compare the minimized flow resistances for laminar and turbulent flows in the T-shaped arrangement (Figure 4.25). It was found that for laminar flow the minimized resistance can be expressed in terms of the two constraints, the tube volume (V) and the total area of the territory (A),

$$R_{\text{lam}} = 4\frac{A^{3/2}}{V^2}. \tag{4.104}$$

The corresponding resistance for turbulent flow is

$$R_{\text{turb}} = 4\frac{A^{7/4}}{V^{5/2}}. \tag{4.105}$$

Equations (4.104) and (4.105) are surprisingly close even though their respective flow regimes are drastically different. As shown in Bejan *et al.* (2000), the R functions before minimization are

$$R_{\text{lam}} = \frac{L_1}{D_1^4} + \frac{L_2}{2D_2^4}, \tag{4.106}$$

$$R_{\text{turb}} = \frac{L_1}{D_1^5} + \frac{L_2}{4D_2^5}. \tag{4.107}$$

These resistances account only for the way in which geometry affects the overall flow resistance of the T construct. Equations (4.104) and (4.105) indicate the role played by global constraints. Flow resistances are smaller when the served territories are smaller, and when the tube volumes are larger.

4.12 Optimal Size of Flow Element

In engineering and natural sciences it has become routine to base the fluid-flow design on the minimization of pumping power (or fluid friction) subject to global space constraints. See, for example, the constraints listed above Equation (4.101). Is there a connection between pumping power savings and the total size (volume, weight) of a flow system? Are these concepts related, or are they a combination of "apples and oranges" invoked solely for the sake of convenience, to give the optimization an economics flavor without engaging in extensive cost analysis and minimization?

We are in a position to answer this basic question on equally basic grounds (Bejan and Lorente, 2002). Consider the simplest model of a flow passage, or a pore and solid combination: the straight tube of inner diameter D and prescribed mass flow rate $\dot{m}$, which is shown in Figure 4.27. Make the additional assumption that this simplest flow system is an element in a greater flow system *with purpose*, for example, an aircraft or a flying animal (insect, bird,

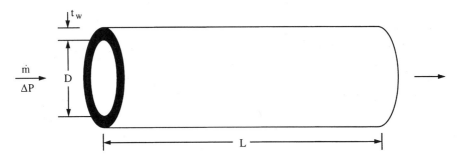

Fig. 4.27. Round duct with specified flow rate (Bejan and Lorente, 2002).

mammal). This assumption is absolutely essential. It means that the purpose of the elemental flow passage is to cause a minimum of "loss," or minimal destruction of useful mechanical power in the greater flow system. In thermodynamics today this means that the best duct is the one that performs its function ($\dot{m}$) while contributing least exergy destruction (irreversibility) to the operation of the greater system (review Sections 3.1 and 3.2).

There are two exergy destruction mechanisms in the configuration of Figure 4.27: the flow with friction through the duct, and the exergy destroyed for the purpose of carrying the duct mass during flight. If the flow regime is laminar and fully developed, the pressure loss per unit of duct length is

$$\frac{\Delta P}{L} = \frac{32\mu}{D^2}U, \tag{4.108}$$

where U is the mean fluid velocity, $\dot{m}/(\rho\pi D^2/4)$. The rate of exergy destruction is the pumping power required to force $\dot{m}$ to flow through the tube,

$$\frac{\dot{W}_1}{L} = \frac{\dot{m}\Delta P}{\eta_p\rho L}, \tag{4.109}$$

where η_p is the pump isentropic efficiency, and ρ is the density of the single-phase fluid.

The exergy destruction rate that is necessary in order to maintain the mass of the duct in flight can be expressed per unit of duct length as (Bejan, 2000, p. 239)

$$\frac{\dot{W}_2}{L} \cong 2\frac{m}{L}gV. \tag{4.110}$$

Here m/L is the duct mass per unit length, and V is the cruising speed. Assume that the competing designs have duct cross-sections that are geometrically similar; that is, the thickness of the duct wall is a fraction of the duct diameter, $t_w = \varepsilon D$, where $\varepsilon \ll 1$, constant. In this case, the duct mass per unit length is $m/L = \rho_w\pi D\varepsilon D$. If many tubes of this type fill a space, then the porosity of that medium is fixed.

The total rate of exergy destruction is the sum of Equations (4.109) and (4.110),

$$\frac{\dot{W}_1 + \dot{W}_2}{L} \cong \frac{c_1}{D^4} + c_2 D^2, \tag{4.111}$$

where $c_1 = 128\,\mu\dot{m}^2/(\pi\eta_p\rho^2)$ and $c_2 = 2\pi\varepsilon\rho_w g V$. We see the competition between the two exergy destruction mechanisms: a large D is attractive from the point of view of avoiding large flow friction irreversibility, but it is detrimental because of the large duct mass. The optimal tube diameter is

$$D_{\mathrm{opt}} = \left(2\frac{c_1}{c_2}\right)^{1/6} = \left(\frac{128\,\mu\dot{m}^2}{\pi^2\eta_p\varepsilon\rho^2\rho_w g V}\right)^{1/6}. \tag{4.112}$$

The optimal diameter increases with the flow duty of the tube as $\dot{m}^{1/3}$, and decreases weakly as the flying speed increases.

Review now the question posed at the start of this section. The optimization of tube diameters in complex flow networks is based routinely on the statements: (I) the minimization of flow resistance, and (II) the constrained (fixed) tube volume. Although (I) can be rationalized as an invocation of the constructal principle (Bejan, 2000), constraint (II) has a less clear theoretical basis. That basis is made clear now by Equation (4.111). The flow resistance (I) is represented by c_1/D^4, and the tube volume (II) by $c_2 D^2$. According to the method of Lagrange multipliers, to minimize (I) subject to fixed (II) is to minimize the sum formed on the right side of Equation (4.111). In other words, all the flow network optimization efforts seen in engineering and physiology can be justified theoretically on the basis of the constructal principle: the global maximization of performance, which led to Equation (4.111) and all the flow structures reported in this chapter.

This simple analysis is just the beginning of what can be done in the direction of optimizing the size of every flow component that works inside a complex flow system. For example, the size of the round tube can be optimized similarly for operation in the turbulent flow regime, and for geometries where the tube wall thickness is not necessarily a fraction of the tube diameter. Fittings, junctions, bends, and all the other geometric features that impede flow and add mass to a bird or aircraft can have their sizes optimized in the same manner.

The flow passages of heat exchangers are next in line, although in their case the exergy destruction process is due to three mechanisms: heat transfer, flow friction, and the airlifting of the heat exchanger mass. The competition between the first two mechanisms was recognized in the past (Bejan, 1982, 1997), and served as the basis for the selection of *shapes*: the length/diameter ratio of flow passages. The competing effect of the heat exchanger mass is new, because it identifies the optimal *size* of every component of this type.

In sum, the method illustrated in this section and in Bejan and Lorente (2002) promises the identification of optimal shapes and sizes for

heat exchangers, as "organs" optimized for the benefit of the greater system
(aircraft, vehicle, animal). If, for example, the elemental tube in Figure 4.27
represents a large blood vessel in a flying bird, then the tube mass per unit
length is $m/L = \rho_B \pi D^2/4$, where ρ_B is the blood density. The thermodynamically optimal tube diameter continues to be given by Equation (4.112), where
$c_2 = (\pi/4)\rho_B gV$.

The main conclusion is that the sizes of flow components can be optimized,
such that the aggregate flow system performs at the highest level possible. We
illustrated this by analyzing a very simple flow organ. This conclusion may
seem counterintuitive. To begin with, the total weight of the aircraft dictates
its power requirement and limits its range. A small total weight is better.
Smaller components in every subsystem of the aircraft may appear to be
preferable.

There is a competing trend, which contradicts the drive toward smaller
sizes. Power and refrigeration systems and their components function less
efficiently when their sizes decrease. Their flow resistances increase when sizes
decrease. In a heat exchanger, for example, the heat-transfer area and the
fluid-flow cross-sections decrease when the total mass and volume decrease.
Larger flow resistances lead to higher rates of exergy destruction and, globally,
to the requirement of installing more fuel on board. More fuel means more
weight.

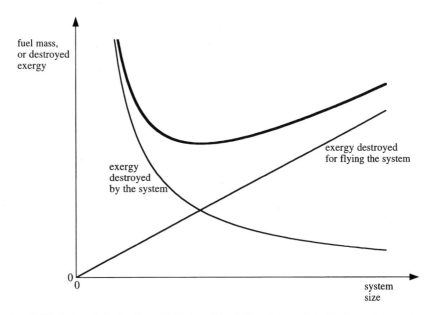

Fig. 4.28. The minimization of the total fuel (food) associated with a flow system:
the fuel destroyed by the flow system plus the fuel installed on board to account for
carrying the system (Ordonez and Bejan, 2003).

This conflict is summarized in general thermodynamic terms in Figure 4.28. The total fuel (or food) required by a flow system is the sum of the fuel required to transport the system and the fuel that must be used in order to produce the exergy that is ultimately destroyed by the system. This tradeoff is basic: we can expect it in every flow system (e.g., vehicle, animal), in every vehicle and living system, no matter how complex.

The theoretical course followed in this chapter and other segments of this book is constructal theory: the idea that it is natural for currents to construct for themselves paths of least resistance. This led to designed tree-shaped flows. Minimal resistance also means minimal irreversibility in purposeful systems with specified internal currents. All the flow systems that have the ability to morph their configurations under constraints progress in steps of geometric form toward better performance—conglomerates of flows that flow more easily. They survive because they change, that is, they project themselves into the future by flowing through beneficially altered configurations.

4.13 Hot Water Distribution Networks

The hot water distribution problem is an interesting way to illustrate the generation tree-shaped flows. It is a combination of two problems, or two functions with their own constraints. One is the fluid mechanics problem of distributing the fluid with minimal resistance, or minimal pumping power, subject to fixed total piping volume or total piping material. The other problem is one of heat transfer, or thermal insulation: the minimization of heat loss to the ambient, subject to fixed total thermal insulation material. The fluid mechanics problem has been addressed in various forms, not only in engineering [e.g., Bejan (2000)] but also in physics and biology [e.g., Thompson (1942), Weibel (1963) and Vogel (1988)].

Heat current distribution networks have also been optimized (Bejan, 2000), but the objective was the maximization of flow access, that is, the minimization of global thermal resistance. In the conceptual development of hot water flow patterns, the objective is just the opposite: it is to prevent the flow of heat from the links of the network to the surrounding territory. This is accomplished by using a finite amount of thermal insulation and distributing it optimally over all the pipes. The same method applies to the optimization of systems for the distribution of cold (chilled) water.

Several flow patterns or rules of network optimization and construction can be selected (Wechsatol et al., 2001). In every case, the total area (A) and the total flow rate of the hot water supply ($\dot{m}$) are fixed. Assume that users are distributed uniformly over the area. Each user receives the same water flow rate $\dot{m}_0 = \dot{m}/n$, where n is the number of users. In the following illustrations, each user occupies an area element of size A/n which is shaped as a square with the side L_0. The elemental length scale is fixed, because it is dictated by the needs of the individual (e.g., size of property). The finite and fixed smallest dimension of tree-shaped flow networks is a characteristic of the constructal designs generated in engineering and in nature (Bejan, 2000).

Three construction rules are shown in Figures 4.29 to 4.31. The simplest is the single stream with no branches (Figure 4.29), which can be coiled so that it covers the given territory. Next is the sequence where each larger construct is shaped as a square (Figure 4.30), and where each new construct is made by grouping four smaller area constructs. The alternative shown in Figure 4.31 is based on the pairing rule: each area construct contains two smaller constructs. A fourth construction rule is discussed in connection with Figure 4.33.

From the fluid mechanics part of the problem, one discovers that in every tree-shaped network there is an optimal distribution of pipe sizes (ratios of successive diameters), for example, Equation (4.101) for fully developed laminar flow. From the heat transfer problem one deduces the optimal distribution of thermal insulation, or the optimal ratio r_0/r_i (outer radius/inner radius) for each insulated pipe. These geometric results are presented in detail in Wechsatol et al. (2001). The thermal insulation part of the problem was pursued in three ways, with similar geometric results: by maximizing the temperature of the hot water received by the most distant user, by minimizing the rate of heat loss from the entire system, and by maximizing the hot water temperature averaged over all the users. The optimal distribution of insulation is such that the optimal r_0/r_i ratios are larger for the smaller ducts. In other words, the insulation shells must be relatively thicker over the smaller pipes, that is, in the regions situated farther from the root of the tree.

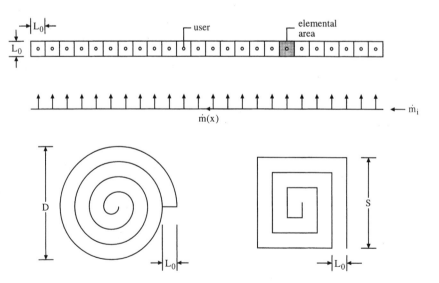

Fig. 4.29. String of hot-water users supplied by a single stream, and round and square territories served by the stream (Wechsatol et al., 2001).

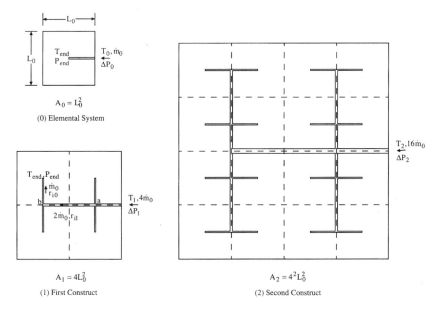

Fig. 4.30. Sequence of square-shaped area constructs obtained by connecting four square constructs (Wechsatol *et al.*, 2001).

The key is to determine which rule of flow pattern construction is superior. The optimized designs showed that the single-stream designs (Figure 4.29) are always inferior to tree constructs. To start with, the water delivered by the tree arrangement of Figure 4.30 is consistently hotter than the water delivered by the coiled string. The coiled string approaches the performance of the tree-shaped construction when the total amount of insulation material increases,

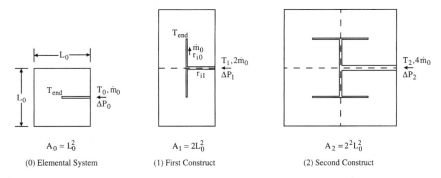

Fig. 4.31. Sequence of area constructs obtained by pairing optimized area constructs (Wechsatol *et al.*, 2001).

and when the dimensionless group N_0 decreases,

$$N_0 = \frac{\pi k L_0}{\dot{m}_0 c_p}. \tag{4.113}$$

In this definition $\dot{m}_0$, c_p, and k are, respectively, the user water flow rate, the water specific heat, and the thermal conductivity of the insulation. By analogy with the number of heat transfer units terminology of heat exchanger design, N_0 may be viewed as "the number of heat loss units" of the insulation; N_0 is small when k is small.

Next, the performance of the tree-shaped network of the pairing sequence (Figure 4.31) is superior to that of the sequence of square areas (Figure 4.30). In Figure 4.32 we reproduce the ratio of the maximized temperatures of the water received by the farthest user in the two constructs sketched above the

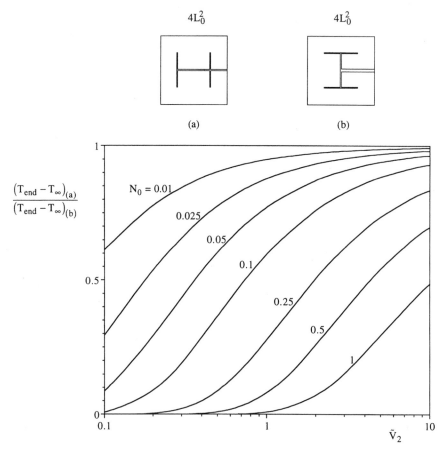

Fig. 4.32. Comparison between the maximized end-user temperatures in areas of size $4L_0^2$ covered by the constructs shown in Figures 4.30 and 4.31 (Wechsatol et al., 2001).

graph. The total area $(4L_0^2)$ and total volume of insulation (V) are the same in both designs. Figure 4.32 shows that the temperature of the hot water received by the end user $(T_{end} - T_\infty)$ in Figure 4.31 (A_2) is consistently higher than in the design of Figure 4.30 (A_1). The temperature T_∞ represents the ambient. The dimensionless volume indicated on the abscissa is defined as $\tilde{V}_2 = V_2/(\pi L_0 r_{i0}^2)$, where V_2 is the total volume of insulation material, and r_{i0} is the inner radius of the insulation shell mounted on the elemental (smallest) tubes. The performance of the sequence of squares (Figure 4.30) approaches the performance of the pairing sequence (Figure 4.31) when the amount of insulating material increases, and when the insulation improves (i.e., when N_0 decreases).

Another useful property of the pairing sequence is that each user receives hot water at the same temperature. The $\dot{m}_0$ stream received by each user passes through the same sequence of insulated tubes.

The comparison illustrated in Figure 4.32 was repeated for a larger system $(16L_0^2)$, on the basis of the same amount of insulation material (Wechsatol et al., 2001). Although the pairing design continues to perform better than the square design, the difference between the two is smaller than in Figure 4.32. This means that as the flow system becomes larger and more complex, the global performance becomes less sensitive to the actual layout of the tubes. More complex designs are more robust. This tendency is general: it is found in other tree-shaped flows, engineered or natural (Bejan, 2000).

Another way to construct the distribution network is to "grow it" optimally, by attaching one new user at a time (Wechsatol et al., 2002a). The growing network is optimized by placing each new user in the best spot that is available for it, that is, in the place where the temperature of the water received by the user after attachment is the highest. Each such step can be executed only after considering all the possible positions for the new user, and by comparing the performance of each possible configuration. The computational work becomes extensive as the network becomes larger and more complex.

Figure 4.33 shows a sequence of steps in the growth of the network. For brevity, the figure shows only the even-numbered steps, where each drawing is the best of all the drawings that could be made at that step. The start of the sequence is not shown: it is the square with four users, shown in the upper-right corner of Figure 4.32, or the A_2 construct of Figure 4.31. As in the flow structures discussed earlier, the one-by-one constructs of Figure 4.33 become better and more robust as they grow larger and more complex. Numerical results (Wechsatol et al., 2002a) show that although the performance of the one-by-one designs is consistently below the level achieved with the pairing construction (Figure 4.31), the gap between the two shrinks as the structures become larger. Once again, optimized tree-shaped flows are robust.

Another way to see that the performance of the structure increases as its complexity increases is shown in Figure 4.34. This figure brings together the one-by-one grown designs of Figure 4.33. This time, the designs are compared

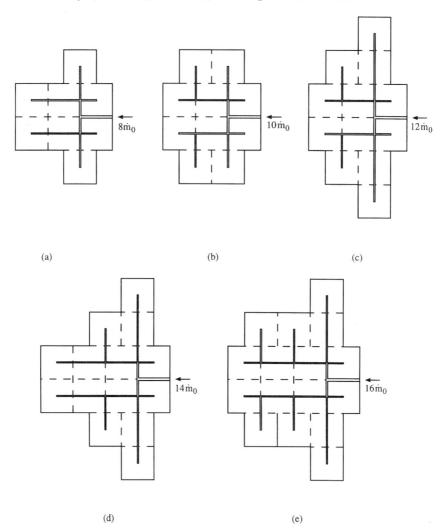

Fig. 4.33. Growing hot water distribution network constructed by adding one new user at a time, and placing it in the best location (Wechsatol *et al.*, 2002a).

on the basis of the same covered territory (A), which means that the L_0 scale of each user in Figure 4.33(e) ($n = 16$) is the shortest. The $n = 16$ structure is the reference. The three designs use the same amount of thermal insulation. The k and $\dot{m}_0$ values are such that the N_0 value based on the L_0 scale of the $n = 16$ case is $N_0 = 0.01$. The dimensionless volume is $\tilde{V}_{16} = V_{16}/(\pi L_0 r_{i0}^2)$, where V_{16} is the total volume of the insulation (the same in all three designs), and L_0 and r_{i0} are the length and radius of the elemental pipes in the $n = 16$ design. Figure 4.34 shows that the temperature of the water received by the end-user increases as the complexity of the structure (n) increases.

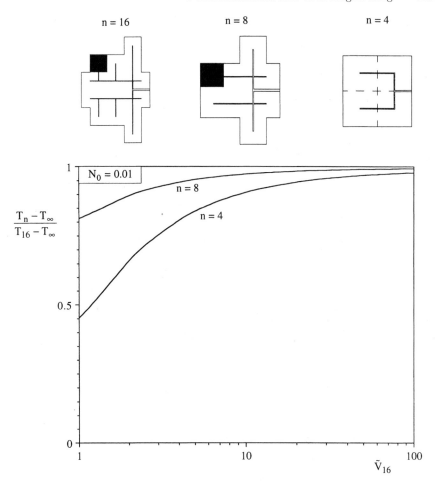

Fig. 4.34. Comparison of the one-by-one designs of Figure 4.26 showing how the performance improves as the complexity (n) increases (Wechsatol *et al.*, 2002a).

A characteristic of the one-by-one growth is the memory that is cemented into each design. What was built prior to the attachment of the newest user is retained. This characteristic is found in natural tree-shaped flows everywhere.

4.14 Minimal Resistance Versus Minimal Flow Length

In this section we explore a method by which the construction of tree-shaped flow structures can be simplified. For illustration, we consider the problem of distributing a stream from one central point to many points situated on a circle. This flow configuration has applications in urban hydraulics and the cooling of electronics. In the latter, a stream of coolant that enters the disc through its center bathes the disc and exits through ports located on the disc perimeter. There are two fundamental problems: the flow architecture for best cooling (minimal overall thermal resistance), and the flow architecture for minimal fluid flow resistance. In this section we consider the fluid flow problem by employing two methods, the minimization of flow resistance, and the much simpler and more direct method of minimizing flow path lengths (Lorente et al., 2002a).

The flow structure covers the disc shown in Figure 4.35. The disc radius is R. The mass flow rate and pressure drop from the center to the periphery are $\dot{m}$ and ΔP. The fluid flows in the Hagen–Poiseuille regime through round tubes of diameter D_i, length L_i and flow rate $\dot{m}_i$, where i indicates the tube level ($i = 0$ near the center, and increasing i values toward the periphery). The flow resistance posed by each tube is

$$\frac{\Delta P_i}{\dot{m}_i} = \frac{128\nu}{\pi} \frac{L_i}{D_i^4}, \tag{4.114}$$

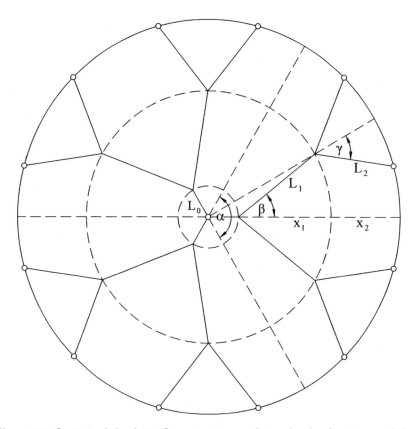

Fig. 4.35. Optimized dendritic flow structure with two levels of pairing and three central tubes (Wechsatol et al., 2002b).

where it has been assumed that the resistance is dominated by the long straight section of the tube, and concentrated losses (e.g., junctions, fittings) can be neglected. For a complete description of losses see Padet (1991). In the example of Figure 4.35, there are only two levels of branching or pairing such that the numbers of tubes of the same level increase from the center to the periphery: $n_0 = 3$, $n_1 = 6$, and $n_2 = 12$. The number of outlets on the disc perimeter is N; in Figure 4.35, for example, N is equal to 12.

The overall flow resistance $\Delta P / \dot{m}$ can be minimized by selecting all the geometric features of the flow structure. These include not only the lengths and angles identified in Figure 4.35, but also the ratios of tube diameters. When the total volume occupied by the tubes is constrained, the optimal tube diameters are sized relative to each other in accordance with Murray's law $D_{i+1}/D_i = 2^{-1/3}$, Equation (4.101). When space is constrained, branchings with only two tributaries pose less resistance than branchings with more tributaries.

The flow architecture of Figure 4.35 was optimized numerically by selecting the angles (α, β, γ) and tube lengths (L_0, L_1, L_2) subject to geometric constraints such as $L_0 + x_1 + x_2 = R$. In Figure 4.35 the flow architecture has two degrees of freedom. The figure shows the optimized layout of tubes. The minimized flow resistance is reported as point 2 $(N = 12)$ in Figure 4.36, where f is a dimensionless flow resistance factor defined by

$$\frac{\Delta P}{\dot{m}} = 8\pi\nu \frac{R^3}{V^2} f. \tag{4.115}$$

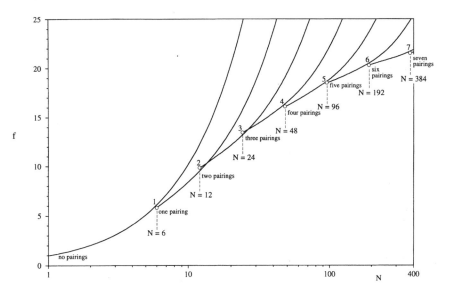

Fig. 4.36. The effect of the number of levels of pairing on the global flow resistance (f) when the number of points on the circle (N) is fixed, and the disc radius and the total tube volume are fixed (Wechsatol et al., 2002b).

In this definition V is the total volume occupied by the tubes. Point 2 is the start of the "two pairings" curve, which was obtained by performing the optimization of structures such as Figure 4.35 for cases with increasing numbers of outlets on the perimeter. The number $N(=4n_0)$ varies discretely as the number of central tubes n_0 increases, in spite of the impression given by the continuous curve shown in Figure 4.36. Every flow structure on that curve has two levels of pairing or bifurcation, that is, two dashed circles, as in Figure 4.35.

Complexity increases when the number of pairing levels increases. This effect is summarized in Figure 4.36. For example, point 3 represents the simplest design with three pairings ($n_0 = 3, n_1 = 6, n_2 = 12, n_3 = 24$), the optimized layout of which is drawn to scale in Figure 4.37a. The numerical details of the optimized structures summarized in Figure 4.36 are reported in Wechsatol *et al.* (2002b).

All the tree-shaped flows developed so far represent design, optimal geometric form with purpose. Which dendritic pattern is better? The answer depends on what is fixed. We rely on the constructal theory statement that the smallest length scale of the flow pattern—the elemental scale—is known and fixed (Bejan, 2000). This length scale is the distance (d) between two adjacent points on the circle. The radius of the circle (R) is also fixed. This means that the number of points on the circle (N) is fixed.

Under these circumstances, formulas such as Equation (4.115) show that the global flow resistance of the tree construct ($\Delta P/\dot{m}$) varies proportionally with f, and the other factors are constant. The flow pattern with less global resistance is the one with the lower f value. The leftmost curve in Figure 4.36 corresponds to purely radial flow (no pairings): one can show that this curve is the line $f = N$. Each of the subsequent $f(N)$ curves is almost a straight line when plotted on a graph with linear scales in f and N.

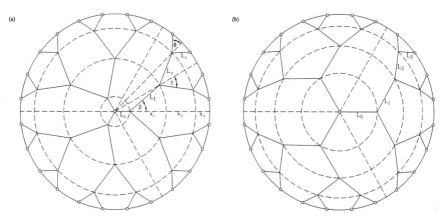

Fig. 4.37. Tree flows between a circle and its center: (a) construct obtained by optimizing every geometric detail, and (b) construct produced by the length-minimization algorithm (Lorente *et al.*, 2002a).

The smallest f corresponds to $N = 1$, or a single radial tube between the center and a point on the circle. The next highest f belongs to $N = 2$. These banal cases fall outside the class of tree-shaped flows: they are point–point flows, and are certainly not relevant to distributing a stream over a disc-shaped body, for example, to provide cooling if the body generates heat.

Figure 4.36 is instructive for several reasons. We read this figure vertically, at $N = $ constant. One conclusion is that pairing is a useful feature if N is sufficiently large (greater than 6). The larger the N value is, the more likely the need to design more levels of pairings into the flow structure. If the number of points on the rim of the structure (N) increases, then the flow structure with minimal flow resistance becomes more complex. Complexity increases because N increases, and because the number of pairing levels increases. Complexity is the mechanism by which the dendritic flow assures its minimal resistance status, that is, its minimal global imperfection. Optimized complexity is the design principle. Optimized complexity must not be confused with maximized complexity. *Optimal distribution of imperfection* is the principle that generates form (Bejan, 2000).

If we think of fixed rims (R) with more and more points (N), then the search for minimal flow resistance between the rim and the center requires discrete changes in the structure that covers the disc. To start with, N has to be large enough for an optimized structure with one or more pairings to exist. These starting N values (6, 12, 24, ...) are indicated with circles in Figure 4.36. As the structures become more complex, these circles describe a nearly smooth curve in the semilogarithmic plot of Figure 4.36. When there are three or more levels of pairing, the circles indicate the *transition* from one type of structure to the next type with one more level of branching.

This transition, or competition between competing flow structures, is analogous to the transition and flow pattern selection in Bénard convection. In the vicinity of each circle in Figure 4.36, the designer can choose between two structures, as both have nearly the same resistance. These choices are illustrated in Figure 4.38, which shows the two dentritic structures that compete in the vicinity of points 3 and 4 of Figure 4.36.

It is much simpler to optimize the layout of the flow structure by minimizing the lengths of the tubes that inhabit every area element (Lorente et al., 2002a). In Figure 4.39 we show how the point-circle flow structure is constructed based on the minimization of lengths. Consider the elemental curvilinear rectangle of radial distance r, angle θ, and radial thickness δ, which is shown in the upper part of Figure 4.39. The area of this element is fixed,

$$A = \frac{1}{2}\theta(2r\delta - \delta^2), \quad \text{constant.} \tag{4.116}$$

The objective is to minimize the length l subject to constraint (4.116),

$$l^2 = a^2 + b^2 = \left[r\left(1 - \cos\frac{\theta}{2}\right) + \delta\cos\frac{\theta}{2}\right]^2 + (r - \delta)^2\sin^2\frac{\theta}{2}. \tag{4.117}$$

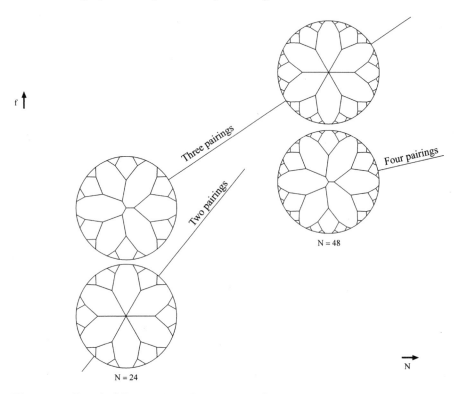

Fig. 4.38. Detail of Figure 4.31: the transition from one tree structure to one with one more level of pairing, as the flow resistance (f) is minimized and the number of points on the rim (N) increases (Wechsatol *et al.*, 2002b).

It is convenient to nondimensionalize Equations (4.116) and (4.117) by using the local radial distance r as length scale,

$$\frac{A}{r^2} = \frac{\theta}{2}(2x - x^2), \tag{4.118}$$

$$\left(\frac{l}{r}\right)^2 = (1 - x)^2 - 2(1 - x)\cos\frac{\theta}{2} + 1, \tag{4.119}$$

where $x = \delta/r$. According to the method of Lagrange multipliers, the problem of finding the extremum of function (4.119) subject to constraint (4.118) is equivalent to finding the extremum of the aggregate function

$$\Phi = (1 - x)^2 - 2(1 - x)\cos\frac{\theta}{2} + 1 + \lambda\theta\left(x - \frac{1}{2}x^2\right). \tag{4.120}$$

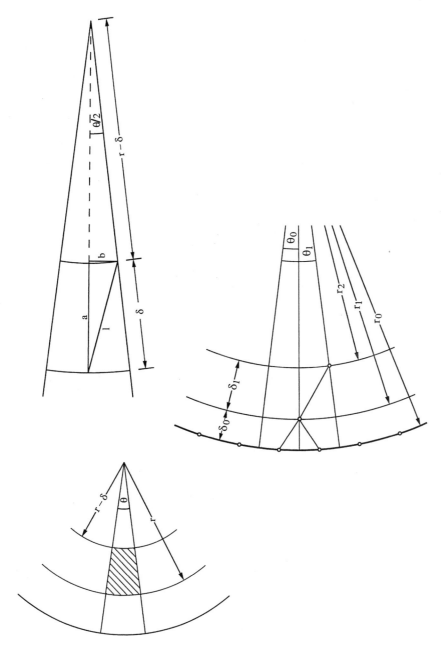

Fig. 4.39. Curvilinear rectangle defined by the intersection of radial lines and concentric circles, and the construction of the minimal-length tree between a circle and its center (Lorente *et al.*, 2002a).

Solving $\partial \Phi / \partial x = 0$ and $\partial \Phi / \partial \theta = 0$, and eliminating the Lagrange multiplier λ, we obtain the relation that pinpoints the optimal aspect ratio of the curvilinear rectangle:

$$\frac{x}{\theta} = \frac{(1-x)^2 \sin \dfrac{\theta}{2}}{(2-x)\left(x - 1 + \cos \dfrac{\theta}{2}\right)}. \tag{4.121}$$

To construct the minimal-length path between the outer circle (radius $r_0 = R$) and the center, we start with the outer circle and approximate it as a string of equidistant points; see the lower part of Figure 4.39. The distance between two consecutive points is d. The angle sustained by the outermost (elemental) rectangle of peripheral length d is $\theta_0 = d/r_0 \ll 1$. The value of θ_0 must be selected at the start of construction, for example, $\theta_0 = 0.1$. Substituting θ_0 for θ in Equation (4.121), we calculate x_0, or the aspect ratio of the elemental rectangle (x_0/δ_0), or the radial thickness of the element $\delta_0 = r_0 x_0$. The radius of the inner circle that borders the elemental rectangle is $r_1 = r_0 - \delta_0$.

The next curvilinear rectangle subtends the angle $\theta_1 = 2\theta_0$, and has the radial position r_1. Equation (4.121) delivers x_1, the aspect ratio x_1/δ_1, and the radial dimension $\delta_1 = r_1 x_1$. The radius of the next circle is $r_2 = r_1 - \delta_1$.

This algorithm can be applied a sufficient number of times, marching toward the center of the circle, and drawing the resulting tree network. The construction must stop at a certain step (i) if $\theta_i > 2\pi$, or $r_i < 0$, whichever occurs first. During the numerical implementation of the algorithm it was

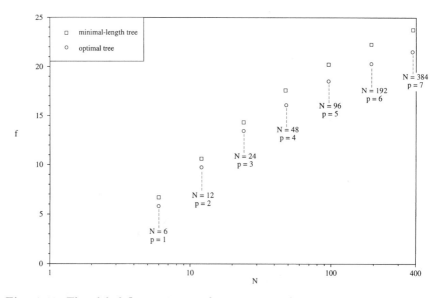

Fig. 4.40. The global flow resistance for constructs obtained by optimizing every geometric detail, and constructs produced by the length-minimization algorithm (Lorente *et al.*, 2002a).

found that thresholds $\theta_i > 2\pi$ and $r_i < 0$ are reached simultaneously in this construction. The construction for $\theta_0 = \pi/12$ is reported in Figure 4.37b. The constructed tree is approximate near the center, because the rectangle approximation breaks down in the limit $r \to 0$.

Figure 4.37b is displayed next to the design based on the resistance minimization method, Figure 4.37a. Visually, there is little difference between the two constructions. The same can be said about the f values of the competing designs. The squares plotted on Figure 4.40 show that, although consistently inferior, the performance of minimal-length structures closely resembles the performance of fully optimized structures. In conclusion, the length-minimization method proposed in Lorente *et al.* (2002a) provides a very effective shortcut to designs that come close to the best designs. The minimal-length designs approach the optimal designs in terms of global performance, and architecturally as well. The closeness documented in Figures 4.37 and 4.40 shows again that optimized tree flow structures are robust.

The design features described in Sections 4.11 to 4.14 are important in general, that is, regardless of what flows through the complex flow structure. For example, structures similar to those of Figure 4.37 can be used in the cooling of electronics. In such cases the disc generates heat at every point, and the "ducts" are blades or fibers with very high thermal conductivity, which are embedded in the heat-generating medium. The heat sink is the center of the disc. In the optimized design the inserts are arranged as a tree, which becomes larger, more complex, and robust as the size of the disc increases (Rocha *et al.*, 2002).

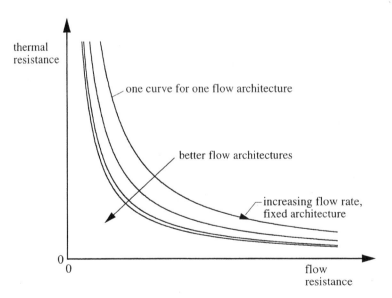

Fig. 4.41. The generation of flow architectures in the pursuit of combined low thermal and fluid resistances (Wechsatol *et al.*, 2003).

The convection cooling of a heat-generating body is a design challenge with at least two objectives: minimal thermal resistance and minimal flow resistance (pumping power). This section outlined the emergence of tree-shaped structures when only one objective is being pursued, minimal flow resistance. Unlike in thermodynamic optimization, where the search is for a unique balance (for minimal irreversibility) between heat-transfer performance and fluid-mechanics performance, in constructal design the search is for flow architectures. On a graph of thermal resistance versus fluid flow resistance (Figure 4.41), there is one curve for each flow architecture subject to specified global constraints. In forced convection configurations the thermal resistance decreases as the pumping power increases. The objective is to "morph" the flow configuration, to optimize the flow architecture so that it is represented by the curve that is situated as close as possible to the origin. The push toward lower and lower curves, or better and better architectures, is a general trend in all flow systems, engineered and natural. This tendency is summarized in the constructal law, as a self-standing principle in the thermodynamics of nonequilibrium (flow) systems (Bejan and Lorente, 2003).

5

Compact Heat Transfer Flow Structures

In this chapter we discuss a series of developments that point in the direction of applying porous media concepts to the description, simulation, and optimization of compact systems with complex flow structures. Compact and miniaturized heat exchangers are primary examples of this trend. Other examples are chemical reactors, fiber filters, brush seals, and the modeling of microsegregation during the solidification of alloys. The new aspect highlighted by these developments is that flow structures are optimized so that flow systems exhibit maximum global performance subject to global constraints. Optimization of flow structure means that in the beginning the flow geometry is free to change. The system is free to morph. Global performance is achieved through the generation of flow architecture (Bejan, 2000). The flow structure becomes a porous medium with purpose, that is, a *designed porous medium*.

We pursue here two geometric ideas that lead to designed porous media. One is the optimization of spacings for fluid flow, or the optimization of the shape of the pore or the fissure, when the flow system has the global objective of maximizing the solid–fluid heat or mass transfer rate per unit volume. The other is the thought that the flows designed to be distributed uniformly through the porous structure are not necessarily the best. Nature shows that nonuniform distribution ("maldistribution") is the norm. This is especially true in flows that connect one point (inlet or outlet) with an entire volume (e.g., river basins, lungs, heat exchangers). Tree-shaped flows are the "maldistributed" patterns that result. These are results of optimization of global performance. They represent *optimal distribution of imperfection* (Bejan, 2000), where the imperfection is due to the resistances encountered by the flow of heat and fluid at every point inside the volume.

These two ideas lead to the proposal that the best flow architecture for a heat exchanger with maximal heat transfer rate density should have tree-shaped flows, where each elemental volume (pore, fissure) is optimized geometrically. Heat exchangers formed by two streams should have two trees, the canopies of which touch in every element that fills the heat exchanger volume

(Section 5.7). Channels form the tree branches, and the interstitial volume that surrounds the channels is visited by diffusion.

5.1 Heat Exchangers as Porous Media

Heat exchangers are a century-old technology based on information and concepts stimulated by the development of large-scale devices such as condensers, boilers, and regenerators for power plants [see, e.g., Bejan (1993, Chapter 9)]. The modern emphasis on heat transfer augmentation, and the more recent push toward miniaturization in the cooling of electronics, have led to the development of compact devices with much smaller features than in the past. These devices operate at much lower Reynolds numbers, where their compactness and small dimensions make them candidates for modeling as fluid-saturated porous media. Such modeling promises to revolutionize the nomenclature and numerical simulation of the flow and heat transfer through heat exchangers.

To illustrate the opportunity that exists for change, consider Zukauskas' (1987) classical chart for the pressure drop in cross-flow through arrays of staggered cylinders; see Figure 5.1, or Bejan (1993, p. 488). The factor f is

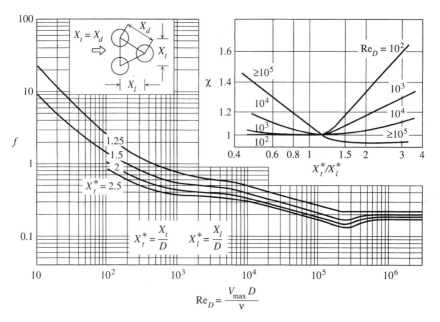

Fig. 5.1. Arrays of staggered tubes: the coefficients f and χ for the pressure drop formula (5.1) (Zukauskas, 1987; from Kakac et al., eds., *Handbook of Single-phase Convective Heat Transfer*, Copyright © 1987 Wiley. This material is used by permission of John Wiley & Sons, Inc.).

for the calculation of the total pressure drop across the bundle,

$$\Delta P = n_1 f \chi \rho V_{\mathrm{max}}^2, \tag{5.1}$$

where n_1 is the number of cylinder rows counted in the flow direction, and V_{max} is the mean velocity through the narrowest cylinder-to-cylinder spacing. The dimensionless factors f and χ are presented in Figure 5.1.

The four curves drawn on Figure 5.1 for the transverse pitch/cylinder diameter ratios 1.25, 1.5, 2, and 2.5 can be made to collapse into a single curve, as shown in Figure 5.2 (Bejan and Morega, 1993). The technique consists of treating the bundle as a fluid-saturated porous medium, and using the volume-averaged velocity U, the pore Reynolds number $U K^{1/2}/\nu$ on the abscissa, and the dimensionless pressure gradient group $(\Delta P/L) K^{1/2}/(\rho U^2)$ on the ordinate. The effective permeability of the bundle of cylinders (K) is estimated based on the Carman–Kozeny model

$$K \cong \frac{D^2 \phi^3}{k_z (1 - \phi)^2}, \tag{5.2}$$

where ϕ is the porosity of the assembly, D is the cylinder diameter, and the empirical constant is $k_z = 100$.

Figure 5.2 shows very clearly the transition between Darcy flow (slope -1) and Forchheimer flow (slope 0). The porous medium presentation of the array

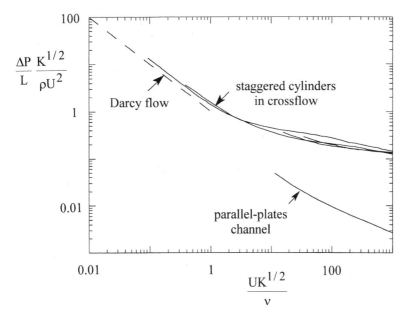

Fig. 5.2. Porous medium representation of the classical pressure-drop data for flow through staggered cylinders and stacks of parallel plates (Bejan and Morega, 1993).

of cylinders leads to a very tight collapse of the curves taken from Zukauskas' chart. The figure also shows the pressure drop curve for turbulent flow through a heat exchanger core formed by a stack of parallel plates. Figure 5.2 extends the curves (Zukauskas' data) reliably into the low Reynolds number limit (Darcy flow), where classical heat exchanger data are not available.

This method deserves to be applied to other configurations. A similar model for three-dimensional flow and heat transfer in a tube bundle viewed as a porous medium was developed by Butterworth (1978). The literature on the calculation of the average permeability of arrays of cylinders was reviewed recently by Papathanasiou (2001).

Another stimulus for progress in this direction is the fact that the heat and fluid flow process can be simulated numerically much more easily if the heat exchanger is replaced at every point by a porous medium with volume-averaged properties. An example is presented in Figure 5.3 (Morega et al., 1995). Air flows from left to right along a hot horizontal surface (the electronics module) and through an array of parallel plate fins of rectangular profile (the heat sink). The aspect ratio of the hot surface is H/L. The Reynolds number Re_L is based on L and the approach velocity. The air flows through and around the heat sink.

The stack volume is treated as a two-dimensional homogeneous and anisotropic porous medium with Darcy flow in the x-direction. The porous medium model has the added advantage that it accounts (in a volume-averaged sense) for the conduction through the board material, longitudinally and transversally. The equations and far-field conditions for the flow and heat transfer in the regions situated outside the stack are in accordance with the formulation of classical forced convection (Bejan, 1984), and are listed in Morega et al. (1995). Inside the $H \times L$ space, the flow is purely longitudinal ($v = 0$), with the velocity

$$u = -\frac{K_x}{\mu}\frac{\partial P}{\partial x}, \tag{5.3}$$

which is constant from $x = 0$ to $x = L$, but may vary in the y-direction. The K_x permeability of a stack with boards spaced equidistantly (spacing $= d$, board thickness $= t$) is a known constant (Bejan, 1984, p. 383):

$$K_x = \frac{d^3}{12(t+d)} = \frac{d^2}{12}\phi, \tag{5.4}$$

where $\phi = d/(d+t)$ is the porosity of the construct. The u velocity is a volume-averaged quantity, which, based on mass conservation, corresponds to the u component immediately outside the stack (in the pure fluid, at $x = 0^-$ and $x = L^+$). The porous medium is impermeable in the transversal direction. The u velocity inside the porous medium is a function of y because it is driven by the pressure difference between the entrance and exit of an individual channel, $-\partial P/\partial x = [P(0,y) - P(L,y)]/L$. This pressure difference is maximum across the channel positioned along the midplane ($y = 0$) and smaller near

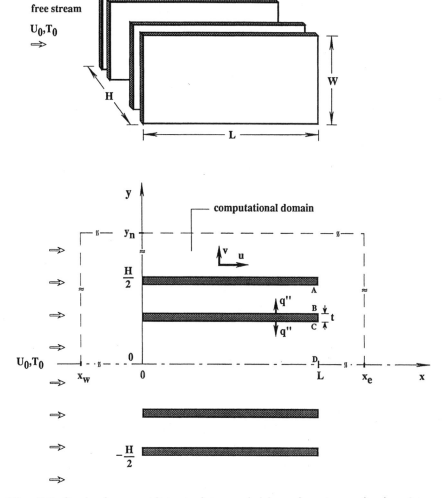

Fig. 5.3. Stack of nonequidistant plates cooled by a free stream (top) and two-dimensional model and computational domain (bottom) (Morega *et al.*, 1995).

the lateral edges $(y = \pm H/2)$. It is assumed that P varies continuously across the entrance and exit planes of the stack $(x = 0, L)$.

The temperature distribution inside the $H \times L$ space is governed by the energy equation

$$\rho c_P u \frac{\partial T}{\partial x} = k_x \frac{\partial^2 T}{\partial x^2} + k_y \frac{\partial^2 T}{\partial y^2} + q'''. \tag{5.5}$$

Local thermal equilibrium is assumed, so that T represents the local temperature of the solid and adjacent fluid. The total heat generation rate of the stack (q') is distributed uniformly over the stack volume, $q''' = q'/HL$. Given the parallel-plates structure of the porous medium (many plates are assumed), the directional thermal conductivities (k_x, k_y) can be estimated based on the parallel resistance and series resistance models:

$$k_x = \phi k + (1 - \phi)k_s, \qquad k_y = \frac{k k_s}{(1 - \phi)k + \phi k_s}, \tag{5.6}$$

where k_s is the thermal conductivity of the solid (the plate material). The temperature T varies continuously across the boundary of the $H \times L$ space. Finally, when we nondimensionalize the energy equation (5.5) using the variables defined in

$$\tilde{x}, \tilde{y} = \frac{(x, y)}{L}, \qquad (\tilde{u}, \tilde{v}) = \frac{u, v}{U_0}, \tag{5.7}$$

$$\tilde{P} = \frac{P}{\rho U_0^2}, \qquad \theta = \frac{T - T_0}{q'/k}, \tag{5.8}$$

we find that the dimensionless temperature inside the stack depends on eight variables, $\theta = \theta[\tilde{x}, \tilde{y}, H/L, \mathrm{Pr}, \mathrm{Re}_L, d/L, t/L(\text{or } \phi), k_s/k]$. The hot spot occurs at or near the exit from the stack. The dimensionless hot-spot temperature depends on a total of six parameters, $\theta_{\mathrm{hot}} = \theta_{\mathrm{hot}}(H/L, \mathrm{Pr}, \mathrm{Re}_L, d/L, t/L, k_s/k)$.

An important question that can be answered numerically by using the above model is this: What is the effect of the conductivity ratio k_s/k on the hot-spot temperature? Several parameters were assumed known: $H/L = 1$, $\mathrm{Pr} = 0.72$, and $t/L = 1/20$. As a first example, we chose an external flow with $\mathrm{Re}_L = U_0 L/\nu = 400$, for which it was shown (Morega et al., 1995) that the optimal number of boards is $n_{\mathrm{opt}} = 5$. This optimum was a numerical test for the forced-convection prediction made later in Equation (5.23). To place the actual channel flow in the Hagen–Poiseuille regime and the porous medium model in the Darcy regime, we chose a larger number of boards, namely, $n = 9$. The other geometric parameters that follow from this choice are $d/L = 0.0688$ and $\phi = 0.579$.

The flow pattern calculated for $\mathrm{Re}_L = 400$ using the porous medium model for the stack is shown in the lower part of Figure 5.4. The view is from above, and only half of the square $H \times L$ is shown. The base of the drawing (the lowest streamline) is the plane of symmetry of the stack $y = 0$. Only the external flow situated in the immediate vicinity of the stack is shown (the computational domain is considerably more extensive). Figure 5.4 shows that when the Reynolds number is large most of the fluid that flows through the stack prefers the channels that are close to the plane of symmetry. The flow through the stack is weaker and more uniform when Re_L is smaller; for example, $\mathrm{Re}_L = 100$ in Figure 5.4.

The patterns of isotherms that correspond to the $\mathrm{Re}_L = 400$ flow are presented in Figure 5.5. We see that the temperature distribution inside the stack is influenced greatly by the thermal conductivity ratio k_s/k. When k_s/k is

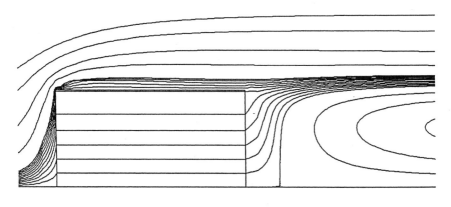

$$\mathrm{Re_L} = 100$$

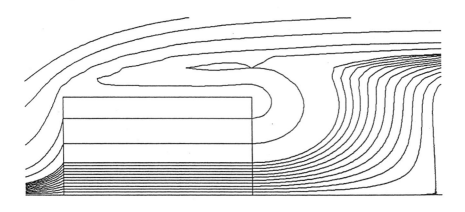

$$\mathrm{Re_L} = 400$$

Fig. 5.4. The flow through and around a stack of rectangular parallel-plate fins attached to a square base and modeled as a porous medium (Morega *et al.*, 1995).

small (Figure 5.5, top), the hot spot occurs near the exit from one of the outer (peripheral) parallel-plate channels, because the flow through that channel is relatively weak (Figure 5.4, bottom). In the opposite extreme, the large k_s/k ratio means that the stack is cooled in the transversal direction by the fluid that flows around the stack. In this limit the hot spot occurs near the exit from the channel that coincides with the plane of symmetry. The hot spot migrates from one position to the other when k_s/k is of order 1.

The additional cooling effect due to transversal conduction is even more evident in Figure 5.6. The hot-spot excess temperature θ_{hot} decreases to about

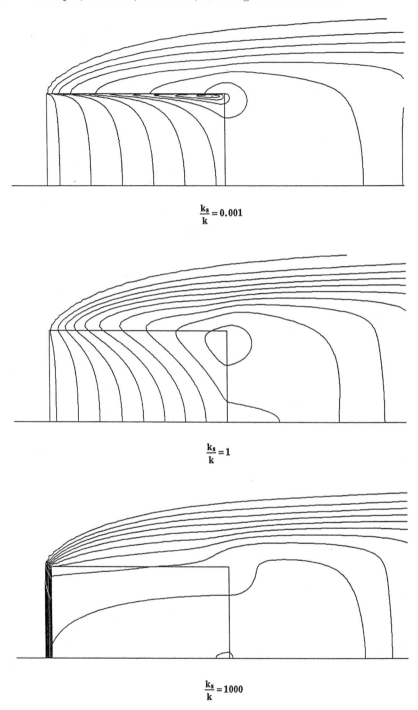

$$\frac{k_s}{k} = 0.001$$

$$\frac{k_s}{k} = 1$$

$$\frac{k_s}{k} = 1000$$

Fig. 5.5. The temperature distribution that corresponds to the $\mathrm{Re}_L = 400$ flow shown in Figure 5.4 (Morega *et al.*, 1995).

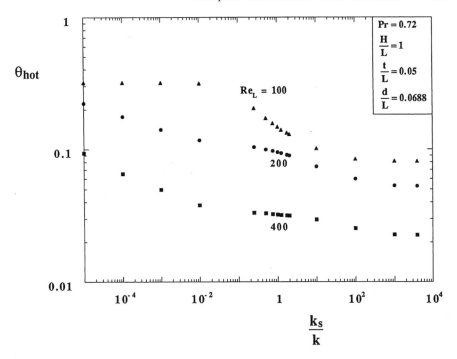

Fig. 5.6. The hot-spot excess temperature obtained by treating the stack as a porous medium (Morega *et al.*, 1995).

one-fifth of its original value as k_s/k increases from 0 to 10^4. Figure 5.6 also summarizes the corresponding conclusions obtained for two additional examples, $\mathrm{Re}_L = 100$ and 200. The θ_{hot} value is larger when Re_L is smaller.

The calculation of θ_{hot} for $k_s/k = 0$ and nine plates was compared with the corresponding results obtained numerically by simulating Hagen–Poiseuille flow in every parallel-plates channel (Morega *et al.*, 1995). The agreement between the two sets of θ_{hot} values is quite good. The loss in the accuracy with which the porous medium model predicts θ_{hot} is balanced by a significant gain in computational speed.

The great interest in heat exchangers as porous media is exemplified by the current work on compactness and heat transfer augmentation by using metallic foams (Boomsma and Poulikakos, 2001, 2002; Boomsma *et al.*, 2003; Angirasa, 2002a,b; Phanikumar and Mahajan, 2002; Bhattacharya *et al.*, 2002).

5.2 Optimal Spacings in Natural Convection

An important application of porous media concepts in engineering is in the optimization of the internal spacings of heat exchangers subjected to overall volume constraints. Packages of electronics cooled by forced convection are examples of heat exchangers that must function inside fixed volumes. The design objective is to install as many components (i.e., heat generation rate) as possible, while the maximum temperature that occurs at a point (hot spot) inside the given volume does not exceed a specified limit. A very basic tradeoff exists with respect to the number of installed components, that is, with respect to the size of the pores through which the coolant flows. This tradeoff is present in natural convection (Bejan, 1984, p. 157, problem 11; Solutions Manual, pp. 93–95) and forced convection (Bejan and Sciubba, 1992).

Imagine the two extremes: numerous components (small pores), and few components (large spacings). When the components and pores are numerous and small, the package functions as a heat-generating porous medium. When the installed heat generation rate is fixed, the hot-spot temperature increases as the spacings become smaller, because in this limit the coolant flow is being shut off. In the opposite limit, the hot-spot temperature increases again because the heat transfer contact area decreases as the component size and spacing become larger. At the intersection of these two asymptotes—the collision between two mechanisms—we find an optimal spacing (or pore size) where the hot-spot temperature is minimal when the heat generation rate and volume are fixed. The same spacing represents the design with maximal heat generation rate and fixed hot-spot temperature and volume. This is the design in which the available volume is used most effectively, that is, to the maximum (Bejan, 2000).

To illustrate in the simplest terms the origin of optimal internal structure, assume that the heat generation rate q is spread almost uniformly over the given volume. Heat-generating devices are mounted on equidistant vertical boards of height H, filling a space of height H, and horizontal dimensions L and W. The configuration is two-dimensional with respect to the ensuing buoyancy-driven flow, as shown in Figure 5.7. The board-to-board spacing D, or the number of vertical parallel-plate channels is allowed to vary, $n = L/D$.

To determine the spacing D that maximizes q is a challenging task. Here we reproduce the back of an envelope method (the *intersection of asymptotes method*) used first in Bejan (1984, p. 157, problem 11). See also Lewins (2003). For simplicity, assume that (i) the flow is laminar, (ii) the board surfaces are sufficiently smooth to justify the use of heat transfer results for natural convection over vertical smooth walls, and (iii) the maximum temperature $T_{\max}$ is representative of the order of magnitude of the temperature at every point on the board surface. The method consists of two steps, where we identify the two possible extremes, the small-D limit and the large-D limit; and then the two asymptotes are intersected to locate the D value that maximizes q.

When D becomes sufficiently small, the channel formed between two boards becomes narrow enough for the flow and heat transfer to be fully developed. According to the first law of thermodynamics we have $q_1 = \dot{m}_1 c_P (T_{\max} - T_0)$ for the heat transfer rate extracted by the coolant from one of the channels of spacing D, where T_0 is the inlet temperature and $T_{\max}$ is the outlet temperature. The mass flow rate is $\dot{m}_1 = \rho D W U$, where the

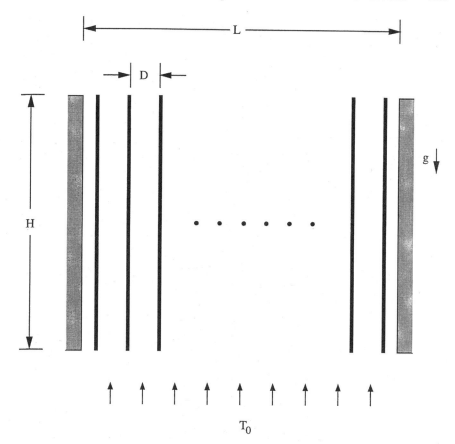

Fig. 5.7. Two-dimensional volume that generates heat and is cooled by natural convection.

mean velocity U can be estimated by replacing $\Delta P/L$ with $\rho g \beta(T_{\max} - T_\infty)$ in the Hagen–Poiseuille flow solution for laminar fully developed flow between parallel plates:

$$\rho g \beta(T_{\max} - T_0) = f \frac{4}{D_h} \frac{1}{2} \rho U^2. \tag{5.9}$$

For laminar flow between parallel plates we have $f = 24/(UD_h/\nu)$ and $D_h = 2D$, hence

$$U = g\beta(T_{\max} - T_0)D^2/(12\nu), \tag{5.10}$$

$$\dot{m}_1 = \rho DWU = \rho W g\beta(T_{\max} - T_0)D^3/(12\nu). \tag{5.11}$$

The rate at which heat is removed from the entire package is $q = nq_1$, or

$$q = \rho c_P WLg\beta(T_{\max} - T_0)^2 D^2/(12\nu). \tag{5.12}$$

In conclusion, in the $D \to 0$ limit the total heat transfer rate varies as D^2.

Consider next the limit in which D is large enough that it exceeds the thickness of the thermal boundary layer that forms on each vertical surface: $\delta_T \sim H \text{Ra}_H^{-1/4}$, where $\text{Ra}_H = g\beta H^3 (T_{\max} - T_0)/(\alpha\nu)$ and $\Pr \geq 1$. In this limit the boundary layers are distinct (thin compared with D), and the center region of the board-to-board spacing is occupied by fluid of temperature T_0. The number of distinct boundary layers is $2n = 2L/D$ because there are two for each D spacing. The heat transfer rate through one boundary layer is $\bar{h} H W (T_{\max} - T_0)$ for which $\bar{h}$ is furnished by the heat transfer solution for boundary layer natural convection, $\bar{h} H/k = 0.517 \text{Ra}_H^{1/4}$; see, for example, Bejan (1984, 1993). The rate of heat transfer extracted from the entire package is $2n$ times larger than $\bar{h} H W (T_{\max} - T_0)$:

$$q = 2\frac{L}{D} H W (T_{\max} - T_0) \frac{k}{H} 0.517 \text{Ra}_H^{1/4}. \qquad (5.13)$$

Equation (5.13) shows that in the large-D limit the total heat transfer rate varies as D^{-1} as the board-to-board spacing changes. The same behavior is exhibited by the overall thermal conductance of the construct, $q/(T_{\max} - T_0)$.

The two asymptotes of the actual (unknown) curve of q versus D show that they intersect above what would be the peak of the actual $q(D)$ curve. It is not necessary to determine the exact form of the actual $q(D)$ relation: the optimal spacing D_{opt} for maximum q can be estimated as the D value where Equations (5.12) and (5.13) intersect [Bejan (1984); see also Bejan (1995a, pp. 202–205)],

$$\frac{D_{\text{opt}}}{H} \cong 2.3 \text{Ra}_H^{-1/4}. \qquad (5.14)$$

This D_{opt} estimate is within 20% of the optimal spacing deduced based on lengthier methods, such as the maximization of the $q(D)$ relation (Bar Cohen and Rohsenow, 1984) and the finite-difference simulations of the complete flow and temperature fields in the package (Anand et al., 1992; Kim et al., 1991). An order of magnitude estimate of the maximum heat transfer rate can be obtained by substituting D_{opt} into Equation (5.13):

$$q_{\max} \lesssim 0.45 k (T_{\max} - T_0) \frac{LW}{H} \text{Ra}_H^{1/2}. \qquad (5.15)$$

The approximate inequality sign is a reminder that the peak of the actual $q(D)$ curve falls under the intersection of the two asymptotes. This result also can be expressed as the maximum volumetric rate of heat generation in the $H \times L \times W$ volume,

$$\frac{q_{\max}}{HLW} \sim 0.45 \frac{k}{H^2} (T_{\max} - T_0) \text{Ra}_H^{1/2}. \qquad (5.16)$$

The assumption that the heat generation rate q is spread on equidistant vertical plates as tall as the volume was made for the sake of illustrating in the simplest terms how internal structure results from the purpose and constraints principle. Optimal internal structure is a characteristic of all systems that share the same purpose and constraints. The generality of this result is supported by several more recent studies in which the stack of H-tall plates was replaced with arrays of internal heating elements of other shapes. This body of work is reviewed in Bejan (2000).

For example, when the fixed volume is filled with many equidistant parallel staggered plates, which are considerably shorter than the volume height ($b \ll H$ in Figure 5.8a), there exists an optimal horizontal spacing (D) between neighboring plates (Ledezma and Bejan, 1997). In laminar natural convection, the optimal spacing scales with the height of the fixed volume, not with the height of the individual plate, for example,

$$\frac{D_{\mathrm{opt}}}{H} \cong 0.63 \left(\frac{Nb}{H}\right)^{1.48} \mathrm{Ra}_H^{-0.19}, \tag{5.17}$$

where N is the number of plate surfaces that face one elemental channel (e.g., $N = 4$ in Figure 5.8a). The dimensionless group (Nb/H) is of the order of 1 and represents the relative contact area present along the boundaries of the elemental channel. The similarities between relations (5.14) and (5.17) are important, because they point to the robustness of the design principle that generates internal spacings. Experiments and numerical simulations show that Equation (5.17) is accurate to within 6% in the range $10^3 \leq \mathrm{Ra}_H \leq 5 \times 10^5$ and $0.4 \leq (Nb/H) \leq 1.2$.

When the volume is heated by an array of horizontal cylinders of diameter D (Figure 5.8b), the optimal cylinder-to-cylinder spacing S scales once again with the overall height of the volume (Bejan *et al.*, 1995). In laminar natural convection the optimal spacing is closely correlated with a formula that resembles Equation (5.14) (Ledezma and Bejan, 1997):

$$\frac{S_{\mathrm{opt}}}{H} \cong 2.72 \left(\frac{H}{D}\right)^{1/12} \mathrm{Ra}_H^{-1/4} + 0.263 \frac{D}{H}, \tag{5.18}$$

where the second term on the right-hand side is a small correction factor. Note the similarities between Equation (5.18) and Equations (5.14) and (5.17). The optimal spacing (5.18) agrees to within 1.7% with numerical simulations and experiments performed in the range $\mathrm{Pr} = 0.72$, $350 \leq \mathrm{Ra}_D \leq 10^4$, and $6 \leq H/D \leq 20$, where $\mathrm{Ra}_D = g\beta D^3(T_{\mathrm{max}} - T_0)/(\alpha\nu)$.

The literature on optimal internal spacings for natural and forced convection was reviewed in Kim and Lee (1996). Additional contributions have been made by Marsters (1975), Farouk and Guceri (1982), Sparrow and Pfeil (1984), Karim *et al.* (1986), Sadeghipour and Kazemzadeh (1992), and Ma *et al.* (1994).

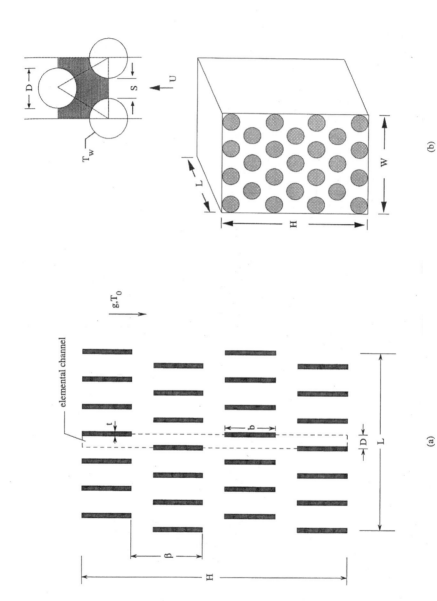

Fig. 5.8. Natural convection cooling of a volume heated uniformly: (a) array of staggered plates, and (b) array of horizontal cylinders.

5.3 Optimal Spacings in Forced Convection

Consider now the analogous problem of optimal spacings in a volume cooled by forced convection (Bejan and Sciubba, 1992; Bejan, 1993). As shown in Figure 5.9, the swept length of each board is L, and the transverse dimension of the entire package is H. The width of the stack W is perpendicular to the plane of the figure. We retain the simplifying assumptions made at the start of Section 5.2. The thickness of the individual board is again negligible relative to the board-to-board spacing D, so that the number of channels is $n = H/D$. The pressure difference across the package, ΔP, is assumed constant and known. We analyze the global heat transfer in the two limits, small D and large D, and intersect the asymptotes.

When D becomes sufficiently small, the channel formed between two boards becomes narrow enough for the flow and the heat transfer to be fully developed. In this limit, the mean outlet temperature of the fluid approaches the board temperature $T_{\max}$. The total rate of heat transfer from the $H \times L \times W$ volume is $q = \dot{m}c_P(T_{\max} - T_0)$, where $\dot{m} = \rho HWU$. The mean velocity through the channel U is known from the Hagen–Poiseuille flow

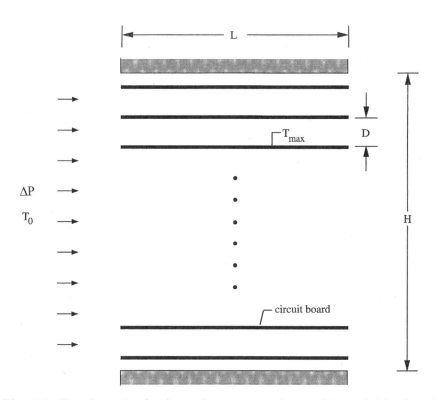

Fig. 5.9. Two-dimensional volume that generates heat and is cooled by forced convection.

solution $U = D^2 \Delta P/(12\mu L)$. The corresponding expression for the total heat transfer rate is

$$q = \rho H W \frac{D^2}{12\mu} \frac{\Delta P}{L} c_P (T_{\max} - T_0). \tag{5.19}$$

In this way we conclude that when $D \to 0$ the total heat transfer rate and the overall thermal conductance vary as D^2.

When D is large enough that it exceeds the thickness of the thermal boundary layer that forms on each horizontal surface, it is necessary to determine the free-stream velocity U_0 that sweeps the boundary layers. Because the pressure drop ΔP is fixed, the balance of forces on the $H \times L \times W$ control volume requires

$$\Delta P H W = 2n\bar{\tau} L W, \tag{5.20}$$

where $\bar{\tau}$ is the wall shear stress averaged over L, namely, $\bar{\tau} = 0.664\, \rho U_0^2 \mathrm{Re}_L^{-1/2}$ for $\mathrm{Re}_L \leq 5 \times 10^5$ (Bejan, 1995a). Equation (5.20) yields

$$U_0 = \left(\frac{\Delta P H}{1.328 n L^{1/2} \rho \nu^{1/2}} \right)^{2/3}. \tag{5.21}$$

The heat transfer rate through the surface of a single board is $q_1 = \bar{h} L W (T_{\max} - T_0)$, where the heat transfer coefficient averaged over L is $\bar{h} = 0.664\,(k/L)\,\mathrm{Pr}^{1/3}\,\mathrm{Re}_L^{1/2}$ when $\mathrm{Pr} \geq 0.5$ (Bejan, 1995a). The total heat transfer rate from the entire package is $q = 2n q_1$ or, after the $\bar{h}$ and U_0 expressions given above are used,

$$q = 1.21\, k H W (T_{\max} - T_0) \left(\frac{\mathrm{Pr}\, L \Delta P}{\rho \nu^2 D^2} \right)^{1/3}. \tag{5.22}$$

In conclusion, in the large-D limit the total heat transfer rate varies as $D^{-2/3}$ as the board-to-board spacing changes.

The intersection of the two $q(D)$ asymptotes, Equations (5.19) and (5.22), yields an estimate for the board-to-board spacing for maximum global thermal conductance (Bejan and Sciubba, 1992; Bejan, 1993):

$$\frac{D_{\mathrm{opt}}}{L} \cong 2.7\, \mathrm{Be}^{-1/4}, \tag{5.23}$$

where Be is the dimensionless group that Bhattacharjee and Grosshandler (1988) and Petrescu (1994) named the Bejan number (see also Furukawa and Yang, 2003),

$$\mathrm{Be} = \frac{\Delta P L^2}{\mu \alpha}. \tag{5.24}$$

Equation (5.23) shows that the optimal spacing increases as $L^{1/2}$ and decreases as $\Delta P^{-1/4}$ with increasing L and ΔP, respectively. Relation (5.23) underestimates by 12% the more exact value obtained by locating the maximum of the actual $q(D)$ curve (Bejan and Sciubba, 1992), and is adequate when the board surface is modeled either as uniform flux or isothermal. It has been shown (Mereu et al., 1993) that relation (5.23) holds even when the board thickness is not negligible relative to the board-to-board spacing.

The manner in which the design parameters influence the maximum rate of heat removal from the filled volume can be expressed as

$$q_{\max} \lesssim 0.6k(T_{\max} - T_0)\frac{HW}{L}\mathrm{Be}^{1/2}, \tag{5.25}$$

which is obtained by setting $D = D_{\mathrm{opt}}$ in Equation (5.19) or (5.22). Once again, the approximation sign is a reminder that the actual $q_{\max}$ value is as much as 20% smaller because the peak of the $q(D)$ curve is situated under the point where the two asymptotes cross. The maximum volumetric rate of heat generation in the $H \times L \times W$ volume is

$$\frac{q_{\max}}{HLW} \lesssim 0.6\frac{k}{L^2}(T_{\max} - T_0)\mathrm{Be}^{1/2}. \tag{5.26}$$

The similarity between the forced convection results and the corresponding results for natural convection is worth noting. The role played by the Rayleigh number Ra_H in the free convection case is played in forced convection by the dimensionless group Be (Petrescu, 1994).

The optimal internal spacings belong to the porous system as a whole, with its purpose and constraints, not to the individual solid element on which heat is being generated. The robustness of this conclusion becomes clear when we look at other elemental shapes for which optimal spacings have been determined. A volume heated by an array of staggered plates in forced convection (Figure 5.10a) is characterized by an internal spacing D that scales with the swept length of the volume L (Fowler et al., 1997):

$$\frac{D_{\mathrm{opt}}}{L} \cong 5.4\,\mathrm{Pr}^{-1/4}\left(\mathrm{Re}_L\frac{L}{b}\right)^{-1/2}. \tag{5.27}$$

In this relation the Reynolds number is $\mathrm{Re}_L = U_\infty L/\nu$. The range in which this correlation was developed based on numerical simulations and laboratory experiments is $\mathrm{Pr} = 0.72$, $10^2 \leq \mathrm{Re}_L \leq 10^4$, and $0.5 \leq (Nb/L) \leq 1.3$.

Similarly, when the elements are cylinders in cross flow (Figure 5.10b) the optimal spacing S is influenced most by the longitudinal dimension of the volume. The optimal spacing was determined based on the method of intersecting the asymptotes (Bejan, 1995b; Stanescu et al., 1996). The asymptotes were derived from the large volume of empirical data accumulated in the literature for single cylinders in cross-flow (the large-S limit) and for arrays with many rows of cylinders (the small-S limit). In the range $10^4 \leq \tilde{P} \leq 10^8$, $25 \leq H/D \leq 200$, and $0.72 \leq \mathrm{Pr} \leq 50$, the optimal spacing is correlated to within 5.6% by

$$\frac{S_{\mathrm{opt}}}{D} \cong 1.59\frac{(H/D)^{0.52}}{\tilde{P}^{0.13}\mathrm{Pr}^{0.24}}, \tag{5.28}$$

where $\tilde{P} = \Delta P D^2/(\mu\nu)$. When the free-stream velocity U_∞ is specified (instead of ΔP), we may transform relation (5.28) by noting that, approximately, $\Delta P \sim (1/2)\rho U_\infty^2$:

$$\frac{S_{\mathrm{opt}}}{D} \cong 1.7 \frac{(H/D)^{0.52}}{\mathrm{Re}_D^{0.26} \, \mathrm{Pr}^{0.24}}. \tag{5.29}$$

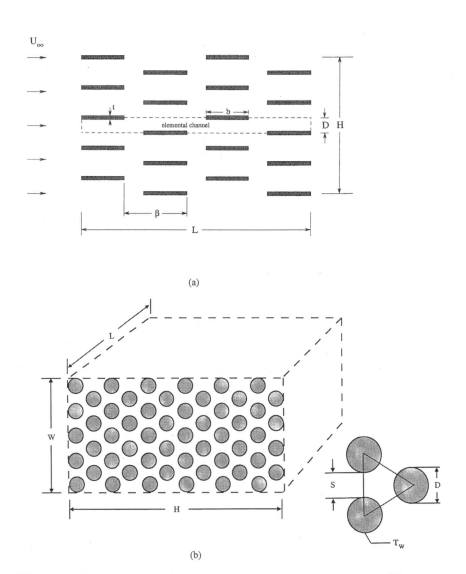

(a)

(b)

Fig. 5.10. Forced convection cooling of a volume heated uniformly: (a) array of staggered plates, (b) array of horizontal cylinders, and (c) square pins with impinging flow.

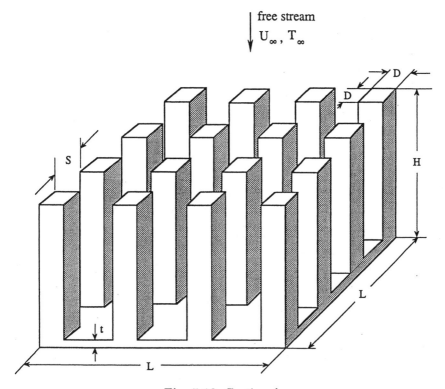

Fig. 5.10. Continued.

This correlation is valid in the range $140 < \mathrm{Re}_D < 14{,}000$, where $\mathrm{Re}_D = U_\infty D/\nu$.

Optimal spacings emerge also when the flow is three-dimensional, as in a dense array of pin fins with impinging flow (Figure 5.10c). The flow is initially aligned with the fins, and later makes a $90°$ turn to sweep along the base plate and across the fins. The optimal spacings are correlated to within 16% by (Ledezma *et al.*, 1996):

$$\frac{S_{\mathrm{opt}}}{L} \cong 0.81 \, \mathrm{Pr}^{-0.25} \mathrm{Re}_L^{-0.32}, \tag{5.30}$$

which has been tested in the range $0.6 < D/L < 0.14$, $0.28 < H/L < 0.56$, $0.72 < \mathrm{Pr} < 7$, $10 < \mathrm{Re}_D < 700$, and $90 < \mathrm{Re}_L < 6000$. Note that the spacing S_{opt} is controlled by the linear dimension of the volume L. Spacings for pin fin arrays in natural convection were optimized by Fisher and Torrance (1998). Additional studies on the optimization of internal spacings for forced convection are reviewed in Kim and Lee (1996).

5.4 Pulsating Flow

Optimal internal spacings are also the answer to maximizing the volumetric rate of heat transfer between a fixed volume (AL, Figure 5.11) and a fluid that flows in pulses through the volume (Rocha and Bejan, 2001). The channels are parallel tubes of radius r_0 and length L. Two periodic flow regimes were

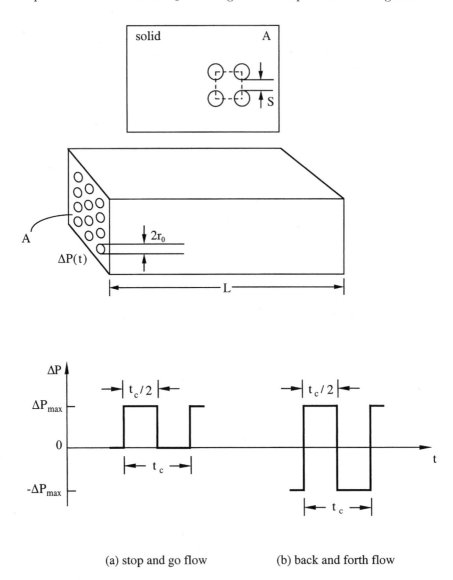

(a) stop and go flow   (b) back and forth flow

Fig. 5.11. Bundle of tubes in a fixed volume connecting two reservoirs of fluid at the same temperature, and two pressure-difference cycles and flows: (a) stop and go, and (b) back and forth (Rocha and Bejan, 2001).

simulated numerically: (a) stop and go flow, and (b) back and forth flow. The flow is driven by the pressure difference $\Delta P(t)$ shown in Figure 5.11. The fluid is Newtonian with $\text{Pr} \gg 1$. The tube wall temperature is assumed uniform (T_w), and the fluid reservoir temperature is T_0 at both ends of each tube.

To maximize the heat transfer performance of the volume as a whole, means to maximize the volumetric heat transfer density associated with one tube and the immediately surrounding solid. If $\bar{q}_1$ is the cycle-averaged heat transfer rate between one tube surface and the fluid that flows through the tube, the volumetric heat transfer rate density is $\bar{q}_1/(\pi r_0^2 L)$, or in dimensionless form,

$$Q = \frac{\bar{q}_1}{\pi r_0^2 L} \frac{\alpha t_c}{2k(T_w - T_0)}. \tag{5.31}$$

In this definition, α, k, and t_c are the fluid thermal diffusivity, fluid thermal conductivity, and pulse period (Figure 5.11).

Another way to see the importance of maximizing Q is to imagine that a large number of tubes (r_0, L) are machined in parallel through the volume of length L and frontal area A. The volume AL is fixed, but the number of tubes (n) and the tube size (r_0) are not. For simplicity, assume that the tube centers form squares, such that $n = A/(2r_0 + S)^2$. Assume further that the spacing between adjacent tubes (S) is a fraction (σ) of the tube diameter $S = \sigma 2r_0$. The porosity of the volume is $\phi = (\pi/4)/(1 + \sigma)^2$. The porosity is fixed when σ is fixed. We are interested in the total heat transfer rate between the fluid that flows through the n tubes and the volume AL. The contribution is $\bar{q}_1$ from one tube, and $n\bar{q}_1$ from all the tubes. The total heat transfer rate per unit volume used is $n\bar{q}_1/(AL)$, or in dimensionless form,

$$\frac{n\bar{q}_1}{AL} \frac{\alpha t_c}{k(T_w - T_0)} = \frac{\pi Q}{2(1 + \sigma)^2}. \tag{5.32}$$

In summary, to maximize the value of Q, which is based on a single tube, is equivalent to maximizing the total heat transfer rate divided by the total volume of the bundle. In other words, the geometric maximization of Q teaches us how to select the dimensions of the tube bundle so that the entire volume AL is reached most effectively by the fluid when the pressure cycle is specified $(t_c, \Delta P_{\max})$.

The volumetric heat transfer density Q was maximized numerically with respect to the tube size $\tilde{r}_0$ by holding $\tilde{L}$ and B fixed, where

$$(\tilde{r}_0, \tilde{L}) = \frac{r_0, L}{(\alpha t_c)^{1/2}}, \qquad B = \frac{\Delta P_{\max} t_c}{\mu}. \tag{5.33}$$

The porosity of the system was set at $\phi = 0.65(\sigma = 0.1)$, because it does not affect the results of geometric optimization. The numerical optimization covered the domain $1 < \tilde{L} < 10^2$ and $10^2 \le B \le 10^6$. At first sight, it is a nice coincidence that the numerical results obtained for $\tilde{r}_{0,\text{opt}}(\tilde{L}, B)$ and $Q_{\max}$ are correlated very tightly when plotted against the groups $\tilde{L}^{1/2}B^{-1/4}$ and

$\tilde{L}^{-1}B^{1/2}$, as shown in Figure 5.12. This tight correlation carries an important message. The optimal tube radius is correlated by

$$\tilde{r}_{0,\text{opt}} = C_r \tilde{L}^{1/2} B^{-1/4}, \tag{5.34}$$

where $C_r = 3.08$ for stop and go flow, and $C_r = 2.70$ for back and forth flow. The standard deviations are respectively, 14 and 21%. The numerical data for the maximum heat transfer rate density are correlated by

$$Q_{\text{max}} = C_Q \tilde{L}^{-1} B^{1/2}, \tag{5.35}$$

where $C_Q = 0.13 \pm 0.21$ for stop and go flow, and $C_Q = 0.26 \pm 0.41$ for back and forth flow. From Equations (5.34) and (5.35) we see that $\tilde{r}_{0,\text{opt}}^2 Q_{\text{max}} =$ constant, which means that at the optimum the thermal conductance per unit length $[\bar{q}_1/L(T_w - T_0)]$ is a constant.

The surprising message hidden in correlations (5.34) and (5.35) is that the time scale t_c does not play a role. The optimized geometry is independent of t_c. Note that Equation (5.34) is the same as

$$\frac{r_{0,\text{opt}}}{L} = C_r \text{Be}^{-1/4}, \tag{5.36}$$

where $\text{Be} = \Delta P_{\text{max}} L^2 / (\mu \alpha)$ [cf. Eq. (5.24)]. The agreement between Equations (5.36) and (5.23) is qualitative as well as quantitative. Note that the optimal tube diameter recommended by Equation (5.36) is

$$\frac{D_{0,\text{opt}}}{L} = 2C_r \text{Be}^{-1/4}, \tag{5.37}$$

where $2C_r$ is either 6.16 or 5.40. If in Equation (5.23) we use the hydraulic diameter $(2D_{\text{opt}})$ in place of D_{opt}, then on the right-hand side of the equation the factor 2.7 is replaced by 5.4. Furthermore, the difference between the factors 6.16 and 5.40 in Equation (5.37) can be attributed to the definition of Be. In back and forth flow, the overall pressure difference is $2\Delta P_{\text{max}}$, that is, twice as much as in stop and go flow. If Be is based on the true overall pressure difference in both flow regimes, then the factor 6.16 must be divided by $2^{1/2}$, and the resulting factor of 4.36 is comparable with 5.40.

Similar agreement is shown by the maximized heat transfer density, Equation (5.35), which can be written as

$$q''' = 2C_Q \frac{k}{L^2} (T_w - T_0) \text{Be}^{1/2}, \tag{5.38}$$

where q''' is the volumetric heat transfer rate, and $2C_Q = 0.26$ for stop and go flow, and $2C_Q = 0.52$ for back and forth flow. Equation (5.38) agrees well with Equation (5.26). The difference between the $2C_Q$ coefficients, 0.26 versus 0.52, can be attributed to the Be definition. If Be for back and forth flow is based on $2\Delta P_{\text{max}}$, which is the true overall pressure difference, then 0.52 is replaced by $0.52/2^{1/2} = 0.36$, which agrees better with the Q_{max} correlation for stop and go flow and the steady-state result (5.26).

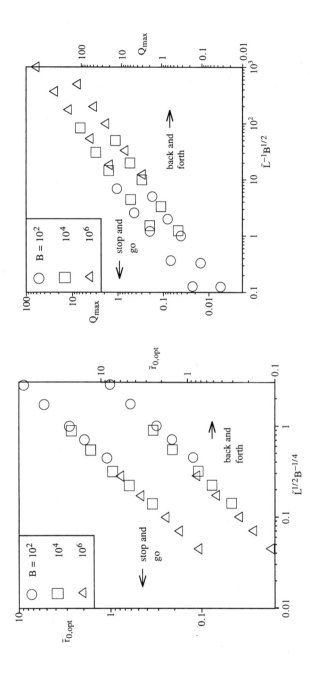

Fig. 5.12. Correlation of the numerical data for the optimal tube size and maximum heat transfer per flow cycle and unit volume in the porous system of Figure 5.11 (Rocha and Bejan, 2001).

5.5 Optimal Packing of Fibrous Insulation

Another class of porous media that owe their structure to the maximization of performance under constraints is the hair (fur) coats of all mammals. In this class, performance means thermal insulation. These complicated and diverse porous structures exhibit some common features. One is the proportionality between the hair strand diameter and the animal body length scale raised to the power 1/2 (Figure 5.13). This allometric law was predicted by minimizing the heat loss from the skin to the ambient in forced convection (Bejan, 1990a) as well as in natural convection (Bejan, 1990b). This theoretical development is reviewed in Nield and Bejan (1999).

The second common feature of fur coats is that the porosity is high and nearly constant ($\phi \sim 0.95$ to 0.99), regardless of animal size (Table 1.1). In this section we show that this feature can also be attributed to design, that is, the maximization of the thermal insulation effect provided by fur as a porous layer containing fibers and air. Consider the unidirectional heat transfer model shown on the right side of Figure 5.14. Fibers are represented by solid plates of thickness D, which are parallel to the skin and are separated by the distance S. We model all the surfaces as diffuse gray with the constant emissivity ε. The overall thickness of the layer is L, and the overall temperature difference is $T_0 - T_L$. If n is the number of plates that fill the layer, then

$$S + D = \frac{L}{n}; \qquad n = \frac{T_0 - T_L}{T_i - T_{i+1}}. \qquad (5.39)$$

Assume that the L-thick layer is the dominant resistance to heat transfer between the skin (T_0) and the ambient (T_L), such that T_L is fixed and

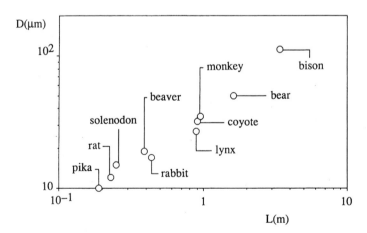

Fig. 5.13. The proportionality between hair strand diameter and animal length scale (Bejan and Lage, 1991).

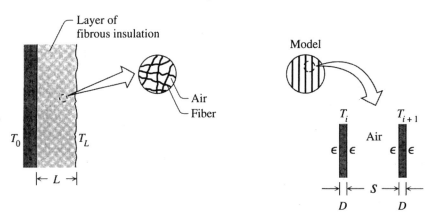

Fig. 5.14. Radiation and conduction model for heat transfer through a layer of fibrous insulation (Bejan, 1993; from Bejan, *Heat Transfer*, Copyright © 1993 Wiley. This material is used by permission of John Wiley & Sons, Inc.).

represents the temperature of the outer plate of the model. The skin surface covered by this insulation is A. The total heat current (q) through the insulating layer is the sum of two currents in parallel,

$$q = q_c + q_r, \qquad (5.40)$$

where q_c accounts for conduction through the air gap of conductivity k,

$$q_c = k\frac{A}{S}(T_i - T_{i+1}) \qquad (5.41)$$

and q_r is the net radiative heat transfer rate from T_i to T_{i+1},

$$q_r = \frac{A\sigma 4T_i^3}{(2/\varepsilon) - 1}(T_i - T_{i+1}). \qquad (5.42)$$

Equation (5.42) is based on the simplifying assumption that $(T_i - T_{i+1})/T_i \ll 1$. Boltzmann's constant is $\sigma = 5.6693 \times 10^{-12}\,W\,cm^{-2}\,K^{-4}$. By eliminating the fiber-to-fiber temperature difference $(T_i - T_{i+1})$ based on Equation (5.39), and using the remaining Equations (5.39) to (5.42), we find that the total heat transfer rate through the porous structure assumes the form

$$\frac{q}{kA(T_0 - T_L)/L} = \left(\frac{D}{S} + B\right)\left(\frac{S}{D} + 1\right). \qquad (5.43)$$

In this expression B is a dimensionless group that accounts for properties and the solid thickness D,

$$B = \frac{\sigma 4T_i^3 D}{k\left(\dfrac{2}{\varepsilon} - 1\right)}, \qquad (5.44)$$

where T_i^3 represents the order of magnitude of T_0^3, or T_L^3.

Equation (5.43) represents the global thermal conductance of the porous structure, in which the air spacing S, or the porosity $\phi = S/(S + D)$, is a degree of freedom in the design of the structure. The conductance is large in the two extremes, small S/D and large S/D. It is minimal when the air spacing has the optimal value

$$S_{\text{opt}} = \frac{D}{B^{1/2}}, \tag{5.45}$$

$$\frac{q_{\text{min}}}{kA(T_0 - T_L)/L} = (1 + B^{1/2})^2, \tag{5.46}$$

where the porosity has the value

$$\phi_{\text{opt}} = \frac{1}{1 + B^{1/2}}. \tag{5.47}$$

The theoretical development reported in this section is the demonstration that there exists an optimal porosity. It happens that when the number B is of order 0.1 or smaller, ϕ_{opt} approaches 1 and is insensitive to the thickness of the solid elements (D). This explains why in the hair coats of all mammals the porosity is nearly constant and close to 1 for all D sizes and, in view of Figure 5.13, for all animal sizes. Indeed, if in Equation (5.44) we use $T_i \approx 300 \, \text{K}$, $\varepsilon \approx 1$ and the conductivity of air at room temperature, $k = 0.025 \, \text{W}/(\text{m K})$, Equation (5.47) yields $\phi_{\text{opt}} = 0.95$ for $D = 0.01 \, \text{mm}$, and $\phi_{\text{opt}} = 0.86$ for $D = 0.1 \, \text{mm}$.

5.6 Optimal Maldistribution: Tree-Shaped Flows

Fluid flow has a mind of its own. It constantly seeks paths of less resistance. It finds such paths, better and better ones, if the flow medium is sufficiently malleable. According to constructal theory, every flow or movement of matter (fluid, solid, goods, people) exhibits economies of scale. The resistance or cost encountered per unit of the mass that moves is much lower when units flow together, as opposed to when they flow individually. In fluid flow, the resistance per fluid packet is much lower when fluid packets flow together through a wide duct—much lower than when each packet flows through its own narrow duct, or through a sequence of connected pores, or through a crowd of slower fluid packets. The wide duct is the highway, and the diffusion of individual fluid packets is the crowd that fills the street.

Both flow regimes are necessary, as the street traffic feeds the highway traffic, and vice versa. Channeled flow coexists with diffusion in every fluid flow class in nature, for example, river basins and deltas, lungs, vascularized tissues, turbulent flow, and Bénard convection (Bejan, 2000). They are arranged optimally, such that each flow channel is surrounded by an optimally sized territory visited by diffusion. The flow pattern is the spatial allocation of diffusion around channels, and channels around diffusion. It is the result of the clash between objective (minimal global resistance) and constraints (space). This method of deducing flow patterns is constructal design. We pursue it further in Section 5.7.

In the present section we use a Darcy porous medium model (Errera and Bejan, 1998) to show how a seepage flow that is originally distributed uniformly over its territory evolves into a tree-shaped flow that has a much lower overall resistance. In the two-dimensional model of Figure 5.15, the surface

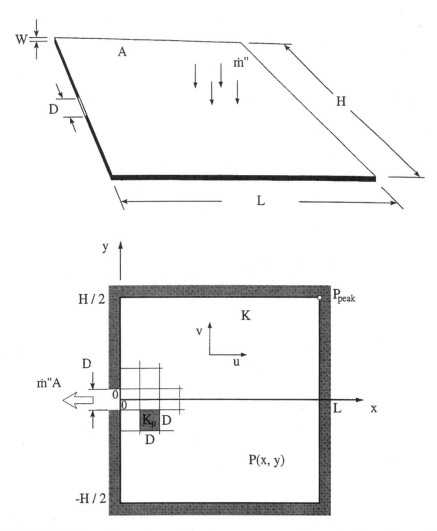

Fig. 5.15. Model of area-to-point flow in a porous medium with Darcy flow and blocks that can be dislodged and swept downstream (Errera and Bejan, 1998).

area $A = HL$ and its shape H/L are fixed. The area is coated with a homogeneous porous layer of permeability K. The small thickness of the K layer, that is, the dimension perpendicular to the plane $H \times L$, is W, where $W \ll (H, L)$.

An incompressible Newtonian fluid is pumped through one of the A faces of the $A \times W$ parallelepiped, such that the mass flow rate per unit area is uniform, $\dot{m}''$ (kg/m^{2}s). The other A face and most of the perimeter of the $H \times L$ rectangle are impermeable. The collected stream $(\dot{m}''\,A)$ escapes through a small port of size $D \times W$ placed over the origin of the (x, y) system. The fluid is driven to that outlet by the pressure field $P(x, y)$ that develops over A. The pressure field accounts for the effect of slope and gravity in a real drainage basin, and the uniform flow rate $\dot{m}''$ accounts for the rainfall.

Another interpretation of this flow configuration comes from the field of heat exchangers. Every stream that leaves a box through one port (e.g., $\dot{m}''\,A$) does so after visiting (bathing, touching, using) every infinitesimal volume of the box (e.g., $\dot{m}''$). The same is true about the earlier flow of the same stream *into* the box, from one inlet $(\dot{m}''\,A)$ to every infinitesimal volume $(\dot{m}'')$. In sum, the flow of one stream through the box is the series arrangement of two flows: from one inlet to an infinity of points spread over A, and from the entire A to the single outlet. Only the latter is shown in Figure 5.15. The two flows share the same A. They are mated on A. After each flow develops its tree structure, the two flows display a structure where the canopy of one tree is mated to the canopy of the other tree at every point inside the box. This is illustrated in three dimensions by the heat exchanger structure developed in Section 5.7.

To determine the flow distribution in Figure 5.15, assume that the flow through the K medium is in the Darcy regime,

$$u = -\frac{K}{\mu}\frac{\partial P}{\partial x}, \qquad v = -\frac{K}{\mu}\frac{\partial P}{\partial y}, \tag{5.48}$$

such that the conservation of mass in every infinitesimal volume element $\Delta x \Delta y W$ requires

$$\frac{\partial u}{\partial x} + \frac{\partial v}{\partial y} - \frac{\dot{m}''}{\rho W} = 0, \tag{5.49}$$

where ρ is the density of the fluid. Combining Equations (5.48) and (5.49), we obtain a Poisson equation for the pressure field,

$$\frac{\partial^2 P}{\partial x^2} + \frac{\partial^2 P}{\partial y^2} + \frac{\dot{m}''\nu}{WK} = 0, \tag{5.50}$$

where $\nu = \mu/\rho$ is the kinematic viscosity. The boundary conditions for solving Equation (5.50) are impermeable boundaries $(u = 0$ or $v = 0)$ all around the domain, except over the sink located at $x = 0$ and $-D/2 < y < D/2$, where $P = 0$.

The resistance encountered by this area-to-point flow is the object of geometric minimization. It is the ratio between the maximal pressure difference (P_{peak}) and the total flow rate ($\dot{m}''\, A$). The location of the point of maximal pressure is not the issue, although in the case of Figure 5.15 its identity is clear. It is important to calculate P_{peak} and to reduce it at every possible turn by making appropriate changes in the internal structure of the $A \times W$ system. Determinism results from invoking a single principle and using it consistently. This is constructal design.

Changes are possible because finite-size portions (blocks) of the system can be dislodged and ejected through the sink. Let us assume that the removable blocks are of the same size and shape (square, $D \times D \times W$). The critical force (in the plane of A) that is needed to dislodge one block is τD^2, where τ is the yield shear stress averaged over the base area D^2. The yield stress and the length scale D are assumed known. They provide a useful estimate for the order of magnitude of the pressure difference that can be sustained by the block. At the moment when one block is dislodged, the critical force τD^2 is balanced by the net force induced by the local pressure difference across the block ΔP, namely, ΔPDW. The balance $\tau D^2 \sim \Delta PDW$ suggests the pressure-difference scale $\Delta P \sim \tau D/W$, which along with D can be used for the purpose of nondimensionalizing Equation (5.50):

$$\frac{\partial^2 \tilde{P}}{\partial \tilde{x}^2} + \frac{\partial^2 \tilde{P}}{\partial \tilde{y}^2} + M = 0, \tag{5.51}$$

$$(\tilde{x}, \tilde{y}) = \frac{(x, y)}{D}, \qquad \tilde{P} = \frac{P}{\tau D/W}, \tag{5.52}$$

$$M = \dot{m}''' \frac{\nu D}{\tau K}. \tag{5.53}$$

We search for the blocks that are dislodged by reasoning as follows. Let s be the direction of the resultant of all the pressure forces that act on the block perimeter $D \times D$. The block does not break away as long as $(\overline{\partial P/\partial s})W < \tau$, which in dimensionless terms is

$$\left(\overline{\frac{\partial \tilde{P}}{\partial \tilde{s}}} \right) < 1. \tag{5.54}$$

The pressure gradient $(\overline{\partial \tilde{P}/\partial \tilde{s}})$ is averaged over the square base area of one block. When this condition is violated, the block is removed and its place is taken by a channel that is considerably more permeable—more conductive for fluid flow—than the original medium (K). The simplest way to implement this change is to assume that the space vacated by the block is also a porous medium with Darcy flow, except that the new permeability (K_p) of this medium is sensibly greater, $K_p > K$. This happens to be the correct assumption when the flow is slow enough (and W is small enough) so that the flow regime in the vacated space is Hagen–Poiseuille between parallel plates. The equivalent K_p value for such a flow is $W^2/12$. The main reason for introducing the K_p assumption in this model is to simplify the calculation of the pressure distribution over the area occupied by the dislodged block. Over that area the pressure is governed by an equation that is the same as Equation (5.51), except that M is replaced with MK/K_p. The pressure varies continuously between the original material K and the vacated domain K_p. When we account for mass continuity across the interface between the K and the K_p domains, the ratio K/K_p emerges as a dimensionless parameter of the system.

The pressure $\tilde{P}$ and the block-averaged pressure gradient [condition (5.54)] increase in proportion with the imposed mass flow rate (M), because Equation (5.54) is linear. When M exceeds a critical value M_c, condition (5.54) is violated and the first block is dislodged. For example, $M_c = 0.00088932$. This value does not depend on K/K_p because in the beginning the entire system is occupied by K material. The first block that breaks away is the one that has the outlet port as one of its four sides. The peak pressure is located in the farthest corners of the A domain, and experiences a drop when the first block is removed at constant flow rate ($M = M_c$). When $K/K_p = 0.1$, this drop occurs from $\tilde{P}_{\text{peak}} = 3.631$ to 2.934. This is in fact the purpose of the change in the internal structure of the area-to-point flow system, that is, the physics principle that we invoke: the resistance to fluid flow is decreased through geometric changes in the internal architecture of the system. This is the constructal law (Bejan, 1996c, 1997).

To generate higher pressure gradients that may lead to the removal of a second block, we must increase the flow-rate parameter M above the first M_c. The removable block is one of the blocks that borders the newly created K_p domain. The peak pressure rises as M increases, and then drops partially as the second block is removed. This process can be repeated in steps marked by the removal of each additional block. In each step, we restart the process by increasing M from zero to the new critical value M_c. During this sequence the peak pressure decreases, and the overall area-to-point flow resistance ($\tilde{P}_{\text{peak}}/M_c$) decreases monotonically.

The key result is that the removal of certain blocks of K material and their replacement with K_p material generate macroscopic internal *structure*. The generalizing mechanism is the minimization of flow resistance, and the resulting structure is deterministic. Every time we repeat this process we obtain exactly the same sequence of images.

For illustration, consider the case $K/K_p = 0.1$, shown in Figure 5.15. The number n represents the number of blocks that have been removed. The domain A is square and contains a total of 2601 building blocks of base size $D \times D$; in other words, $H = L = 51D$. The pressure field equations were solved with the finite-element method. In most cases the variation of $\tilde{P}$ over the edges of the blocks located near the interfaces between the two porous domains was quite smooth, almost linear. Details of the numerical work are given in Errera and Bejan (1998).

The curves plotted in Figure 5.16 are unique and reproducible. They show the evolution of the critical flow rate and peak pressure. The curves appear ragged because of an interesting feature of the erosion model: every time that a new block is removed, the pressure gradients redistribute themselves and blocks that used to be "safe" are now ready to be dislodged even without an

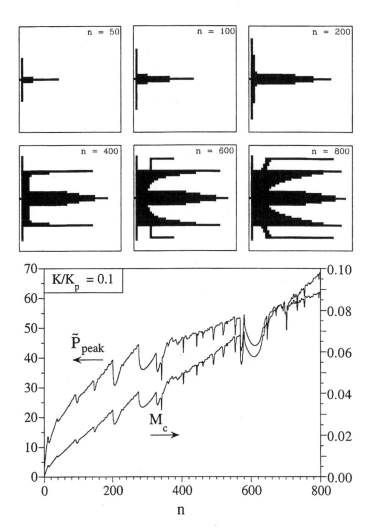

Fig. 5.16. The evolution of the structure of the system of Figure 5.15 when $K/K_p = 0.1$: the growth of the K_p domain and the variations in the critical flow rate and peak pressure (Errera and Bejan, 1998).

increase in M. The fact that the plotted M_c values drop from time to time is due to restarting the search for M_c from $M = 0$ at each step n.

The shape of the high-permeability domain K_p that expands into the low-permeability material K is that of a tree. New branches grow in order to channel the flow collected by the low-permeability K portions. The growth of the first branches is stunted by the fixed boundaries (size, shape) of the A domain. The older branches become thicker; however, their early shape (slenderness) is similar to the shape of the new branches.

The slenderness of the K_p channels and the interstitial K regions is dictated by the K/K_p ratio, that is, by the degree of dissimilarity between the two flow paths. Highly dissimilar flow regimes $(K/K_p \ll 1)$ lead to slender channels (and slender K interstices) when the overall area-to-point resistance is minimized.

The availability of two dissimilar flow regimes $(K_p \neq K)$ is a necessary precondition for the formation of deterministic structures through flow-resistance minimization. The "glove" is the high-resistance regime (K), and the "hand" is the low-resistance tree (K_p): these two flow modes work together toward minimizing the overall resistance.

The optimized maldistribution of point-area and point-volume flows, and the discovery of dendritic flow patterns based on this principle, were reported most recently by Ordonez *et al.* (2003), Borrvall and Petersson (2002), and Borrvall *et al.* (2002). The effect of root dendrites on plant transpiration was documented by Lai and Katul (2000).

5.7 Dendritic Heat Exchangers

In this section we rely on the main ideas of this chapter to develop the constructal architecture that gives a heat exchanger the ability to pack maximum heat transfer rate in a fixed volume (Bejan, 2002). For simplicity consider the heat exchanger between two streams with the same capacity rate (Figure 5.17). Each stream carries a single-phase fluid (e.g., gas or liquid) with constant properties and a Prandtl number of order 1 or greater than 1. More general (unbalanced) heat exchangers for single-phase and/or two-phase flow can be configured by using the method described here (see the end of this section).

5.7.1 Elemental Volume

In the simplest description, every flow passage can be modeled as a two-dimensional space formed between two parallel plates. Two such passages are shown in counterflow in Figure 5.17a, and in crossflow in Figure 5.17b. A sandwich of only two passages is called *elemental volume*, and is labeled (0). The elemental volume is the geometric feature with the smallest scale in the considerably larger and more complex structure that is the heat exchanger. The envisioned structure contains a large number of elemental volumes, which

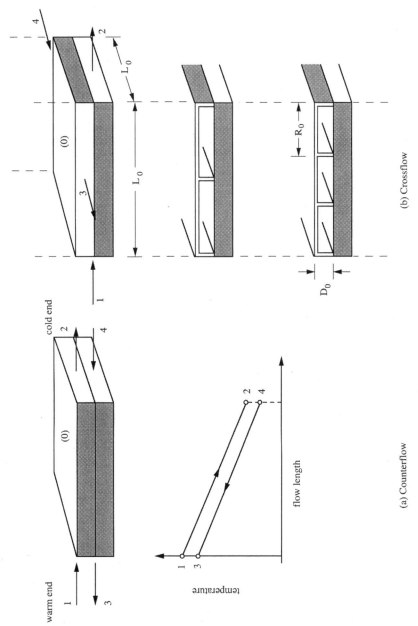

(a) Counterflow

(b) Crossflow

Fig. 5.17. Elemental volumes containing two flow passages: (a) counterflow; (b) crossflow (Bejan, 2002).

are arranged in vertical stacks indicated by dashed lines in Figure 5.17b. The challenge is to find the flow architecture, that is, the best way to connect and assemble the elemental volumes into a much larger device that receives and discharges only two streams, one hot and the other cold. The challenge is to bathe the entire volume with the two streams and, later, to reconstitute the streams before they are discharged.

The proposed architecture has several new features, and several old ones. An old feature is the crossflow arrangement chosen for each elemental volume (Figure 5.17b). It is known that the best arrangement from a thermodynamic optimization point of view is the counterflow. In the counterflow configuration, the entropy generation rate for a fixed heat transfer area is the smallest, and vanishes when the area becomes sufficiently large [e.g., Bejan (1997, p. 611)]. The need to stack many elemental volumes together makes the use of counterflows at the smallest scales difficult, if not impossible. This is especially true when the passage dimension (the spacing between plates) is small. It is difficult to distribute many streams of type 1 to one side of the assembly, while having to collect many streams of type 3 from the same side.

This difficulty is eliminated by the crossflow arrangement, where each of the four sides of the assembly of elemental volumes is devoted to distributing or collecting a single stream that flows through many elemental slits in parallel. Counterflows are not abandoned, however. They will be used in all the subsequent (larger) scales of construction (the first construct, the second construct, ...), where the use of counterflows is not as difficult as at the elemental level.

Another old feature is the solid material into which the flow structure is machined. The material must be one with high thermal conductivity, such as, aluminum or silicon. The outer surfaces of the entire heat exchanger are insulated with respect to the ambient.

One new feature is that most of the heat transfer between the two streams occurs in the elemental volumes, across small separating surfaces of size L^2, as shown in Figure 5.17. The two streams will come in contact by counterflow at larger scales, in the larger ducts that distribute the streams to (and later collect the streams from) the elemental volumes. The following labeling convention is used for the ports of a volume, regardless of the size of the volume:

1. The inlet of the warm stream,
2. The outlet of the warm stream,
3. The outlet of the cold stream, and
4. The inlet of the cold stream.

Taken together, ports 1 and 3 constitute the "warm side" of the volume. Ports 2 and 4 constitute the "cold side."

The key feature is that the dimensions of every flow passage are such that the flow is laminar, and the flow length matches the thermal entrance length $L_0 \cong X_T$. The reason for selecting this feature is explained in Section 5.3 and

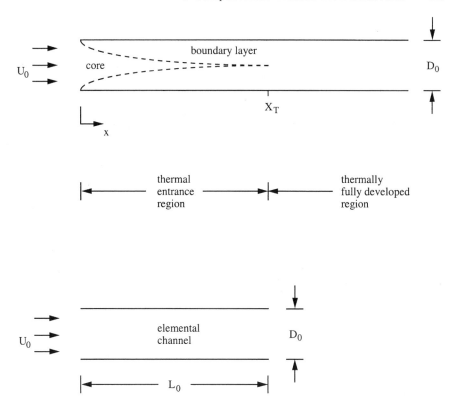

Fig. 5.18. The flow length of the elemental channel (L_0) must match the thermal entrance length of the laminar flow (X_T) (Bejan, 2002).

Figure 5.18. Let D_0 and U_0 be the plate-to-plate spacing and mean velocity of one elemental passage. The entrance is the region where thermal boundary layers are present. The thermal entrance length X_T is the approximate longitudinal position where the thermal boundary layers have just merged. Downstream of X_T the temperature distribution across the channel has a fully developed profile. Said another way, the stream must travel a certain distance (X_T) before it is penetrated fully by the diffusion of heat from or to the wall. Although the value of X_T is not a precise number, its order of magnitude is certain [e.g., Bejan (1995a, pp. 122–132)],

$$\left(\frac{X_T/D_h}{\mathrm{Re}_{D_h}\,\mathrm{Pr}}\right)^{1/2} \cong 0.2, \qquad (5.55)$$

where $\mathrm{Re}_{D_h} = U_0 D_h/\nu$, with ν as the kinematic viscosity of the fluid and D_h as the hydraulic diameter, $D_h = 2D_0$. Selecting the elemental flow length such that $L_0 \cong X_T$, we find from Equation (5.55) that

$$\frac{L_0}{D_0} \cong 0.16 \frac{U_0 D_0}{\nu} \Pr > 1. \tag{5.56}$$

If D_0 is the smallest dimension that can be manufactured, then L_0 is dictated by Equation (5.56) when U_0 is known. The elemental velocity U_0 is known as soon as the volume and macroscopic streams of the entire heat exchanges are specified. Let V and $\dot{m}_h$ be the total volume and the mass flow rate of the hot streams that must be distributed throughout V. According to the constructal method (Bejan, 2000), the volume V is filled by interconnected elemental volumes. The order of magnitude of the cross-sectional area of V is $V^{2/3}$. The stream $\dot{m}_h$ perfuses with the mean velocity U_0 through the $V^{2/3}$ area, therefore, from mass conservation, $U_0 \approx \dot{m}_h/(\rho V^{2/3})$.

Here is the reason for choosing an elemental length that matches the thermal entrance length. The key to packing maximum heat transfer per unit volume is the observation that every infinitesimal packet of fluid must be used for the purpose of transferring heat. Fluid flow regions that do not work must be avoided. Flow regions that have worked too much, and have become ineffective, must be eliminated.

With reference to the top drawing of Figure 5.18, the designer has two choices, and both are not the best. If L_0 is made shorter than X_T, then the fluid that occupies the core of the duct does not participate in the heat transfer enterprise. Such fluid must not be allowed to leave the elemental channel without having interacted thermally with the walls. In the other extreme, when L_0 is made longer than X_T, all the fluid has interacted with the walls. This fluid is so saturated with heating or cooling from the wall that it can accommodate further heating or cooling only by overheating, that is, by changing its bulk temperature in the downstream direction.

This extreme (the fully developed regime) must be eliminated. It is important to note that the decision to avoid the thermally fully developed flow regime contradicts current practice in microscale heat exchangers, where laminar fully developed flow is seen as the key to maximizing the heat transfer density. We return to this important difference in Section 5.8.

The best choice is in between, $L_0 \approx X_T$, because in this configuration all the fluid of the channel cross-section "works" in a heat transfer sense. The fluid leaves the channel as soon as it completes its mission. In this configuration the elemental volume is used to the maximum for the purpose of transferring heat between the stream and the walls.

The elemental channel may have a variety of cross-sectional shapes that are represented by the spacing D_0. Several designs are sketched in Figure 5.17b. The elemental duct of cross-section $D_0 \times L_0$ could be a bundle of parallel channels machined or etched into high-conductivity wall material. The ribs between the machined channels serve as fins, and provide mechanical strength for the elemental assembly. The ribs are not new features. New is the idea that if the spacing between ribs is R_0, and if D_0 is smaller than or equal to R_0, then the elemental duct $D_0 \times L_0$ must be designed in accordance with Equation (5.56). The important dimension of the smallest cross-section ($D_0 \times R_0$) is the smaller of the two dimensions, namely, D_0. The elemental volume geometry must be optimized by selecting L_0, D_0, and U_0, such that Equation (5.56) is respected as closely as possible. This holds for both sides of the heat transfer surface, and, consequently, the elemental surface must be a square of side L_0.

5.7.2 First Construct

A large number (n_1) of elemental volumes can be assembled into a larger system in the manner shown in Figure 5.19. This larger system is called a *first*

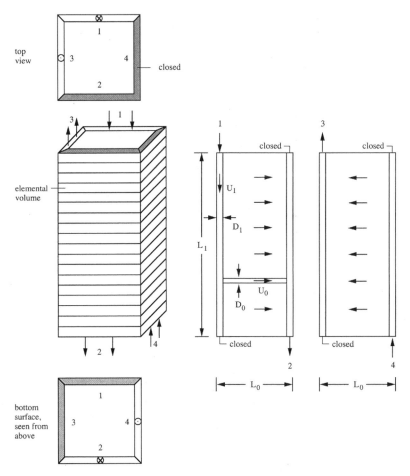

Fig. 5.19. First construct containing a large number of elemental volumes (Figure 5.17) stacked in the D_0 direction (Bejan, 2002).

construct (Bejan, 2000). It is a parallelepiped of size $L_1 \times L_0 \times L_0$, where, if the solid elemental walls are thin in comparison with D_0, then $L_1 = 2n_1 D_0$. This assembly has a fluid jacket of height L_1 and spacing D_1. The jacket is divided into four vertical parallel-plate channels of cross-section $D_1 \times L_0$, one channel for each side of the $L_0 \times L_0$ square base. Each $D_1 \times L_0$ channel is closed at one end. The cross-section of the $D_1 \times L_0$ channel may have ribs for mechanical stability and additional heat transfer (e.g., Figure 5.17b), but such details are not essential as long as D_1 is the smaller dimension of the $D_1 \times L_0$ cross-section.

The streams enter and exit the first construct through slits: two slits at the warm end of the construct and two at the cold end. These ports are paired together such that one end of the stack becomes the "warm side," and the other end becomes the "cold side." Seen from the outside, the first construct resembles the counterflow arrangement shown in Figure 5.17a. This is an important step in the construction. The first construct is the first module (the smallest scale) where the crossflows of the elemental volumes are organized in such a way that they give birth to counterflows of a larger scale. Stream 1 enters vertically through a D_1-wide slit, flows horizontally through the stack of n_1 elemental channels of spacing D_0, and continues vertically through another D_1 channel to exit 2. The cold stream follows a similar path, as shown on the right side of Figure 5.19.

In this simplest description of the first construct, the jacket spacing D_1 and the elemental spacing D_0 were assumed constant. In such a design the horizontal flow through the elements that form the core of the first construct would be distributed nonuniformly. Specifically, the velocity U_0 through the top and bottom regions of the stack would be larger than through the mid-height regions. This flow can be made uniform by using (i) tapered D_1 channels that reach $D_1 = 0$ at their closed ends, and (ii) smaller D_0 spacings in the top and bottom portions of the stack (Bejan, 2002). Past experience with the optimal tapering of channels and fins shows that, if implemented, such structural details can improve the overall performance by changes on the order of 10% [e.g., Bejan (2000, pp. 62–65)]. These improvements may not be warranted in view of the difficulties associated with manufacturing first constructs with smoothly varying D_1 and D_0.

5.7.3 Second Construct

The assembly of a number (n_2) of first constructs is called a *second construct*. An example with $n_2 = 5 \times 5$ is illustrated in Figure 5.20. The key to this arrangement is the counterflows formed between streams 1 and 3, and 2 and 4. In other words, when two of the first constructs shown in Figure 5.19 are fused, their lateral surfaces of size $L_1 \times L_0$ are fused in such a way that the surface with outlet 3 is fused with the surface with inlet 1. Similarly, the surface with the cold inlet 4 is fused with the number 2 surface of another first construct. This construction is illustrated in Figure 5.21.

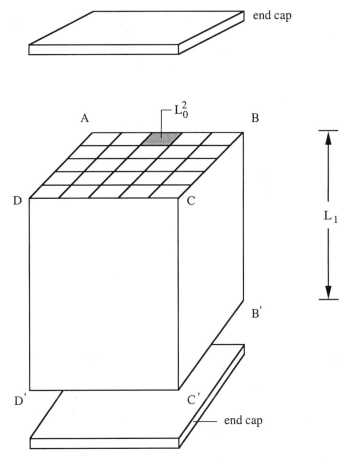

Fig. 5.20. Second construct containing a number of first constructs (Bejan, 2002).

The second construct is a parallelepiped of height L_1 and base $n_2 L_0^2$. The base is not necessarily square. If the total volume allotted to the second construct is V_2, and if the shape of the V_2 parallelepiped is also specified, then several internal architectural features $(n_1, n_2, \ldots)$ can be selected such that the size and shape of the second construct match the size and shape of the available space.

The left side of Figure 5.22 shows how the top side of the second construct looks when it is viewed from above. The top surface is pierced by counterflows of fluid sheets 1 and 3. The right side shows the corresponding view of the bottom surface, when the bottom is also viewed from above. Note the vertical alignment of corners A and A', B and B', and so on. The bottom surface is pierced by counterflows of fluid sheets 2 and 4.

The top and bottom end plates have the purpose of reorganizing the streams. They connect all the fluid sheets of one type (e.g., sheet 1,

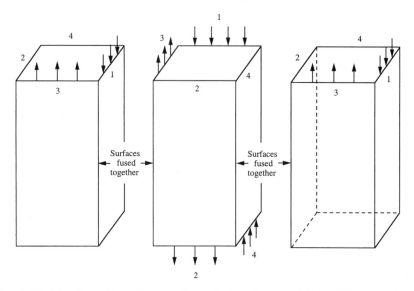

Fig. 5.21. The formation of counterflows during the assembling of the second construct (Bejan, 2002).

Figure 5.22a) to a single inlet or outlet that supplies or discharges the stream of that type. Another feature is that the flow network that connects all the streams of type 1 is placed in balanced counterflow with the flow network that connects all the streams of type 3. Similarly, at the bottom end of the second construct, the flow network that connects the streams of type 2 is placed in balanced counterflow with the network connecting the streams of type 4. Tree-shaped flows in balanced counterflow are a prevailing flow structure in subskin vascularized tissues (Weinbaum and Jiji, 1985; Bejan, 2001). The purpose of the intimate thermal contact between the streams in counterflow is to minimize the leakage of heat (an enthalpy current) along the counterflow, from the warm end to the cold end. The counterflow provides thermal insulation in the flow direction: this insulation effect has its origin in the minimization of thermal resistance in the direction perpendicular to the streams (Bejan, 1979b, 1982). This special feature, and the fact that the streamwise leakage of heat vanishes as the thermal contact between streams becomes perfect, is the reason why the balanced counterflow is the best arrangement from the point of view of minimizing heat transfer irreversibilities.

Each of the four end-plate networks connects one point (source or sink) to n_2 points distributed almost uniformly over the top or bottom areas. The n_2 'points' are thin slits of size $D_1 \times L_0$. There are several flow structures that effect such point-area connections. The most effective are the tree-shaped networks. One example is shown in Figures 5.23 and 5.24: note the superposition of two tree-shaped flows in counterflow, namely, streams 2 and 4.

The two trees are machined into the plate that caps one end of the second construct. For example, at the bottom end (A'B'C'D' Figure 5.22b) the end plate is a sandwich of three layers (α, β, γ). The inner layer (α) makes contact with the A'B'C'D' surface, and has parallel rows of perforations that communicate with the slits of type 2 and 4 of the A'B'C'D' surface. Rows that communicate with slits of type 2 alternate with rows that communicate with slits of type 4. The traces of these perforations are indicated with dots and crosses on the A'B'C'D' frame shown in Figure 5.24. The rows are parallel to the diagonal of the A'B'C'D' square. On the outer side of layer α, the orifices

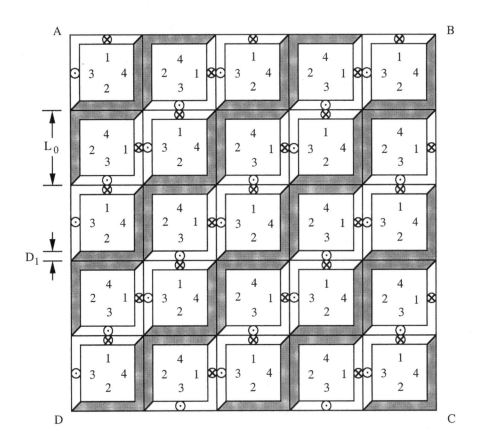

(a)

Fig. 5.22. (a) The top surface and (b) the bottom surface of the second construct. Both surfaces are viewed from above (Bejan, 2002).

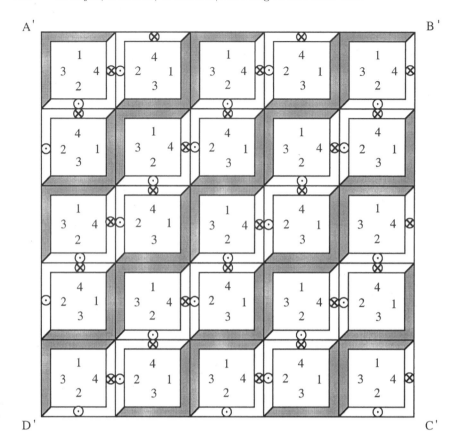

(b)

Fig. 5.22. Continued.

communicate with channels that collect the fluid of only one type (e.g., type 2), and separate it from the fluid of the competing type (e.g., type 4).

The next layer (β) caps the α channels. Layer β has only two rows of orifices, which connect all the α channels of one type and place their fluid in counterflow with the fluid of the second type. The orifices of the two channels of layer β are projected on the plane A'B'C'D' in Figure 5.24. The α and β channels of the same fluid type are arranged in a dendritic pattern, where the branches (or tributaries) are perpendicular to the stem. Dendritic patterns of other types can be machined into the α and β layers. Finally, the function of layer γ is to cap the β channels. One of the two ports machined into the

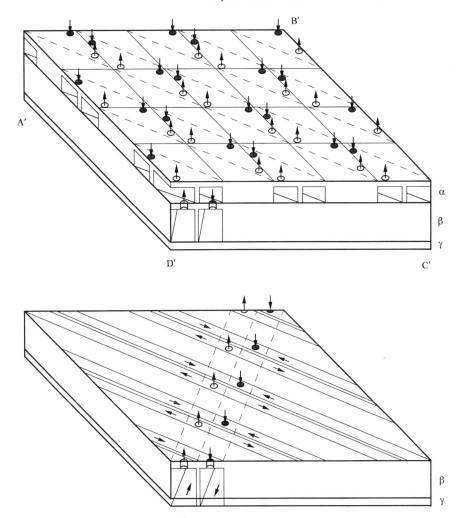

Fig. 5.23. Counterflow tree-shaped networks in the α and β layers of the second-construct end plates (Bejan, 2002).

γ layer allows the total stream of type 2 to leave the second construct. The other port is the inlet of the total stream of fluid 4.

The second construct is an arrangement that unites the large number of ministreams of fluid and delivers to the outside just two outflowing macroscopic streams (2 and 3). At the same time, the architecture of the second construct allows the two inflowing streams (1 and 4) to spread through the entire volume, and bathe every single elemental volume in practically the same way (namely, as in Figure 5.18). At scales of construction that are larger than those of the elemental volume, every stream and flow channel is placed in counterflow with a stream and channel of the same size. These

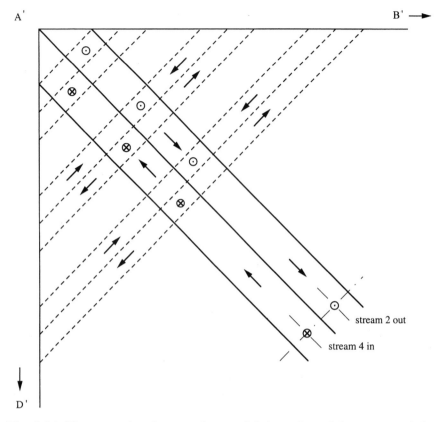

Fig. 5.24. The connections between the α and β channels, and the two ports (inlet, outlet) of the cold end of the second construct (Bejan, 2002).

counterflows force the two macroscopic streams to exchange heat at every scale, all the way to the largest scale. Note the counterflow formed between the two channels of the β layer. By making these channels relatively deep, we can maximize the thermal contact between the two streams.

Larger third constructs can be made by assembling more than four second constructs. In every case, the rule is to guide streams 1 and 3 in counterflow into the third construct, and into the second constructs that are present in the third construct. Details of the third construct and the optimization of all channel spacings for minimal global flow resistance are presented in Bejan (2002).

In conclusion, the pursuit of minimal overall flow resistance between a point (inlet, outlet) and an entire volume led to tree-shaped flows, or flows that are not distributed uniformly through the volume. They are optimal arrangements of small and large streams, with diffusion-penetrated volumes between them (Section 5.6). What may appear as flow "maldistribution" is in fact the optimal or near optimal flow structure for distributing a stream to a volume, or collecting a stream from a volume.

In this section we put these ideas together and described the conceptual design of a heat exchanger that takes maximum advantage of the high heat transfer density promised by the use of small-scale channels with laminar flow. The heart of the design is the smallest (elemental) channels, which fill most of the volume allocated to the heat exchanger. Each elemental channel has a length that matches the thermal entrance length of the stream that flows through it. This feature gives the channel the ability to transfer heat at a maximum rate per unit volume.

The rest of the design has the purpose of organizing and connecting the elemental channels so that they fill the heat exchanger volume effectively. The objective is minimum pumping power and maximum thermal contact between hot and cold streams. This leads to a sequence of constructs of increasing size, such that each stream bathes the heat exchanger volume as a tree-shaped flow. Later, the same stream is reconstituted by flowing as a tree or as a river basin. Each stream has a flow architecture similar to two trees matched canopy to canopy. The two streams, and their associated ramifications, are placed in counterflow at every length scale except the smallest. There the elemental streams are oriented in crossflow, to facilitate assembly.

The maximum heat transfer rate per unit volume, which built into this design, requires emphasis. It is well known that small channel sizes (D_0) and laminar flow lead to high heat transfer coefficients, $h \approx k/D_0$. This feature is also present in the constructal design described in this section. If the channel is long so that the laminar flow is fully developed, the channel acquires a longitudinal temperature difference associated with the bulk enthalpy increase experienced by the stream. The longitudinal temperature difference, which in the small-scale heat exchanger literature is recognized as an additional "thermal flow" resistance (Kraus and Bar-Cohen, 1995), is avoided in the present design.

The thermal-entrance operation of each elemental channel (Figure 5.18) offers an additional benefit. If the heat transfer coefficient associated with laminar fully developed flow is h, then according to laminar boundary layer theory the heat transfer coefficient averaged over the entrance (elemental) length is $2h$. The maximum heat transfer density of the proposed design is due to two effects: the elimination of the longitudinal flow thermal resistances, and the doubling of the heat transfer coefficient that would have been offered by laminar fully developed flow.

The order of magnitude of the heat transfer coefficient in the elemental volumes (k/D_0) justifies the continued push toward miniaturization. Smaller elemental spacings (D_0) lead to greater heat transfer densities. In this direction, the optimization rule described in Equation (5.56) reaches an important limitation: L_0 decreases faster than D_0 because it is proportional to D_0^2. When the slenderness ratio L_0/D_0 becomes smaller than the order of 10, or when $(U_0 D_0/\nu)\mathrm{Pr} < 10^2$, the theory of Section 5.3 must be replaced by one that holds in the limit of decreasing Be values, where boundary layers are absent.

5.8 Constructal Multiscale Structure for Maximal Heat Transfer Density

Two ideas attain greater clarity as we pursue "designed" porous structures. One is the generation of organized flow structures with multiple length scales, which are organized hierarchically. This development was amply illustrated by the tree-shaped flow paths optimized in this chapter. The other idea is the very meaning of the concepts of porous medium and REV. At the start of this book we pointed out that distinctions between porous media, coarse porous media, and channels with flowing fluid depend on the observer—the distance between the analyst and the analyzed. The study of designed porous structures forces the analyst to get 'into the structure', and to optimize its architectural features. Examined up close, the designed structure is a multiscale, nonuniformly distributed flow system, heat exchanger, and the like. Seen from a distance, or when miniaturization packs enough solid and fluid space features into the visible volume, the multiscale flow structure is better described as a designed porous medium. In this section, we give one more example of this kind.

A key result of constructal theory is the prediction of optimal spacings for the internal flow structure of volumes that must transfer heat and mass to the maximum. This idea holds for both forced and natural convection, and is outlined in Sections 5.2 and 5.3. Optimal spacings have been determined for several configurations, depending on the shape of the heat transfer surface that is distributed through the volume: stacks of parallel plates, bundles of cylinders in crossflow, and arrays of staggered plates (e.g., Figure 5.25). In each configuration, the reported optimal spacing is a single value, that is, a *single length scale* that is distributed uniformly through the available volume.

Is the stack of Figure 5.25 the best way to pack heat transfer into a fixed volume? It is, but only when a single length scale is to be used, that is, if the structure is to be *uniform*. The structure of Figure 5.25 is uniform, because it does not change from $x = 0$ to $x = L_0$. At the most, the geometries of single-spacing structures vary periodically, as in the case of arrays of cylinders and staggered plates.

Bejan and Fautrelle (2003) pointed out that the structure of Figure 5.25 can be improved if more length scales $(D_0, D_1, D_2, \ldots)$ are available. The technique consists of placing more heat transfer in regions of the volume HL_0 where the boundary layers are thinner. Those regions are situated immediately downstream of the entrance plane $x = 0$. Regions that do not work in a heat transfer sense must either be put to work, or eliminated. In Figure 5.25, the wedges of fluid contained between the tips of opposing boundary layers are not involved in transferring heat. They can be involved if heat-generating blades of shorter lengths (L_1) are installed on their planes of symmetry. This new design is shown in Figure 5.26.

Each new L_1 blade is coated by Blasius boundary layers [e.g., Bejan (1995a, p. 49)] with the thickness $\delta(x) \cong 5x(Ux/\nu)^{-1/2}$. Because δ increases as $x^{1/2}$,

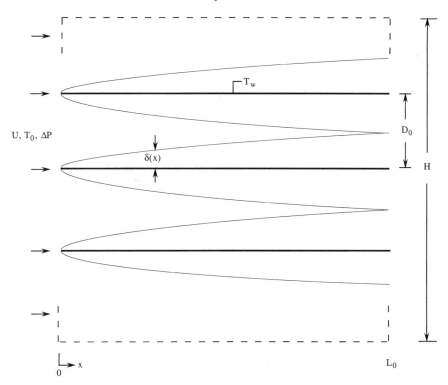

Fig. 5.25. Optimal package of parallel plates with one spacing (Bejan and Fautrelle, 2003).

the boundary layers of the L_1 blade merge with the boundary layers of the L_0 blades at a downstream position that is approximately equal to $L_0/4$. The approximation is due to the assumption that the presence of the L_1 boundary layers does not significantly affect the downstream development $(x > L_0/4)$ of the L_0 boundary layers. This assumption is made for the sake of simplicity. The order-of-magnitude correctness of this assumption is clear, and it comes from geometry: the edges of the L_1 and L_0 boundary layers must intersect at a distance of order

$$L_1 \cong \frac{1}{4}L_0. \tag{5.57}$$

Note that by choosing L_1 such that the boundary layers that coat the L_1 blade merge with surrounding boundary layers at the downstream end of the L_1 blade, we once more invoke the optimal packing principle of Sections 5.2 and 5.3, and Figure 5.25. We are being consistent as constructal designers and, because of this, every structure with merging boundary layers will be optimal, no matter how complicated.

The wedges of isothermal fluid (T_0) remaining between adjacent L_0 and L_1 blades can be populated with a new generation of even shorter blades,

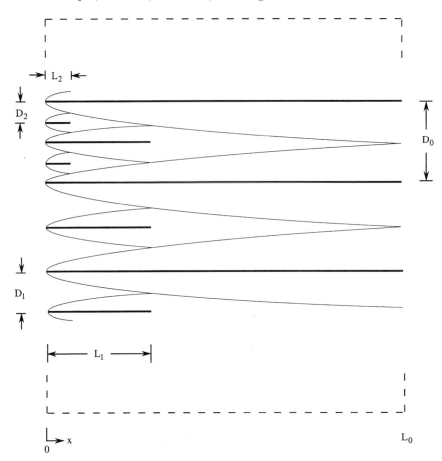

Fig. 5.26. Optimal multiscale package of parallel plates (Bejan and Fautrelle, 2003).

$L_2 \cong L_1/4$. Two such blades are shown in the upper-left corner of Figure 5.26. The length scales become smaller (L_0, L_1, L_2), but the shape of the boundary layer region is the same for all the blades, because the blades are all swept by the same flow (U). The merging and expiring boundary layers are arranged according to the algorithm

$$L_i \cong \frac{1}{4}L_{i-1}, \qquad D_i \cong \frac{1}{2}D_{i-1} \qquad (i = 1, 2, \ldots, m), \qquad (5.58)$$

where we show that m is finite, not infinite. In other words, as in all the constructal tree structures (Bejan, 2000), the image generated by the algorithm (5.58) is not a fractal. It is a Euclidean image [cf. Bejan (1997, p. 765)]. The sequence of decreasing length scales is finite, and the smallest size (D_m, L_m) is known, as we show in Equation (5.72).

To complete the description of the sequential construction of the multiscale flow structure, we note that the number of blades of a certain size increases as the blade size decreases. Let $n_0 = H/D_0$ be the number of L_0 blades in the uniform structure of Figure 5.25, where

$$D_0 \cong 2\delta(L_0) \cong 10 \left(\frac{\nu L_0}{U} \right)^{1/2}. \tag{5.59}$$

The number of L_1 blades is $n_1 = n_0$, because there are as many L_1 blades as there are D_0 spacings. At scales smaller than L_1, the number of blades of one size doubles with every step,

$$n_i = 2n_{i-1} \qquad (i = 2, 3, \ldots, m). \tag{5.60}$$

Two conflicting effects emerge as the structure grows in the sequence started in Figure 5.26. One is attractive: the total surface of temperature T_w installed in the HL_0 volume increases. The other is detrimental: the flow resistance increases, the flow rate driven by the fixed ΔP decreases, and so does the heat transfer rate associated with a single boundary layer. The important question is how the volume is being used: what happens to the heat transfer rate *density* as complexity increases?

5.8.1 Heat Transfer

The total heat transfer rate from the T_w surfaces to the T_0 fluid can be estimated by summing up the contributions made by the individual blades. The heat transfer rate through one side of the L_0 blade is equal (in an order of magnitude sense) to the heat transfer rate associated with a laminar boundary layer [cf. the Pohlhausen solution for Prandtl numbers of order 1, e.g., Bejan (1995a)]

$$\frac{\bar{q}_0'' }{\Delta T} \frac{L_0}{k} \cong 0.664 \left(\frac{UL_0}{\nu} \right)^{1/2}. \tag{5.61}$$

Here $\bar{q}_0''[W/m^2]$ is the L_0-averaged heat flux, $\Delta T = T_w - T_0$, and k is the fluid thermal conductivity. There are $2n_0$ such boundary layers, and their combined contribution to the total heat transfer rate of the package of Figure 5.26 is

$$q_0' = 2n_0\bar{q}_0''L_0 \cong 1.328k\Delta T n_0 \left(\frac{UL_0}{\nu} \right)^{1/2}. \tag{5.62}$$

The same calculation can be performed for any group of blades of one size, L_i. Their total heat transfer rate $q_i'[W/m]$ is given by a formula similar to Equation (5.62), in which n_0 and L_0 are replaced by n_i and L_i. The heat transfer rate of all the blades is the sum

$$q' = \sum_{i=0}^{m} q_i' \cong 1.328 k \Delta T n_0 \left(\frac{UL_0}{\nu}\right)^{1/2} S, \tag{5.63}$$

where S is the dimensionless geometric parameter

$$S = 1 + \frac{n_1}{n_0}\left(\frac{L_1}{L_0}\right)^{1/2} + \frac{n_2}{n_0}\left(\frac{L_2}{L_0}\right)^{1/2} + \cdots + \frac{n_m}{n_0}\left(\frac{L_m}{L_0}\right)^{1/2} = 1 + \frac{m}{2}. \tag{5.64}$$

This analysis confirms the anticipated trend: the total heat transfer rate increases monotonically as the complexity of the structure (m) increases.

5.8.2 Fluid Friction

It is necessary to evaluate the flow resistance of the multiscale structure, because the velocity U that appears in Equation (5.63) is not specified. The pressure difference ΔP is specified, and it is related to all the friction forces felt by the blades. We rely on the same approximation as in the case of heat transfer, and estimate the friction force along one face of one blade by using the solution for the laminar boundary layer [e.g., Bejan (1995a)],

$$\tau_i \cong C_{fi}\frac{1}{2}\rho U^2, \qquad C_{fi} = \frac{1.328}{(UL_0/\nu)^{1/2}} \tag{5.65}$$

Here τ_i and C_{fi} are the averaged shear stress and skin friction coefficient, respectively. The total force felt by the blades of size L_i is

$$F_i = 2n_i\tau_i L_i \cong 1.328\rho(\nu L_i)^{1/2}n_i U^{3/2}. \tag{5.66}$$

The total force for the multiscale package is

$$F = \sum_{i=0}^{m} F_i \cong 1.328\rho(\nu L_0)^{1/2}n_0 U^{3/2} S. \tag{5.67}$$

This force is balanced by the longitudinal force imposed on the control volume $\Delta P H = F$, which combined with Equation (5.67) and the D_0 formula (5.59), yields the order of magnitude of the average velocity of the fluid that permeates the structure:

$$U \cong 2.7 \left(\frac{\Delta P}{\rho S}\right)^{1/2}. \tag{5.68}$$

This result confirms the second trend that we anticipated: the flow slows down as the complexity of the structure (S or m) increases.

5.8.3 Heat Transfer Rate Density: The Smallest Scale

Putting together the results of the heat transfer and fluid flow analyses, we find how the structure performs globally when its constraints are specified $(\Delta P, \Delta T, H, L_0)$. Eliminating U between Equations (5.63) and (5.68) yields the dimensionless global thermal conductance,

$$\frac{q'}{k\Delta T} \cong 0.36 \frac{H}{L_0} \mathrm{Be}^{1/2} S^{1/2}, \tag{5.69}$$

where the Bejan number is based on ΔP and L_0,

$$\mathrm{Be} = \frac{\Delta P L_0^2}{\mu\alpha}. \tag{5.70}$$

In this expression μ and α are the viscosity and thermal diffusivity of the fluid. The alternative to using the global conductance is the heat transfer rate density $q''' = q'/HL_0$. Both quantities increase with the applied pressure difference (Be) and the complexity of the flow structure (S). In conclusion, in spite of the conflicting effects of S in Equations (5.63) and (5.68), the effect of increasing S is beneficial from the point of view of packing more heat transfer in a given volume. Optimized complexity is the route to maximal global performance in a morphing flow architecture (Bejan, 2000).

How large can the factor S be? The answer follows from the observation that the geometry of Figure 5.26 is valid when boundary layers exist, that is, when they are distinct. To be distinct, boundary layers must be slender. Figure 5.26 makes it clear that boundary layers are less slender when their longitudinal scales (L_i) are shorter. The shortest blade length L_m below which the boundary layer heat transfer mechanism breaks down is

$$L_m \sim D_m. \tag{5.71}$$

In view of Equations (5.58), this means that $L_0 \sim 2^m D_0$. Finally, by using Equations (5.53) and (5.68) we find the smallest scale, which occurs at the level m given by

$$2^m \left(1 + \frac{m}{2}\right)^{1/4} \sim 0.17\,\mathrm{Be}^{1/4}. \tag{5.72}$$

In view of the order-of-magnitude character of the analysis based on Equation (5.71), the right side of Equation (5.72) is essentially $(\mathrm{Be}/10^3)^{1/4}$. Equation (5.72) establishes m as a slowly varying monotonic function of $\mathrm{Be}^{1/4}$. This function can be substituted in (5.69) to see the complete effect of Be on the global heat transfer performance,

$$\frac{q'}{k\Delta T} \cong 0.36 \frac{H}{L_0} \mathrm{Be}^{1/2} \left(1 + \frac{1}{2}m\right)^{1/2}. \tag{5.73}$$

In conclusion, the required complexity (m) increases monotonically with the imposed pressure difference (Be). More flow means more length scales, and smaller smallest scales. The structure becomes not only more complex but also finer. The monotonic effect of m is accompanied by diminishing returns: each new length scale (m) contributes to global performance less than the preceding length scale $(m - 1)$. If the construction started in Figure 5.26 is arbitrarily continued ad infinitum [arbitrarily so, because this limit is valid only if Be is infinite, cf. Equation (5.73)], then the resulting image is a fractal and m and the heat transfer density (5.73) tend to infinity.

Forced convection was used in Bejan and Fautrelle (2003) only for illustration, that is, as a flow mechanism on which to build the multiscale structure. A completely analogous multiscale structure can be deduced for laminar natural convection. The complete analogy that exists between optimal spacings in forced and natural convection was described by Petrescu (1994). In brief, if the structure of Figure 5.25 is rotated by 90° counterclockwise, and if the flow is driven upward by the buoyancy effect, then the role of the overall pressure difference ΔP is played by the difference between two hydrostatic pressure heads, one for the fluid column of height L_0 and temperature T_0, and the other for the L_0 fluid column of temperature T_w. If the Boussinesq approximation applies, the effective ΔP due to buoyancy is (Bejan, 1995a, p. 186)

$$\Delta P = \rho g \beta \Delta T L_0, \tag{5.74}$$

where $\Delta T = T_w - T_0$, β is the coefficient of volumetric thermal expansion, and g is the gravitational acceleration aligned vertically downward (against x in Figure 5.25). By substituting the ΔP expression (5.74) into the Be definition (5.70) we find that the dimensionless group that replaces Be in natural convection is the Rayleigh number $\mathrm{Ra} = g\beta\Delta T L_0^3/(\alpha\nu)$. Other than the Be $\to$ Ra transformation, all the features that are due to the generation of multiscale blade structure for natural convection should mirror, at least qualitatively, the features described for forced convection in this section.

Finally, the hierarchical multiscale flow architecture constructed in this section is a theoretical comment on fractal geometry. Fractal structures are generated by assuming (postulating) certain algorithms. In much of the current fractal literature, the algorithms are selected such that the resulting structures resemble flow structures observed in nature. For this reason, fractal geometry is descriptive, not predictive (Bejan, 1997; Bradshaw, 2001). Fractal geometry is not a theory.

Contrary to the fractal approach, the constructal method used in this section generated the construction algorithms [Equations (5.58) and (5.60)], including the smallest-scale cutoff, Equation (5.72). The algorithms were generated by the constructal principle of optimization of global performance subject to global constraints (Bejan, 2000). This principle was invoked every time the optimal spacing between two blades was used. Optimal spacings were assigned to all the length scales, and were distributed throughout the volume. With regard to fractal geometry and why (empirically) some fractal structures happen to resemble natural flow structures, the missing link has been the origin of the algorithm (Nottale, 1993). Constructal theory delivers the algorithm as an optimization result, from the constructal principle.

5.9 Concluding Remarks

Research marches forward, even in an apparently mature field such as convection in porous media. The drive for innovation is provided by the premium that we all put on performance and finite resources (e.g., space). Better devices, better numerical models, and more comprehensive theoretical views—they all keep the field alive. In this chapter, we reviewed recent developments that bring porous media in close contact with the design of small-scale flow structures of high complexity. The results represent designed porous media. The flow architecture of designed porous media has purpose: the maximization of system performance at the global level, subject to global constraints.

Heat exchangers can be modeled as fluid-saturated porous media in the limit of small Reynolds numbers (Section 5.1). The spacings (pores) of volumes that generate heat can be optimized so that the volumetric density of heat transfer is maximized. This opportunity is present in many configurations in natural and forced convection (Sections 5.2 and 5.3). The same optimal internal structure is recommended by the maximization of heat transfer density in a porous space cooled by pulsating flow (Section 5.4). The optimized geometry is robust relative to changes in the flow direction (stop and go, back and forth, steady) and the frequency of pulsations. Robustness was also a feature in all the internal structures optimized for steady flow in Sections 5.2 and 5.3.

The experience with the optimization of geometry for steady-state heat transfer showed that the constructal optimization principle—optimal flow structure for best global performance—works in all the applications in which it has been invoked (Bejan, 2000). Optimal spacings and flow paths (e.g., trees) have been developed for both simple and complicated geometries. Similarly, the principle that in this chapter was demonstrated for relatively simple designs can be expected to apply in more complex designs of porous media permeated by flows.

Constructal heat exchangers have also been described by Tondeur et al. (2000). Dendritic microchannel structures for cooling heat-generating volumes have been reported by Bejan and Errera (2000), Bejan (2000), Chen and Cheng (2002), Pence (2002), Lorente et al. (2002a), and Wechsatol et al. (2002b). Tree-shaped fins for augmenting condensation have been studied experimentally by Belghazi et al. (2002). Understanding the formation of dendritic structures is essential to theoretical progress on rapid solidification [e.g., Bejan (2000), Dupuoy et al. (1993, 1997), Dupouy and Camel (1998), Goyeau et al. (1999), Jaluria (2001), and Mat and Ilegbusi (2002)]. Tree-shaped inserts with high conductivity for conduction have been optimized numerically by Xia *et al.* (2002) and Rocha *et al.* (2002). The constructal optimization of conduction in a three-dimensional porous medium was proposed by Ait Taleb *et al.* (2002).

In Section 5.8, we outlined a new concept for generating a multiscale flow structure that maximizes the heat transfer density installed in a fixed volume (Bejan and Fautrelle, 2003). The method consists of exploiting every available flow volume element for the purpose of transferring heat. In laminar forced convection, the working volume has a thickness that scales with the square root of the length of the streamwise heat transfer surface. Larger surfaces are surrounded by thicker working volumes. The starting regions of the boundary layers are thinner. They are surrounded by fresh flow that can be put to good use: heat transfer blades with smaller and smaller lengths can be inserted in the fresh fluid that enters the smaller and smaller channels formed between existing blades.

The number of scales of the multiscale flow structure (m) increases slowly as the flow becomes stronger. The flow strength is accounted for by the pressure difference maintained across the structure (Be). Boundary layers become thinner as Be increases, and this means that more small-scale heat transfer blades can be inserted in the interstitial spaces of the entrance region of the complex flow structure.

Two trends compete as the number of length scales increases. The structure becomes less permeable, and the flow rate decreases. At the same time, the total heat transfer surface increases. The net and most important result is that the heat transfer *density* increases as the number of length scales increases. This increase occurs at a decreasing rate, meaning that each new (smaller) length scale contributes less to the global enterprise than the preceding length scale. There exists a characteristic length scale below which heat transfer surfaces are no longer lined by boundary layers. This smallest scale serves as the cutoff for the algorithm that generates the multiscale structure.

Brod (2003) devised tree-shaped networks for distributing polymer melts. As the optimization criterion, he used the minimization of travel time, or residence time for the melt. He achieved this by selecting the tree tube diameters such that the local apparent shear stress is uniform throughout the system.

Another example is the optimization of the size and spacing of the holes in the wall of a cooled high-temperature turbine blade (Cizmas and Bejan, 2001). The design of holes in the gas distributor plate in gas–solid fluidized beds controls the flow distribution and performance (Hepbasli, 1998a,b). Nearly uniform flow injection can be achieved by using a tree-like piping injection system (Cheng *et al.*, 2000). Porous materials with tree-like flow channels are also formed by chemical dissolution (Daccord and Lenormand, 1987). The multiscale properties of such porous media have been used as the basis of a permeability model (Yu and Cheng, 2002) and wall roughness in microchannels (Chen and Cheng, 2003). Such opportunities deserve to be considered in future studies, in the wide and always expanding arena of engineered and natural flow systems.

6

Living Structures

The development of complex cellular systems such as the vertebrates requires the availability of large amounts of oxygen for the metabolic needs of the cells. The respiratory and circulatory systems are the specialized and hierarchically organized flow systems that meet this need. The respiratory system carries oxygen from the air to the pulmonary veins, whereas in the circulatory system oxygen is carried by the blood and is delivered to the cells by the thinnest vessels, the capillaries. A countercurrent of oxygen and carbon dioxide crosses the capillary wall, such that carbon dioxide is extracted from the blood stream and is discharged from the lungs into the atmosphere during exhalation.

6.1 Respiratory System

The respiratory system consists of the entire pathway for airflow from the mouth or nose to the alveoli (Figure 6.1). The air that enters the trachea continues through two main ducts called bronchi, one for the left lung, and the other for the right lung. Each bronchus branches off into a tree network of small ducts called bronchioles. At the end of each bronchiole is situated an air sack called the alveolus. The alveoli are the final branching of the respiratory tree—the air spaces with the smallest length scale.

The tree structure can also be described in the reverse direction, from small to large, from the alveoli to the trachea, from the tips of the smallest tree branches to the trunk. In this direction, the branches coalesce. Instead of bifurcations, we speak of pairings. We see the bronchial tree as the result of a construction, the principles of which form the subject of constructal theory (Bejan, 2000).

The respiratory system has the following functional components.

(1) *Conducting region*: This consists of air ducts that guide and distribute air into the lung. It comprises the pathway from the mouth or nose to the terminal bronchioles. This region also performs functions of humidification and temperature regulation (conditioning the air), removal of particles,

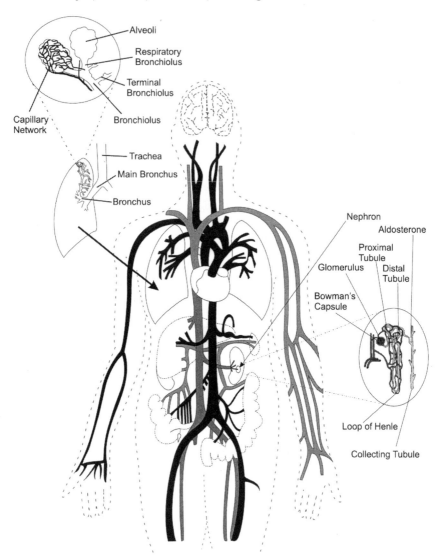

Fig. 6.1. Human circulatory system, composed of systemic circulation (which serves the body except the lungs) and pulmonary circulation (blood carried to and from the lungs) (Courtesy of P.S.C. Neves).

antibacterial and immunological defense, the sense of smell (via specialized olfactory epithelium), and the production of sound (the voice).

(2) *Respiratory region:* This is the volume where the gas exchange takes place and, unlike the conduction region, this volume is not covered by a protective mucus layer. The airways that participate in gas exchange with capillary blood are the respiratory bronchioles and the alveoli (alveoli ducts included).

The alveoli act as the primary gas exchange units of the lungs. It is estimated that an adult has approximately 600 million alveoli with a surface area for gas exchange of about $75\,m^2$, which are perfused by more than $2000\,km$ of capillaries (see Section 6.2). Oxygen reaches the blood by rapidly diffusing through a very thin barrier between the alveolar space and the capillary endothelium. Carbon dioxide follows the reverse course to reach the alveoli.

6.1.1 Airflow Within the Bronchial Tree

During heavy work, more than $1.5 \times 10^{-3}\,m^3$ of tidal air could be inhaled and exhaled with each breath. During normal breathing turbulent flow may be found at the entrance but never in the smallest airways. During periods of rest, the tidal volume of air inhaled and exhaled with each breath may be three or more times smaller than during exercise.

In order to optimize its function (see Section 4.11), the airway tree exhibits 23 levels of bifurcations after the trachea (Weibel, 1963). Along a streamline, the air ducts become smaller in diameter and length, as compared to early branches (Hinds, 1999). If the flow within the bronchial tree is laminar and fully developed, minimum resistance is achieved when the diameter of the nth bronchiole is given by Equation (4.101),

$$\frac{D_n}{D_0} = 2^{-n/3}, \tag{6.1}$$

whereas in view of Equation (4.102) its length is

$$\frac{L_n}{L_0} = 2^{-n/3}, \tag{6.2}$$

where D_0 and L_0 are the diameter and the length of the trachea, respectively. The flow resistance posed by nth bronchial tube is proportional to $L_n D_n^{-4}$,

$$\frac{\Delta P_n}{\dot{m}_n} = \frac{128\nu 2^{-n/3} L_0}{\pi 2^{-4n/3} D_0^4} = 2^n \frac{128\nu L_0}{\pi D_0^4}. \tag{6.3}$$

At the nth bifurcation, the bronchial tree has 2^n tubes and the total flow resistance

$$2^{-n} \frac{\Delta P_n}{\dot{m}_n} = \frac{128\nu L_0}{\pi D_0^4}. \tag{6.4}$$

This means that the overall resistance of the bronchial tree to laminar airflow does not change from one bifurcation level to another. Therefore, if ΔP is the average pressure difference that drives the airflow $\dot{m}$ into and out of the lungs, the overall resistance of the respiratory tree (trachea and the bronchial tree with n bifurcations) is

$$\frac{\Delta P}{\dot{m}} = (n+1)\frac{128\nu}{\pi} \frac{L_0}{D_0^4}. \tag{6.5}$$

The overall resistance of the optimized respiratory tree is controlled by the trachea morphological parameter L_0/D_0^4. Reis *et al.* (2004) showed based on Bejan's constructal theory that among the flow architectures that could bring oxygen from outside the body to a surface alveolus in the lung, the air tree with successive dichotomous branching and ending with a smallest-scale volume with diffusion is the architecture that performs best. They found that the best oxygen access to the alveolar tissues is provided by a tree flow structure with 23 levels of bifurcation. In addition, there exists a dimensionless length scale defined as the ratio of the square of the first airway diameter to its length, which is constant for all individuals of the same species (constructal law).

Andrade *et al.* (1998) have investigated the influence of the Reynolds number on the flow through a cascade of bifurcations. They used direct simulations based on the 2-D Navier–Stokes equation. They found that for a fully symmetric tree with $n \geq 3$ bifurcations, at low Re the flow was equally distributed within the ducts corresponding to the same level of branching. They also found that the flow distribution becomes significantly heterogeneous at higher Reynolds numbers (e.g. Re $\geq$ 4800), when inertial effects become relevant. They suggested that the flow asymmetry due to inertial effects could be compensated by the structural asymmetry of the lungs.

6.1.2 Alveolar Gas Diffusion

After being transported through the bronchial tree, the air is brought into contact with the capillary bed that surrounds each alveolus. The oxygen diffuses first through the alveolar air and then through the tissue before reaching the red blood cells within the capillaries.

With the objective of studying respiratory abnormalities, Kulish *et al.* (1999, 2002) and Kulish and Lage (2001) have presented a model for the diffusion process in the lungs. The lung diffusing capacity is a lumped parameter estimated from the measurements of the concentration of carbon monoxide (CO) in the inhaled and the exhaled gas. This measurement is used to indicate abnormalities in the diffusion process. However, because diffusion occurs at the alveolar scale, the parameter that is most effectively related to possible abnormalities is the effective diffusivity D_{eff}, which Kulish *et al.* (2002) use in order to account for the gas diffusivities of all constituents present within the alveolar representative elementary volume (REV: see Section 1.1). If P denotes the REV-averaged partial pressure of the gas, then the diffusive process is represented by

$$\frac{\partial P}{\partial t} = D_{\text{eff}} \nabla^2 P. \tag{6.6}$$

The lung diffusing capacity D_L as a lumped parameter is obtained from the equation (Kulish *et al.*, 2002)

$$\frac{V_A}{P_{\text{ref}}} \frac{dP}{dt} = -D_L P, \tag{6.7}$$

the integral form of which is known as the Krogh equation

$$P = P_0 \exp\left[-\left(\frac{D_L P_{\text{ref}}}{V_A}\right) t\right]. \tag{6.8}$$

Here $P_{\text{ref}} \sim 9.51 \times 10^4\,\text{Pa}$ is the total pressure of the inhaled air, and $V_A \sim 4.93 \times 10^{-3}\,\text{m}^3$ is the alveolar volume, that is, the inspired volume plus the residual lung volume. Equation (6.8) allows the lung diffusing capacity to be written as

$$D_L = \frac{V_A}{P_{\text{ref}}} \ln\left(\frac{P_0}{P}\right). \tag{6.9}$$

By assuming values for D_{eff}, Kulish *et al.* (2002) have solved Equation (6.6) iteratively so that P matches the experimental value T_L in Equation (6.9). In this way they found that $D_{\text{eff}} = 2.68 \times 10^{-7}\,\text{m}^2/\text{s}$. They have also established a criterion for the validity of the lumped capacity approach; namely,

$$P_{\text{ref}} \frac{L_K D_L}{A D_{\text{eff}}} < 1, \tag{6.10}$$

where L_K is the diffusion length scale in the REV.

6.1.3 Particle Deposition

During inhaling, the gas mixture and particles transported through the airway system may deposit on the duct walls. Deposition depends on particle size, breathing pattern, and air duct geometry. Once deposited, particles are retained for varying times. The residence times depend upon the particle characteristics, location within the airway system, and type of clearance mechanism involved. A review of physiological events involved during the removal of lung particles may be found in Lehnert (1990).

The importance of understanding particle deposition and retention in the airway system encompasses the following.

(1) Assessing the health hazards of particles: micrometric particles ($>1\,\mu\text{m}$) may be captured in the conducting region of the airway system, and subsequently cleared by the action of the mucus layer. Submicrometric particles may reach the gas-exchange surface of the alveolar region, and if their size is less than 50 nm they may have exceptional toxicity (Donaldson *et al.*, 1998).

A number of occupational diseases are directly linked to deposition and retention of inhaled particles such as coal, silica, and asbestos (Douglas, 1989). Deposition of inhaled allergens may cause acute bronchoconstriction for bronchitis and asthma sufferers (Gonda, 1997), as well as tissue injuries and lung diseases such as bronchial carcinoma and even lung cancer (Spencer, 1985).

(2) For an optimal delivery and evaluation of therapeutic effects of pharmaceutical aerosol particles by inhalation (inhalation route drug delivery) in order to reach the predetermined sites of action in the lung (Ganderton, 1999): inhalation of drug aerosol deposited directly to airway target areas results in a reduction of the adverse reactions in the therapy of asthma, bronchitis, and other respiratory disorders (Smith and Bernstein, 1996).

Deposition of aerosol particles in the respiratory system follows mechanisms similar to those already described in Section 4.5 for filters. However, some specific features of the respiratory system should be stressed.

(1) Interception deposition occurs when particles that follow air streamlines touch the airway walls and become attached to them. This mechanism is especially important in the deposition of fibrous particles (particles with an elongated form) in the smaller airways.

(2) Inertial impaction deposition is characteristic of large particles, and occurs when particles are unable to adjust to the abrupt streamline changes in the neighborhood of airway walls. Therefore, inertial impaction is most effective at or near the tracheal bifurcation, and also plays a role at subsequent bifurcations. Inertial impaction is the predominant deposition mechanism for particles larger than $1\,\mu m$ in the tracheobronchial region (Kim et al., 1983).

(3) The settling of particles due to gravity is very effective for heavy particles transported at low velocities, that is, in the smaller airways. This mechanism of deposition is dominant when particles are larger than $0.5\,\mu m$, after the tracheobronchial region where $v < 300\,\text{cm}^3/\text{s}$ (Hinds, 1999).

(4) Diffusion deposition is most important for submicrometric particles, especially in the airways far from the trachea, where the residence times are relatively long and the distance to airway walls relatively small.

(5) Electrostatic deposition is important only in cases with charged particles, when the airway walls are charged with an opposite charge, or when the concentration of charged particles is high enough to produce a repulsive force between particles. This force drives the particles against the walls.

Many studies related to particle deposition in the airway system have been reported in the literature [e.g., ICRP (1994), Kim and Fisher (1999), Comer et al. (2000), Oldham et al. (2000), Zhang et al. (2002) and Goo and Kim (2003)]. The ICRP model predicts particle deposition in the airway system for male and female adults at three levels of exercise (sitting, light exercise, and heavy exercise). Hinds (1999), based on data predicted with the ICRP model for adults at these three levels of exercise, reported deposition fractions for three regions, from mouth/nose to larynx (Φ_{mn-1}), from trachea to terminal bronchioles (Φ_{t-tb}), and the region beyond the terminal bronchioles (Φ_{btb}):

$$\Phi_{mn-1} = \left[1 - \left(0.5 - \frac{0.5}{1 + 5.29 \times 10^{-3} r_p^{2.8}}\right)\right]\left[\frac{1}{1 + \exp[6.84 + 1.183\ln(2r_p)]}\right.$$

$$\left. + \frac{1}{1 + \exp[0.924 - 1.885\ln(2r_p)]}\right], \tag{6.11}$$

$$\Phi_{t-tb} = \frac{1.76 \times 10^{-3}}{r_p}\{\exp[-0.234(\ln(2r_p) + 3.4)^2]$$

$$+ 63.9\exp(-0.819(\ln(2r_p) - 1.61)^2)\}, \tag{6.12}$$

$$\Phi_{btb} = \frac{7.75 \times 10^{-3}}{r_p}\{\exp[-0.416(\ln(2r_p) + 2.84)^2]$$

$$+ 19.11\exp(-0.482(\ln(2r_p) - 1.362)^2)\}. \tag{6.13}$$

In these expressions r_p is the particle radius expressed in microns, and should range between 0.0005 and 50 μm. The deposition fraction for the entire airway system (Φ) is the sum of the three regional deposition factions (Φ_{mn-1}, Φ_{t-tb}, Φ_{btb}). The mass total of particles with radius r_p deposited per minute ($\dot{m}_p$) can be obtained from

$$\dot{m}_p = \frac{8}{6}\pi c_p \dot{V}_{ia}\rho_p r_p^3\Phi, \tag{6.14}$$

where c_p is the concentration of particles of density ρ_p and $\dot{V}_{ia}$ is the volume inhaled per minute.

6.2 Blood and the Circulatory System

The circulatory system has three distinct parts (Figure 6.1): pulmonary, coronary, and systemic circulation. The movement of blood from the heart to lungs and back to the heart again is known as pulmonary circulation. Coronary circulation consists of the movement of blood through the tissues of the heart, to which it supplies nourishment. The systemic circulation is the major part of the circulatory system, and supplies blood to the rest of the body tissues through a complex (dendritic) system of channels. The systemic circulation comprises two flows: the flow oriented from a point (the heart) to a volume (the body) through the distributing fluid trees, and the flow oriented from the tissues to the heart through the collecting fluid trees (Bejan, 2000). The blood carries oxygen and nutrients that are vital to the body cells, and also carbon dioxide and other waste products that are generated by the cells. This is essential for the functioning, growing, and repairing of the body.

In order to feed the body cells, the circulatory system carries oxygen- and nutrient-rich blood to the capillaries. A countercurrent of carbon dioxide and other wastes crosses the capillary wall in the direction of the blood stream. The capillaries are the smallest elements of the circulatory system. The blood inside the capillaries performs two important functions: furnishing oxygen and nutrients to the tissues, and removing carbon dioxide and waste products. The total number of capillaries in the body has been estimated to be on the order of 40 to 50 billion (Guyton, 1991).

In order to optimize the exchange of substances, capillaries are arranged in tree networks, as is shown in Figure 6.2. These networks cover the entire body.

The structure of the capillary network is tailored to the tissues in order to meet specific requirements. Tissues with low metabolic activity have reduced networks, as compared to those with heavy metabolic requirements. For example, a dense connective tissue has a poor capillary network as compared to the cardiac muscle.

Another modification in the structure is the presence of arteriovenous shunts that directly connect the arterial and venous systems. During sweating, for example, blood can flow rapidly and the nourishing function is neglected.

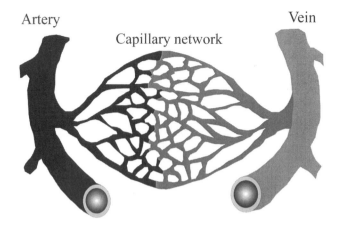

Fig. 6.2. Capillary network.

These shunts can be observed in the capillaries of the skin, enabling cutaneous blood flow to increase during intensive physical exercise or in warm environments, in order to promote the dissipation of heat from the body (see also Bejan, 2001).

6.3 Biomembranes: Structure and Transport Mechanisms

The exchange of substances between the blood and the cells occurs through two different types of biomembranes, the cell membrane (also called the plasma membrane) and the capillary wall. For the organism to function properly, these biomembranes have to allow selective and specific permeation of substances, according to the needs of the body cells. Basically, membranes separate a feed solution into two parts (Figure 6.3): the retained part (the feed that does not cross the membrane) and the permeated part (the feed that crosses the membrane).

6.3.1 Cell Membrane

All biological cell membranes have a great deal in common with regard to organization and composition (Sackmann, 1995; Läuger, 1970). Cell membranes have an average thickness of about seven nanometers and their main components are phospholipids and proteins (Figure 6.4). The key to understanding their functioning lies in the understanding of these specialized parts.

Phospholipids are amphipathic molecules, that is, molecules that contain hydrophilic and hydrophobic regions. The hydrophilic region is attracted to water, and the hydrophobic region is repelled by water. When mixed with

Feed solution

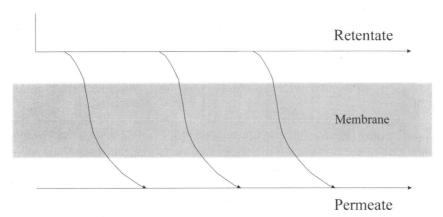

Fig. 6.3. Membrane separation: feed mixture, retained and permeated.

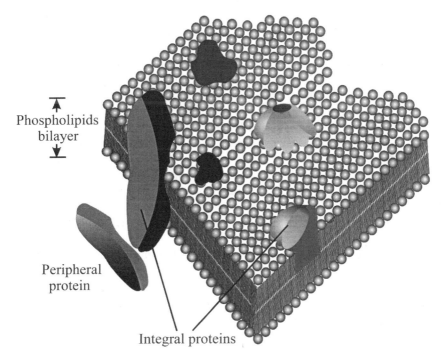

Fig. 6.4. Cell membrane or plasma membrane showing the phospholipid bilayer and membrane proteins (courtesy of P.S.C. Neves).

water, phospholipids line up in double-layered spheres known as a phospholipid bilayer. Phospholipids appear together with some cholesterol, which is a relatively flat and rigid molecule. This combination provides the membrane with the required flexibility. The phospholipid bilayer is selectively permeable, being highly permeable to gases, small organic molecules, and water. It is much less permeable to small polar solutes (urea, glucose, and amino acids), and impermeable to large biological molecules such as adenosine triphosphate (ATP).

Proteins associated with the cell membranes can be distinguished into types: (i) integral proteins, namely, proteins that extend completely through the phospholipid bilayer (transmembrane proteins) and proteins that extend only partially; and (ii) peripheral proteins, that is, proteins that are not connected directly with the cell membrane, but are associated with integral proteins and frequently function as enzymes.

In cells, there is no convective transport of solute and solvent because the total pressure gradient across the cell membrane is null. Therefore, the transport of solutes and solvents through the cell membrane occurs by a combination of the following mechanisms.

(1) *Diffusion, or simple diffusion*:   This mechanism is the most important means of transporting nutrients, oxygen, carbon dioxide, and metabolic end-products across the cell membrane. Diffusion in complex structures such as membranes depends upon the diffusion process in the fluid phase, and on the structure of the membrane. For neutral membranes (i.e., membranes without electrical charge) the diffusive transport of substances is driven by the chemical potential gradient, which is generally proportional to the concentration gradient [cf. Equation (1.41)]. However, when membranes are electrically charged, both the electrical potential and concentration gradients drive the transport of substance (see Sections 4.8 and 6.4). This overall gradient is known as the electrochemical potential gradient (Guggenheim, 1933).

(2) *Osmosis*:   This is the main mechanism for water transfer to and from living cells. Figure 6.5 shows a membrane permeable to the solvent but not permeable to the solute. Osmosis is the process by which the solvent moves to the side containing the solution, until it reaches the equilibrium that corresponds to the equality between the chemical potentials on the two sides of the membrane. The flow of solvent generates a pressure difference between the solvent and the solution. This difference is known as the osmotic pressure. Every particle of solute that is retained in solution contributes to the osmotic pressure, independently of its size. Bloodstream concentrations of salt produce an osmotic pressure of about 3×10^5 Pa. Protein concentration in blood serum is about $1.5 \, mol \, m^{-3}$, and the corresponding osmotic pressure is about 4×10^3 Pa. Thus, as osmotic pressure depends on the number of molecules present, the contribution of small molecules is more significant. If the osmotic pressure is created by protein concentration, then it is known as colloid osmotic pressure, or simply oncotic pressure.

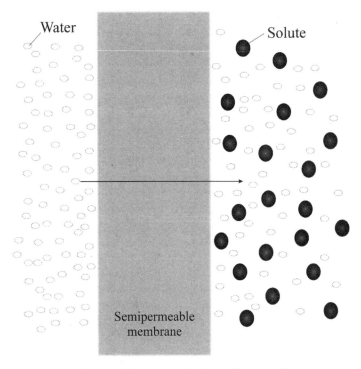

Fig. 6.5. Movement of solvent by osmosis.

A solution separated from another solution by a semipermeable membrane can have three osmotic states. When a cell is surrounded by a solution with a higher concentration of solute or osmolarity, it is said to be in a hypertonic solution. Osmotic equilibrium requires solvent to flow out of the cell, thus leading to the shrinkage of the cell. In a hypotonic solution, a solution with a lower osmolarity surrounds the cell and the equilibrium requires the osmosis of solvent into the cell. As a consequence, the cell swells. In an isotonic solution the cell and the surrounding solution have the same osmolarity. Osmotic equilibrium between the cell and the surrounding solution exists, and there is no exchange of solvent by osmosis. Isotonic solutions are used for maintaining tissues and cells alive outside the body, and for preventing dehydration of patients during and immediately after surgical operations.

(3) *Facilitated diffusion*: This mechanism enables some ions and small polar molecules to cross the cell membrane much more easily than by simple diffusion. It involves transmembrane proteins known as *channels* and *carriers*. Solutes that cross the cell membrane by either channel or carrier proteins are transported passively; that is, the transport occurs down an electrochemical potential gradient.

Channel proteins have a "donut" shape, and allow the solute to diffuse through them as shown in Figure 6.6. Solutes may be neutral or may carry an electrical charge. As mentioned in (1), neutral solutes diffuse only under concentration gradients (namely, from regions of high concentration to regions of lower concentration), but electrically charged solutes are affected by both concentration gradients and electrical potential gradients.

The transport of charged solutes across the channels contributes to building up the membrane potential. In general, the inside of the cell is negative with respect to the outside. The latter is the membrane equilibrium potential, which is about −90 mV (Guyton, 1991). In nerve and muscle cells the resting potential can change very quickly, in order to make possible the fast propagation of nerve signals from neuron to neuron, for the prompt contraction of muscle fibers. This quick change in the membrane potential is known as active potential. The evolution of the active potential is possibly due to the presence of voltage-gated sodium and potassium channels, which become active when the membrane potential is less negative than the equilibrium potential.

In general, cells contain a concentration of potassium ions as much as 20 times higher than that of the extracellular fluid. On the other hand, the extracellular fluid contains a concentration of sodium ions about 10 times higher than that inside the cell. As the voltage-gated sodium channels become active, the rapid flux of ions into the cell increases the membrane potential in the positive direction. This is known as the depolarization of the cell membrane. Voltage-gated potassium channels become active more slowly than sodium channels. In fact, they become fully active only after the sodium channels have become inactive. This allows a rapid loss of charged ions from the cell,

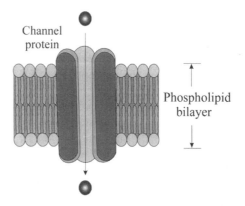

Fig. 6.6. Channel proteins exhibit a "donut" shape.

and thus the reestablishment of the normal negative equilibrium potential of the cell.

In conclusion, the transfer of ions through the membrane leads to changes in the electrical potential of the membrane. The electrical current (I) of ions and the change in electrical potential of the membrane ψ with respect to the initial conditions state ψ_0 are related by

$$I = \frac{\zeta_M}{t}(\psi - \psi_0), \qquad (6.15)$$

where ζ_M is the membrane capacitance, t is the time corresponding to the change in the electrical potential of the membrane, and the subscript 0 indicates the initial state.

Carrier-mediated transport operates in sequence as shown in Figure 6.7: (i) recognition and binding of the target solute; (ii) translocation of the target solute through the membrane (i.e., change of transporter configuration, from "ding" to "dong"); (iii) release of the solute by the carrier protein; and (iv) return of the carrier protein to its original condition. There are some basic features that characterize this transport mechanism. Carriers are specific; that is, there are different carrier proteins for different solutes. These proteins can transport only a single solute (in which case they are known as uniport), or two different solutes in the same direction (symport), or in opposite directions (antiport). Carrier-mediated transport presents a nonlinear relationship between the solute flux and the concentration of solute, as is shown in Figure 6.8. A small concentration produces a linear relationship between the flux and the concentration of solute, as in the case of simple diffusion. At high concentrations the solute flux is independent of concentration; that is, all the carrier sites are occupied or become occupied as soon as they appear.

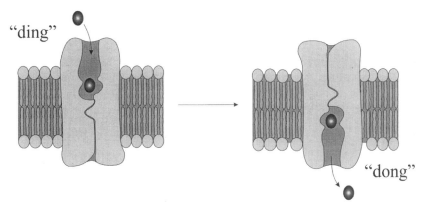

Fig. 6.7. Carrier-mediated transport: the carrier binds to the solute and experiences a change from "ding" to "dong."

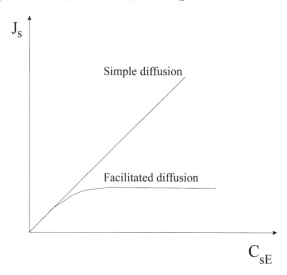

Fig. 6.8. The solute flux J_s versus the extracellular concentration of solute C_{sE} (intracellular concentration constant).

The probability Θ for a carrier to be occupied (i.e., for the carrier to engage a solute) is given by (Freitas Jr., 1999)

$$\Theta = \frac{C_c}{\lambda_{d-a} + C_c} \tag{6.16}$$

with

$$\lambda_{d-a} = \frac{\lambda_d}{\lambda_a}, \tag{6.17}$$

where C_c is the solute concentration available to be carried across the membrane (m^{-3} or $\mathrm{mol\,m}^{-3}$, depending on whether the concentration is defined on a molecular or molar basis), λ_a is the solute-carrier association rate constant ($\mathrm{m^3\,s^{-1}}$ or $\mathrm{m^3\,mol^{-1}\,s^{-1}}$), and λ_d is the solute-carrier complex dissociation rate constant (s^{-1}).

(4) *Ion pumps*: The proper functioning of cells requires differences in the concentration of potassium ions (high inside the cell, and low outside) and sodium ions (high outside the cell, and low inside). As was mentioned already, cells contain a concentration of potassium ions much higher than that in the extracellular fluid, whereas in the extracellular fluid the concentration of sodium ions is higher than inside the cell. The reestablishment of the necessary concentration differences requires the transport of potassium

ions into the cell and sodium ions out of the cell, against the existing electro-chemical potential gradient. Such transport requires specific carrier proteins and the use of cellular energy. This is known as the primary active transport (Figure 6.9). The high amount of energy provided by the hydrolysis of adenosine triphosphate (ATP) into adenosine diphosphate (ADP) on the cytoplasmatic side of the membrane is enough to pump two potassium ions into the cell and three sodium ions out of the cell. This ionic transport is known as Na–K ATPase (Ewart and Klip, 1995). Other ions such as calcium, magnesium, and chloride can also be transported by specific carrier proteins, in a similar process. The mechanism of action of Na–K ATPase has been discussed recently by Tsong (2002) and Tsong and Chang (2003).

The separation of potassium and sodium ions accomplishes several vital functions. It establishes a net charge across the cell membrane, the exterior of the cell being positively charged with respect to the inner part, and preserves the cell membrane potential (Läuger, 1991). Furthermore, the accumulation of sodium ions outside the cell draws water out of the cell, thus maintaining the osmotic balance and preventing the cell from swelling and bursting. The ion gradients established by the primary active transport may permit the transport of an uncharged solute such as glucose or amino acid into the cell against the respective concentration gradient. This process is known as the secondary active transport.

The functioning of the cells requires higher sodium ion concentration outside the cell than inside the cell. Under this gradient of concentration, some of these ions bind to a specific carrier protein that is transported into the cell. This allows the uncharged solute to bind to the same carrier. Through a change in the carrier configuration (from "ding" to "dong"), both the sodium ion and the uncharged solute cross the membrane and are released into the cell. The carrier protein reverts to its original configuration, and the sodium ion is pumped out of the cell by the Na–K ATPase pump described earlier.

The minimum energy required to pump ions of solute from one side of the cell membrane of concentration $C_{s,l}$ to the other side with the concentration $C_{s,h}$ is the minimum energy required for primary active transport. It is given by

$$\Delta G = K_{Bt} T \ln \frac{C_{s,h}}{C_{s,l}} + \frac{z F \Delta \psi}{N_A}, \qquad (6.18)$$

where K_{Bt} is the Boltzmann constant, z is the ion number (i.e., $+1$ for Na^+, -1 for Cl^-, $+2$ for Ca^{2+}, and zero for neutral solutes), F is Faraday's constant, $\Delta \psi$ is the variation of the electrical potential across the membrane, and N_A is Avogadro's constant.

The knowledge obtained from the Na–K ATPase and other similar processes allowed the development of new products and techniques of great importance in biomedicine. Artificial ion-gated membranes (Burgmayer and Murray, 1982) and voltage-gated membranes (Nishizawa et al., 1995) have been reported in the literature. The electroporation technique found

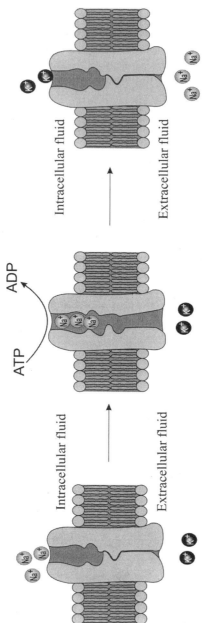

Fig. 6.9. Primary active transport.

widespread use in genetic studies. This technique permits the insertion of molecules such as DNA into cells by using a short and intense pulse of energy, such as a laser pulse, to open the cellular pores.

(5) *Endocytosis and exocytosis*: These mechanisms are shown in Figure 6.10. They contribute little to the total transfer of small solutes across the cell membrane, but may be significant in the transfer of large particles and molecules such as proteins (Lauffenberger and Linderman, 1993). When the solute approaches the cell membrane, a portion of the cell cytoplasm surges forward to surround the solute, and forms a vesicle containing the extracellular fluid and the associated solute. The vesicle then drifts further into the body of the cell. This is known as endocytosis. The opposite process is exocytosis. A vesicle containing fluid and its associated solute drifts in the direction of the cell membrane, where it forms a temporary opening through which the vesicular contents are released into the extracellular fluid.

Inflammations and other health problems can stimulate the local accumulation of fluid in the intracellular space. This fluid accumulation is known as edema, and can arise as a result of the increase of cell-membrane diffusivity to sodium, or due to a malfunction of the sodium-ion pump. In both cases, there is an accumulation of sodium ions within the cell, and extra water flux due to the increase in osmotic pressure.

6.3.2 Capillary Wall

Capillaries are typically 2 to 6 μm in radius, and less than 1 mm long (Guyton, 1991). The walls of these vessels are composed of small pores having diameters of about 6 to 7 nm (nanoporous structures), and consist of a single

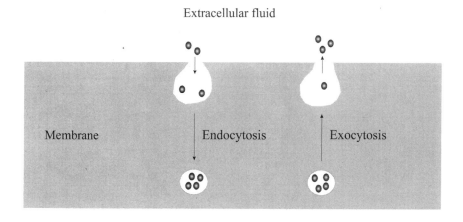

Fig. 6.10. Endocytosis and exocytosis.

layer of endothelial cells cut by intercellular clefts set on a basement membrane, as is shown in Fig. 6.11. Endothelial cells serve as regulators of ion and molecule transport. They can be classified into three types (Renkin, 1988): continuous, fenestrated, and discontinuous. Continuous endothelium forms a continuous ring that is connected by tight junctions to the basement membrane. It is present in the capillaries of the brain, heart, and lungs. Fenestrated endothelium is present in the capillaries of the intestines, glands, and kidney glomerulus. It is characterized by endothelial cells pinched together at numerous points forming openings or gaps called fenestrae (literally, windows). In some cases, a fenestra contains a membrane with selective properties. Discontinuous endothelium is characterized by large openings or gaps of up to 0.2 μm, and is present in the liver, bone marrow, and spleen.

Gaps or openings present in fenestrated and discontinuous endothelia contribute to the transport of ions such as sodium and chloride. Some endothelial cells also have a number of vesicles that can provide paths for transporting proteins (large molecules).

The single layer of endothelial cells and the basement membrane form the capillary wall, which is typically 0.5 μm thick. Because of the absence of smooth muscle cells, capillaries have a very limited ability to contract or dilate.

Along the capillary network (Figure 6.2), pressure varies from about 4 kPa at the arterial end to about 1.7 kPa at the venous end. As a result, blood moves along the capillaries, while exchanging essential substances between the blood and the tissue surrounding the capillaries. Diffusion is the

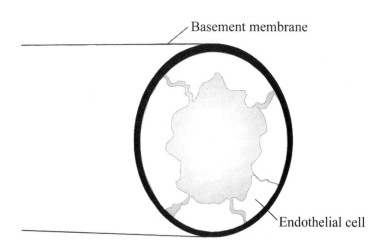

Fig. 6.11. Capillary wall structure: endothelial cell and basement membrane.

most important means of transporting nutrients, oxygen, carbon dioxide, and metabolic end-products across the capillary wall. This is mainly due to certain factors:

(1) relatively high diffusivity of the capillary wall to the majority of these substances (typically $10^{-11} - 10^{-9}\,\mathrm{m^2s^{-1}}$), together with a small thickness ($\sim 0.5\,\mu\mathrm{m}$);
(2) numerous branches and tributaries of the capillary network, which offer a very large surface area for diffusion; and
(3) very small blood velocity along the capillary network ($\sim 0.06\,\mathrm{cm\,s^{-1}}$), allowing long residence times for blood inside the capillaries.

There are many large molecules (proteins), the sizes of which are larger than the pore diameter of the capillary wall. Such molecules are retained on one side of the membrane. The resulting difference in protein concentration between the blood and the tissue surrounding the capillary induces an oncotic pressure difference of about 2.6 kPa. This is offset, in part, by a hydrostatic pressure difference between the capillary blood and the interstitial space, which produces a convective flow (filtration) of fluid across the capillary wall. Recall that the pressure difference varies along the capillary from about 4.7 kPa at the arterial end to about 1.7 kPa at the venous end. Thus, near the arterial end of the capillary, the hydrostatic pressure difference between the blood and the tissue exceeds the oncotic pressure difference. As a result, there is a net flow of fluid from the capillary into the tissue surrounding the capillary. However, at the venous end the difference in oncotic pressure exceeds the difference in hydrostatic pressure, and this produces a net flow of fluid to the capillary. The end result of these outflows and inflows along the capillary wall is a net flow of fluid from the blood into the tissue surrounding the capillary. The fluid not reabsorbed by the capillary is collected in the interstitial space, and under normal circumstances it is drained completely by the lymphatic system. The lymphatic system is a unidirectional flow system that removes the excess fluid from the tissues and returns it to the blood. In a number of diseases, the accumulation of interstitial fluid (edema) can occur in many places in the body. This can be due to the malfunction of the lymphatic system, the increase of the filtration flow from the capillary, and the retention of sodium in the interstitial space. The latter causes osmosis of water into the tissue.

Large particles and molecules are transferred across the endothelium by vesicles (see endocytosis and exocytosis, Figure 6.10). In addition, vesicles can also transport water and small solvent, but they contribute very little to the total transfer of such substances (Lauffenberger and Linderman, 1993). The absence of transmembrane proteins in the capillary wall prohibits the existence of mechanisms that involve both channel and carrier proteins, that is, the mechanisms of facilitated diffusion and ion pump.

6.4 Transport of Neutral Solutes Across Membranes

The transport of small solutes across biomembranes occurs as combinations of several distinct processes, such as diffusion, osmosis, and convection. The interdependence of these processes is likely to occur, and this recommends the use of nonequilibrium thermodynamics for the purpose of describing membrane processes (Kondepudi and Prigogine, 1998).

Consider a thermodynamic system composed of a solute and a solvent, which are separated by an isothermal membrane. The solute and the membrane (e.g., capillary wall) are not charged electrically. In order to obtain the general form of the mass fluxes and the respective driving forces, we start from the expression for the entropy production inside a unit control volume, Equation (1.66),

$$\dot{S}_{gen}''' = -\frac{\mathbf{j}_v \nabla P_v}{C_v T} - \frac{\mathbf{j}_s \nabla P_s}{C_s T} \qquad (6.19)$$

in which

$$\mathbf{j}_v = C_v \mathbf{v}_v \qquad \mathbf{j}_s = C_s \mathbf{v}_s. \qquad (6.20)$$

Here C is the concentration. The subscripts v and s stand for solvent and solute, respectively. Assuming that the solvent and solute form an ideal solution with the pressure P, according to Raoult's law the respective pressures are

$$P_v = \frac{C_v}{C}P, \qquad P_s = \frac{C_s}{C}P, \qquad (6.21)$$

where, in addition,

$$C = C_v + C_s. \qquad (6.22)$$

Then, assuming constant concentration C, Equations (6.19) and (6.21) yield

$$\dot{S}_{gen}'' = -\frac{\mathbf{v}}{T} \cdot \nabla P - \mathbf{v}_{sv} \cdot R\nabla C_s, \qquad (6.23)$$

where $\mathbf{v}$ represents the average volumetric flux of solution $(\mathbf{v}_v C_v + \mathbf{v}_s C_s)/C$, $\mathbf{v}_{sv}$ the volumetric flux of solute with respect to solvent $(\mathbf{v}_s - \mathbf{v}_v)$, and R the ideal gas constant.

For small concentrations and pressure gradients, the fluxes across the membrane can be expressed as linear functions of the forces

$$\mathbf{v} = -\frac{L_{11}}{T}\nabla P - L_{12}R\nabla C_s, \qquad (6.24)$$

$$\mathbf{v}_{sv} = -\frac{L_{21}}{T}\nabla P - L_{22}R\nabla C_s, \qquad (6.25)$$

where L_{11}, L_{12}, L_{21}, and L_{22} are phenomenological coefficients. According to Onsager's Reciprocity Theorem, the cross-phenomenological coefficients L_{12} and L_{21} are equal. Therefore, Equations (6.24) and (6.25) can be rewritten as

$$\mathbf{v} = \frac{L_{11}}{T}(-\nabla P + \sigma RT\nabla C_s), \qquad \cdot \qquad (6.26)$$

$$\mathbf{v}_{sv} = \frac{L_{11}}{T}\left(\sigma\nabla P - \frac{L_{22}}{L_{11}}RT\nabla C_s\right), \qquad (6.27)$$

with

$$\sigma = -\frac{L_{12}}{L_{11}} = -\frac{L_{21}}{L_{11}}. \qquad (6.28)$$

The coefficient σ is a measure of the restrictions of the pores of the membrane to the entry of a particular solute. It is also known as the Staverman reflection coefficient (Staverman, 1951). It varies between 0 and 1, as shown in Figure 6.12. When the reflection coefficient is equal to 1, the molecules of the solute are completely barred from crossing the pores of the membrane. This means that the solute is completely "reflected" by the membrane, or that the membrane is completely selective to the solute. When the reflection coefficient is zero, the molecules of solute may cross the pores of the membrane freely. Reflection coefficients for different microvessels and solutes may be found in Michel (1997).

The total transfer of solute across the membrane ($\mathbf{J}_s$) is given by

$$\mathbf{J}_s = (\mathbf{v} + \mathbf{v}_{sv})C_{sM}, \qquad (6.29)$$

where C_{sM} is the average concentration of solute within the membrane. By substituting Equations (6.12) and (6.27) into (6.29) we obtain

$$\mathbf{J}_s = C_{sM}(1 - \sigma)\mathbf{v} + C_{sM}L_{11}\left(\sigma^2 - \frac{L_{22}}{L_{11}}\right)R\nabla C_s. \qquad (6.30)$$

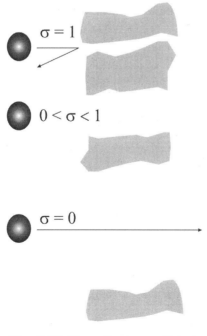

Fig. 6.12. The reflection coefficient.

The unknown phenomenological coefficient L_{11} may be determined by considering the limiting situation of a nonselective membrane ($\sigma = 0$). Then Equation (6.26) reads

$$\mathbf{v} = -\frac{L_{11}}{T}\nabla P. \tag{6.31}$$

This equation represents the Darcy flow (cf. Section 1.3) across the capillary wall—the process of filtration—which leads to identifying

$$\frac{L_{11}}{T} = \frac{K}{\mu}, \tag{6.32}$$

where μ is the dynamic viscosity of the solution. The hydraulic permeability of the membrane K is known in biomedical literature as the filtration coefficient. Experimental measurements of the capillary permeability reported in the literature reveal that K varies between 10^{-23} and $10^{-17}\,\mathrm{m}^2$, with the lower and higher scale values corresponding to capillaries with continuous and discontinuous endothelial cells, respectively. For example, the glomeruli of the kidney capillary wall (fenestrated capillary) has a permeability of about $10^{-19}\,\mathrm{m}^2$.

If Darcy flow is absent ($\mathbf{v} = 0$), then Equation (6.30) becomes

$$\mathbf{J}_s = C_{sM}L_{11}\left(\sigma^2 - \frac{L_{22}}{L_{11}}\right)R\nabla C_s. \tag{6.33}$$

This corresponds to pure diffusion of solute, or Fickian diffusion. By comparing this expression with Fick's law, we make the identification

$$\left(\sigma^2 - \frac{L_{22}}{L_{11}}\right) = -\frac{D_M}{L_{11}RC_{sM}}, \tag{6.34}$$

where D_M is the solute diffusivity, or the solute diffusion coefficient. For biomembranes, the diffusivities of solutes vary in a wide range, from 10^{-19} to $10^{-9}\,\mathrm{m}^2\mathrm{s}^{-1}$, with the lower and higher values corresponding to proteins and small solutes (ions), respectively. The substitution of Equations (6.32) and (6.34) into (6.26) and (6.30) yields

$$\mathbf{v} = -\frac{K}{\mu}(\nabla P - \sigma\nabla\Pi), \tag{6.35}$$

$$\mathbf{J}_s = C_{sM}(1 - \sigma)\mathbf{v} - D_M\nabla C_s, \tag{6.36}$$

where Π is the osmotic pressure, or $RT\nabla C_s$. Equations (6.35) and (6.36) are known as the Kedem–Katchalsky equations (Kedem and Katchalsky, 1958). The first term on the right side of (6.35) represents the primary force that drives the filtration, and the second term stands for the additional contribution to the filtration arising from the osmotic pressure. On the right side of Equation (6.36), the two terms represent the solute transport arising from filtration, and the solute flux due to diffusion, respectively. The application of (6.35) and (6.36) is not restricted to natural biomembranes. These relations are also applicable to synthetic uncharged membranes, which are often used in a wide variety of biomedical applications.

Parameters K, D_m, and σ completely characterize the functioning of a membrane as a system for solute and solvent transport. Analytical equations are available in the literature for defining these parameters in relation to the geometric characteristics of the membrane and the properties of the solute. However, as the membrane structure is too complex to allow the derivation of these parameters, they are usually determined by fitting experimental data to the transport equations. Models are particularly useful in membrane design, or whenever the available experimental data are few. A summary of relevant expressions for calculating the permeability K is presented in Chapters 1 and 4. Expressions for calculating the reflection coefficient σ for special geometric arrangements are listed in Table 6.1.

The diffusion of solute molecules through membranes is a less efficient process than free diffusion in a bulk medium, because of the following factors.

(1) A solute molecule that is not much smaller than the size of the pores cannot be distributed evenly over the whole pore cross-sectional area. The closest that its center can get to the pore wall is a distance equal to the radius of the solute. The actual space that is available to the large solute is bounded by the dashed line in Figure 6.13a. This effect is known as steric exclusion. As a consequence, the equilibrium concentration in the bulk solution is higher than in the "mouth" of the pore (Figure 6.13b).

(2) Molecules that are inside the pores feel the hydrodynamic drag produced by the solvent over the surface of the molecule (Figure 6.14). In this way, the mobility of the solute molecules decreases.

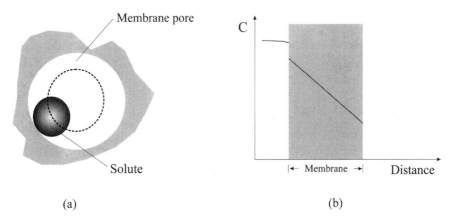

(a)   (b)

Fig. 6.13. Concept of steric exclusion: (a) large solute inside a membrane pore; (b) schematic of concentration profile across the membrane.

Table 6.1. Reflection coefficients (σ) for special geometric arrangements

Geometry	Reflection coefficient	Validity domain	Source
Capillary tubes with diameter d; spherical solute with diameter d_s	$\sigma = 1 - \left(1 - \dfrac{d_s}{d}\right)^2 \left[2 - \left(1 - \dfrac{d_s}{d}\right)^2\right]$ $\left[1 - \dfrac{2}{3}\left(\dfrac{d_s}{d}\right)^2 - 0.163\left(\dfrac{d_s}{d}\right)^3\right]$	if $\dfrac{d_s}{d} > 1$, then set $\dfrac{d_s}{d} = 1$	Anderson and Quinn (1974)
Narrow capillary fissures with width W; spherical solutes with diameter d_s	$\sigma = 1 - 2\left(1 - \dfrac{d_s}{W}\right)^2 + \left(1 - \dfrac{d_s}{W}\right)^4$	if $\dfrac{d_s}{W} > 1$, then set $\dfrac{d_s}{W} = 1$	Bean (1972); Curry (1974)
Random fiber arrangement with diameter d_c and solidity α; spherical solute with diameter d_s	$\sigma = 1 - 2\left\{\exp\left[-\alpha\left(\dfrac{2d_s}{d_c} + \dfrac{d_s^2}{d_c^2}\right)\right]\right\}$ $+ \left\{\exp\left[-\alpha\left(\dfrac{2d_s}{d_c} + \dfrac{d_s^2}{d_c^2}\right)\right]\right\}^2$		Ogston et al. (1973); Curry and Michel (1980)

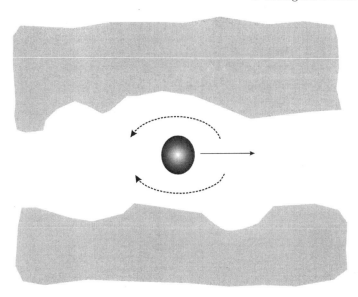

Fig. 6.14. Hydrodynamic drag produced by the movement of solvent through solute in the pore.

(3) The tortuosity of the pores leads to additional effects that reduce the diffusion rate of solute. In tortuous pores the solute molecule has to follow a path that is longer than the thickness of the membrane, therefore lowering the effective gradient of concentration along the path.

In summary, because of (1) to (3) the diffusion coefficient in membranes (D_M) is lower than the bulk diffusion coefficient (D). For dilute solutions, the bulk diffusion coefficient can be predicted from the Stokes–Einstein equation (Bird *et al.*, 1960)

$$D = \frac{RT}{3\pi\mu d_s \mathrm{N_A}},\tag{6.37}$$

where $\mathrm{N_A}$ is Avogadro's constant and d_s is the diameter of the solute. The relation between the two diffusion coefficients is (Renkin, 1954; Deen, 1987)

$$D_M = \frac{f_{dm} f_{pt} D}{T^*},\tag{6.38}$$

where the solute partition coefficient f_{pt} represents the fraction of pore volume available to the solute molecule, and accounts for steric exclusion. The coefficient f_{dm} accounts for the decrease in the solute mobility, and T^* is the tortuosity. Expressions for calculating the coefficients f_{pt} and f_{dm} in particular geometric arrangements are listed in Table 6.2.

The biomembrane structure is too complex to allow the derivation of the diffusivity based on Equation (6.38). For this reason, methods for determining the membrane diffusivity from appropriate experimental data are often used, which are described next.

Table 6.2. The coefficients f_{pt} and f_{dm} in special geometric arrangements

Geometry	Coefficient	Validity domain	Source
Cylindrical capillary tubes with diameter d; and spherical solutes with diameter d_s	$f_{dm} = 1 - 2.1\dfrac{d_s}{d} + 2.09\left(\dfrac{d_s}{d}\right)^3 - 0.95\left(\dfrac{d_s}{d}\right)^5$	$\dfrac{d_s}{d} < 0.4$	Beck and Schultz (1970); Paine and Scher (1975)
	$f_{dm} = \dfrac{1 - 1.25(d_s/d)\ln(d_s/d) - 1.54(d_s/d)}{(1 - d_s/d)^2}$	$\dfrac{d_s}{d} < 0.2$	Brenner and Gaydos (1977)
	$f_{pt} = \left(1 - \dfrac{d_s}{d}\right)^2$	$\dfrac{d_s}{d} < 1$	Ferry (1936)
Narrow capillary fissures with width W; spherical solutes with diameter d_s	$f_{dm} = 1 - \dfrac{d_s}{W} + 0.42\left(\dfrac{d_s}{W}\right)^3 + 2.1\left(\dfrac{d_s}{W}\right)^4 - 1.7\left(\dfrac{d_s}{W}\right)^5$		Curry (1974)
	$f_{pt} = \left(1 - \dfrac{d_s}{W}\right)^2$		Ferry (1936)
Random fiber arrangement with diameter d_c and solidity α; spherical solutes with diameter d_s	$f_{dm} = \exp\left[-\alpha^{0.5}\left(1 + \dfrac{d_s}{d_c}\right)\right]$		Ogston et al. (1973); Curry and Michel (1980)
	$f_{pt} = \exp\left[-\alpha\left(\dfrac{2d_s}{d_c} + \dfrac{d_s^2}{d_c^2}\right)\right]$		

(1) *Unsteady diffusion*: In most applications, the transport across membranes can be modeled as one-dimensional, because membranes are very thin. If J_{sx} is the total flux of solute normal to the plane of the membrane (the x-direction), and if Darcy flow is absent ($\mathbf{v} = 0$), then Equation (6.36) reads (see also Sections 1.6 and 4.8)

$$J_{sx} = -D_M \frac{\partial C_s}{\partial x}. \tag{6.39}$$

As long as chemical reactions in the membrane are negligible, according to Equation (1.40) the variation of the solute concentration within the membrane is given by

$$\phi \frac{\partial C_s}{\partial t} = D_M \frac{\partial^2 C_s}{\partial x^2}, \tag{6.40}$$

where ϕ is the membrane porosity. For the membrane shown in Figure 6.15 the boundary conditions are

$$C_s = \text{constant} \quad \text{at } x = 0 \text{ and } x = W; \quad \frac{\partial C_s}{\partial x} = 0 \quad \text{at } x = \frac{W}{2}; \tag{6.41}$$

and Equation (6.40) has the solution (Gates and Newman, 2000)

$$C_s = C_{s0} + \sum_{n=1}^{\infty} \beta_i \exp\left(-\frac{n^2 \pi^2 D_M}{\phi W^2} t\right) \cos\left(\frac{n\pi x}{W}\right). \tag{6.42}$$

At extended times, the Fourier series converges, and a plot of $\ln(C_s - C_{s0})$ versus time has a slope equal to $\pi^2 D_M / W^2$. The coefficient D_M may be obtained from the slope of the curve $\ln(C_s - C_{s0})$ versus t, by plotting the experiment concentration data against time.

(2) *Steady diffusion*: In the steady state, Equation (6.40) reduces to $\partial^2 C_s / \partial x^2 = 0$, and the solute flux is given by

$$J_{sx} = D_M \frac{C_{s0} - C_{sW}}{W}. \tag{6.43}$$

Under the same conditions, the mass conservation during transportation can be written in the form

$$\left\{\begin{array}{c} \text{rate of solute} \\ \text{leaving side 0} \end{array}\right\} = \left\{\begin{array}{c} \text{rate of solute transported} \\ \text{through the membrane} \end{array}\right\} = \left\{\begin{array}{c} \text{rate of solute} \\ \text{entering side W} \end{array}\right\}. \tag{6.44}$$

Let A_M be the surface area of the membrane normal to the direction of solute transport, and V_M the membrane volume (Figure 6.15). Equation (6.44)

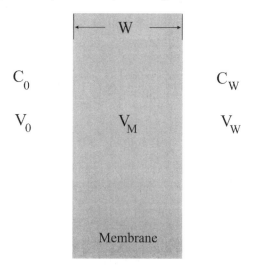

Fig. 6.15. Membrane with thickness W surrounded by volumes V_0 and V_W.

expresses two equalities

$$-V_0 \frac{\partial C_{s0}}{\partial t} = A_M D_M \frac{C_{s0} - C_{sW}}{W}, \qquad (6.45)$$

$$A_M D_M \frac{C_{s0} - C_{sW}}{W} = V_W \frac{\partial C_{sW}}{\partial t}. \qquad (6.46)$$

Generally, the volume of the cell membrane is much smaller than the volume of the confining fluid. If the volumes of fluid V_0 and V_W on the two sides of the membrane are approximately equal, then (6.45) and (6.46) yield

$$V_0 \frac{\partial (C_{s0} - C_{sW})}{\partial t} = -2A_M D_M \frac{C_{s0} - C_{sW}}{W}. \qquad (6.47)$$

By integrating Equation (6.47) subject to the initial conditions $C_{s0} = C_{t0}$ and $C_{sW} = 0$ leads to

$$C_{s0} = 0.5C_{t0} \left[1 + \exp \left(-\frac{2A_M D_M}{W V_0} t \right) \right]. \qquad (6.48)$$

Under the above assumptions, Equation (6.48) provides the basis for a simple method of determining the membrane diffusivity D_M. The characteristic time for diffusion in the membrane is $t_{DM} = W^2/D_M$ (Bird et al., 1960), and according to (6.48) the characteristic time for changes in concentration on the two sides of the membrane is $(t_0, t_w) = WV_0/(2A_M D_M)$. Combining these two relations yields

$$(t_0, t_w) = \frac{V_0}{2V_M} t_{DM}.$$ (6.49)

When $(V_0, V_W) \gg V_M$, as was assumed, we see that $(t_0, t_w) \gg t_{DM}$.

6.5 Transport of Charged Solutes Across Membranes

In the cell membrane Darcy flow is absent. The passive transport of solutes is due to the combined effect of the concentration gradients and the electrical potential gradient across the cell membrane. The solute flux can be obtained by the well-known Nernst–Planck equation [cf. Equation (4.86)].

For the x component of the solute transport, and for the one-dimensional case, Equation (4.86)) can be written as

$$J_{sx} = -D_M \left(\frac{\partial C_s}{\partial x} + \frac{z\mathrm{F}}{RT} C_s \frac{\partial \psi}{\partial x} \right).$$ (6.50)

The solute diffusivity D_M can be related to the mobility v through the Einstein relation [cf. Equation (4.80)], such that Equation (6.50) becomes

$$J_{sx} = -v \left(RT \frac{\partial C_s}{\partial x} + z\mathrm{F}C_s \frac{\partial \psi}{\partial x} \right).$$ (6.51)

The integration of Equation (6.51) subject to the boundary conditions $C_s = C_{s0}$(at $x = 0$) and $C_s = C_{sW}$(at $x = W$) (see also Figure 6.15) yields

$$J_{sx} = vRT \frac{C_{s0} - C_{sW} \exp(z\mathrm{F}\Delta\psi/RT)}{\int_0^w \exp(z\mathrm{F}\psi/RT)\,\mathrm{d}x}.$$ (6.52)

This equation describes the flux of ions across the cell membrane. It is an important tool for the detailed analysis of this process.

6.5.1 Membrane Potential

The diffusion of a certain ion under a concentration gradient generates an electric gradient that cannot be maintained without energy consumption (see Section 6.2). When concentration and electric gradients for an ion solute balance each other, there is no net flux of ions ($J_{sx} = 0$). Because the integral shown in Equation (6.52) is nonzero for all finite values of ψ, it can be disregarded in this case, and reduces to equation

$$C_{sW} = C_{s0} \exp\left(-\frac{zF\Delta\psi}{RT}\right).$$ (6.53)

The above relation represents the equilibrium concentration of a particular ion as a function of the electric potential in the membrane ($E = \Delta\psi$). The equilibrium membrane potential E reads, after using (6.53),

$$E = \frac{RT}{zF} \ln \frac{C_{s0}}{C_{sW}}.$$ (6.54)

The membrane potential E can be regarded as an electrical measure of the diffusion strength of a particular ion that arises from unequal concentrations on the two sides of the membrane. Equation (6.54) is known as the Nernst equation.

In living cells, the ions of potassium, sodium, and chloride are exchanged through the cell membrane (cf. Section 6.3). In the steady state, the ion concentration is constant in time, and the flux of ions across the membrane is zero,

$$J_{sx}^{K^+} + J_{sx}^{Na^+} + J_{sx}^{Cl^-} = 0.$$ (6.55)

In view of Equation (6.52), the above equation becomes

$$\frac{C_{s0}^{K^+} - C_{sW}^{K^+} \exp\left(\dfrac{F\Delta\psi}{RT}\right)}{\int_0^w \exp\left(\dfrac{F\psi}{RT}\right) dx} + \frac{C_{s0}^{Na^+} - C_{sW}^{Na^+} \exp\left(\dfrac{F\Delta\psi}{RT}\right)}{\int_0^w \exp\left(\dfrac{F\psi}{RT}\right) dx}$$

$$+ \frac{C_{s0}^{Cl^-} - C_{sW}^{Cl^-} \exp\left(-\dfrac{F\Delta\psi}{RT}\right)}{\int_0^w \exp\left(-\dfrac{F\psi}{RT}\right) dx} = 0.$$ (6.56)

In order to evaluate the integrals appearing in the denominator of Equation (6.56), it is common to assume that the membrane potential is a linear function of the distance across the membrane,

$$\psi = \frac{x}{W}\Delta\psi.$$ (6.57)

The integration of Equation (6.56) subject to (6.57) and the boundary conditions $C_s = C_{s0}$(at $x = 0$) and $C_s = C_{sW}$(at $x = W$) yields (see also Figure 6.15)

$$E^{K,Na,Cl} = \frac{RT}{F} \ln \left(\frac{P_M^{K^+} C_{sW}^{K^+} + P_M^{Na^+} C_{sW}^{Na^+} + P_M^{Cl^-} C_{s0}^{Cl^-}}{P_M^{K^+} C_{s0}^{K^+} + P_M^{Na^+} C_{s0}^{Na^+} + P_M^{Cl^-} C_{sW}^{Cl^-}} \right). \tag{6.58}$$

The coefficient P_M is defined as the ratio between the solute diffusivity and the membrane thickness (D_M/W), and is known in biomedical literature as the solute permeability. Note that the solute permeability is not an intrinsic property of the membrane, because it depends on the membrane thickness. Equation (6.58) is known as the Goldman–Hodgkin–Katz equation, and is a generalization of the Nernst equation (6.54). This equation is based on the assumption that the membrane potential is a linear function of the distance across the membrane. Although this assumption is not completely valid for all cases, it serves as an useful approximation for predicting the potential of many membranes.

In the methods considered above, the processes of generating and maintaining the concentration gradients across the membrane were not considered. As mentioned in Section 6.3, the reestablishment of the necessary concentration gradients for continued diffusion of ions requires the use of specific carrier proteins and cellular energy (active transport). In membranes there are fluxes of ions due to both active and passive transport. Therefore, in the steady state, the net flux of ions across the membrane is

$$J_{sx}^{active} + J_{sx}^{passive} = 0. \tag{6.59}$$

The active transport has to be included in the membrane potential model when sodium and potassium ions diffuse down their concentration gradients, and the reestablishment of concentration gradients is made by active transport through the Na–K ATPase pump. Assuming that only sodium and potassium ions are involved in the process, in the steady state the equations accounting for the sodium and potassium fluxes read

$$J_{sx}^{Na^+,active} + J_{sx}^{Na^+,passive} = 0, \tag{6.60}$$

$$J_{sx}^{K^+,active} + J_{sx}^{K^+,passive} = 0. \tag{6.61}$$

On the other hand, the equation accounting for the transport ratio due to Na–K ATPase is (Mullins and Noda, 1963)

$$\frac{J_{sx}^{Na^+,active}}{n^{Na^+}} + \frac{J_{sx}^{K^+,active}}{n^{K^+}} = 0, \tag{6.62}$$

where n^{K^+} and n^{Na^+} are the number of potassium ions pumped into the cell and sodium ions pumped out of the cell, respectively. From Equations (6.60) to (6.62) we also conclude that

$$J_{sx}^{\mathrm{Na}^+,\mathrm{passive}} = -\frac{n^{\mathrm{Na}^+}}{n^{\mathrm{K}^+}} J_{sx}^{\mathrm{K}^+,\mathrm{passive}}. \tag{6.63}$$

The substitution of Equation (6.52) for sodium and potassium ions into (6.63) yields

$$E^{\mathrm{K},\mathrm{Na}} = \frac{RT}{\mathrm{F}} \ln \left[\frac{(n^{\mathrm{Na}^+}/n^{\mathrm{K}^+})P_M^{\mathrm{K}^+} C_{sW}^{\mathrm{K}^+} + P_M^{\mathrm{Na}^+} C_{sW}^{\mathrm{Na}^+}}{(n^{\mathrm{Na}^+}/n^{\mathrm{K}^+})P_M^{\mathrm{K}^+} C_{s0}^{\mathrm{K}^+} + P_M^{\mathrm{Na}^+} C_{s0}^{\mathrm{Na}^+}} \right]. \tag{6.64}$$

This equation represents the equilibrium membrane potential in the presence of active transport, and is known as the Mullins–Noda equation. Equation (6.64) can be compared with (6.58), which is based on neglecting active transport, if only sodium and potassium ions are considered. The net effect of the active transport is to change only the permeability of the ions, because of the factor $n^{\mathrm{Na}^+}/n^{\mathrm{K}^+}$, which is equal to 3/2 (see Figure 6.9). This result explains why passive transport-based models match data from real cell membranes well, despite the inaccuracy in the assumption on which the models are based.

6.5.2 Electrical Equivalent Circuit

Equation (6.51) can be rewritten for electrical current by multiplying each term by the ion number and the Faraday constant

$$I = z\mathrm{F}\upsilon \left(RT\frac{\partial C_s}{\partial x} + z\mathrm{F}C_s\frac{\partial \psi}{\partial x} \right). \tag{6.65}$$

The electrical current density is defined as positive in the opposite direction of the flux, in order to be consistent with the usual convention of electric circuit theory and the second law of thermodynamics. Equation (6.65) can be integrated between the two sides of the membrane,

$$\frac{RT}{z\mathrm{F}} \int_0^W \frac{d\ln C_s}{dx} dx + \int_0^W \frac{d\psi}{dx} dx = \int_0^W \frac{I}{z^2\mathrm{F}^2\upsilon C_s} dx. \tag{6.66}$$

If the current I is constant along the membrane (i.e., the current I is independent of x), Equation (6.66) becomes

$$\Delta V_M = IR_M. \tag{6.67}$$

This states that the current I is driven by a force (ΔV_m),

$$\Delta V_M = \Delta\psi - E \tag{6.68}$$

through a membrane the resistance of which is

$$R_M = \int_0^W \frac{1}{z^2\mathrm{F}^2\upsilon C_s} dx. \tag{6.69}$$

This resistance depends, among other factors, on the concentration C_s and mobility of the solute v. Equation (6.67) is analogous to Ohm's law for electrical circuits. However, this equation is a generalization of Ohm's law to the case of ions responding to electrochemical gradients, instead of electrical gradients only.

In practice, Equation (6.67) is a very important tool in the analysis of the transport of ions through charged membranes. Based on this equation, Alan Lloyd Hodgkin and Andrew Fielding Huxley, winners of the Nobel Prize in Physiology or Medicine in 1963 for their studies on the nature of the voltage-gated channels in the squid axon, were able to determine experimentally the membrane resistance of the squid axon. In fact, Equation (6.67) relates three quantities two of which have to be determined experimentally in order to calculate the third. With this objective, Hodgkin and Huxley (Nobel Lectures, 1973)

> ... introduced two electrodes into the giant nerve fibre of the squid. One served to clamp the voltage in predetermined steps, the other to measure the current produced during activity. Calculation gave the third quantity, the resistance of the membrane,..., was the one which the experiments were designed to measure..

6.6 The Kidney and the Regulation of Blood Composition

The kidney is more than an organ for excretion. In addition to removing wastes, it regulates continuously and within narrow limits the chemical composition of the blood. If excess water, sodium ions, calcium ions, and so on are present, the excess is quickly transferred to the urine. Conversely, the kidneys step up their reclamation of these same substances when they are present in the blood in insufficient amounts. So the right levels are maintained to keep the body healthy.

The human kidneys are bean-shaped organs and receive 20 to 25% of the total arterial blood pumped by the heart. Inside the kidneys, the blood is filtered through very fine networks of tubes called nephrons (Figure 6.1). Each kidney has about 1 million nephrons. The nephron is a tube closed at one end and open at the other. Waste products in the blood move across from the bloodstream (the capillaries) into the urine-carrying tubes (the tubules) inside the nephron. As the blood passes through the blood vessels of the nephron, all unwanted waste is removed. Any chemicals needed by the body are kept or returned to the blood stream by the nephrons. Capillary networks are located inside the glomerulus, in the vicinity of the proximal tubule and the distal tubule.

Nephric filtrate collects within Bowman's capsule, and then flows into the proximal tubule. Here the glucose, amino acids, large amounts of uric acid, and about 70% of inorganic salts are reabsorbed by active transport. As these solutes are removed from the nephric filtrate, a large volume of the water follows them by osmosis. As the fluid flows into the descending segment of the loop of Henle, water continues to leave by osmosis because the interstitial fluid is very hypertonic. This water flow is caused by the active transport of sodium ions out of the fluid that flows through the tubule, as the fluid moves up the ascending segment of the loop of Henle. In the distal tubules more sodium is reclaimed by active transport, and still more water follows by osmosis. Final adjustment of the sodium and water content of the body occurs in the collecting tubules.

6.6.1 Kidney Failure and Dialysis

For patients with kidney failure, dialysis is a treatment that cleans the blood by means of semipermeable membranes. The following are required in order to achieve full blood cleaning (Daugirdas et al., 2000).

(1) *Depuration of toxic substances accumulated in the body*: This occurs through the processes of diffusion, convection (filtration), and adsorption, and can be used to remove substances such as cytokines and anaphylatoxins. These processes are described in Sections 1.1 and 6.3.

(2) *Hydroelectrolytic patient reequilibrium*: This process involves the removal and uptake of solute by diffusion and convection (filtration). The coefficients used to describe the hydroelectrolytic reequilibrium are called clearance and dialysance. Both coefficients express the ratio between the solute passing through the semipermeable membrane and the concentration. Clearance accounts for the concentration of a given substance in the blood inlet of a device known as a dialyzer (also called an artificial kidney). Dialysance accounts for the difference between concentrations in the blood and the dialysate at the dialyzer inlet.

(3) *Acid-base balance*: Due to the lack of renal function, patients are able to balance the metabolic acidosis. During dialysis the correction of acidosis can be made by administering a buffer substance into the vein, by postdilution or added to the dialysate. Various substances are used, for example, acetate and bicarbonate.

Two dialysis processes are available, peritoneal dialysis and hemodialysis, both working on the same principle. The main difference is that in peritoneal dialysis the blood is cleaned inside the patient's body, whereas in hemodialysis it is cleaned externally by a machine.

The peritoneal cavity in the patient's abdomen is lined by a thin membrane that surrounds the intestines and other internal organs. In the peritoneal dialysis this cavity is filled with dialysis fluid (the liquid into which contaminants are passed during dialysis), which enters the body through a soft plastic tube (catheter). Waste products and excess fluid from the blood pass into the dialysis fluid. In the end, this fluid is eliminated from the patient's body through the catheter.

The hemodialysis machine pumps blood from the patient's body through the dialyzer. The dialyzer has two spaces separated by a thin membrane. This membrane is shaped into small hollow-fiber tubes or into parallel plates. Blood flows on one side of the membrane, and dialysis fluid flows on the other. The wastes and the fluid excess, which patients accumulated between treatments, pass from the blood through the membrane into the dialysis fluid, which is later eliminated. The cleaned blood is returned to the patient's body.

Dialyzer membranes can be of several types:

(1) Cellulose derived from wood products or processed cotton;
(2) Substituted cellulose: cellulose acetate that has acetate bound to the free hydroxyl groups, or modified cellulose that has tertiary amino compound bound to the groups; and
(3) Synthetic thermoplastics such as polyacrylonitrile, polymethylmethacrylate, polyamide, and polysulfone.

Cellulose membranes are cheaper, but the presence of free hydroxyl groups along their surfaces seems to be connected to some side effects (Gutch et al., 1999). Synthetic membranes are thought to be more biocompatible than cellulose types, but their higher hydraulic permeability may increase the risk of backfiltration and consequent blood contamination (Gutch et al., 1999).

6.6.2 Pumping Blood Through Semipermeable Membranes

Water and small solutes such as glucose readily cross a semipermeable membrane, and bigger solutes such as red cells, white cells, and some proteins are held back. The retained part induces a differential of osmotic pressure, and water flows from the side of low solute concentration to the side of high solute concentration. The flow may be stopped, or even reversed by applying external pressure (e.g., piston motion) on the retained solute side (Freitas, Jr., 1999).

For processing a volume rate $\dot{V}$ through a membrane having a volume V_m and porosity ϕ_m, the piston has to act during the time t_{pm},

$$t_{pm} = \frac{\phi_m V_m}{\dot{V}}. \tag{6.70}$$

To oppose the osmotic backflow, the piston must move with the frequency ν_{pm},

$$\nu_{pm} = \frac{1}{t_{pm}} \geq \frac{\dot{V}}{\phi_m V_m}, \tag{6.71}$$

or, according to Darcy's law,

$$\nu_{pm} \geq \frac{K_m \Delta P}{\mu \phi_m L_m^2}. \tag{6.72}$$

In order for small solutes to be able to cross the membrane, the fluid flow should be slow enough to allow solutes to line up with the membrane pores. If the solutes have the mass m_s and radius r_s (which is comparable to but smaller than the pore radius of the membrane) each pore will be able to hold at least n_p molecules in a row,

$$n_p = \frac{L_m}{2r_s}. \tag{6.73}$$

Each stroke of the piston will push a number of molecules (n_m) through each pore,

$$n_m = \frac{\dot{V} C_s V_p t_{pm}}{\phi_m V_m}. \tag{6.74}$$

Under the action of the piston, the small solutes (particles) experience rotation around randomly oriented axes. The root mean square angle of rotation (γ) can be obtained from (Einstein, 1956)

$$\gamma = \left(\frac{K_B T t_{rt}}{4\pi \mu r_s^3} \right)^{0.5}, \tag{6.75}$$

where K_B is the Boltzmann's constant, T the temperature, and t_{rt} the time available for solute rotation (for ballistic rotation, $t_{rt} = m_s / 15\pi \mu r_s$). Assuming that the time available for solute rotation is much higher than the solute diffusion time, the layer formed by the rotating target solute in the neighborhood of the pores has a thickness that approximately matches the membrane thickness, and

$$t_{rt} = t_{pm} \frac{n_p}{n_m}. \tag{6.76}$$

To allow solutes to line up with the pores, the volume processing rate of the membrane should be small, and ν_{pm} should not be very large. By substituting Equations (6.70) and (6.73) to (6.75) into Equation (6.76) we obtain

$$\nu_{pm} \leq \frac{K_B T L_m}{8\pi \mu \gamma^2 C_s V_p r_s^4} \tag{6.77}$$

Equations (6.72) and (6.77) constitute the bounds for the frequency of the power stroke for best membrane performance.

The application of an external pressure on the system and the transport of small solutes and water through the membrane pores generate heat. Thus the temperature in the system will increase. Blood is composed of solutes that are very sensitive to temperature: proteins may experience degrading after reaching a certain temperature level. The first law of thermodynamics defines the maximum pressure $P_{\max}$ that can be applied to the system,

$$P_{\max} < \rho_{bl} C_V (T_{den} - T_{bl}), \tag{6.78}$$

where C_V is the specific heat of blood, T_{den} is the temperature above which solutes are degraded, and T_{bl} is the blood temperature in the human body ($\sim$309.5 K).

7

Drying of Porous Materials

7.1 Introduction

Drying is the process of thermally removing volatile substances (e.g., moisture) to yield a solid product. Mechanical methods for separating a liquid from a solid are not considered drying. When a wet solid is subjected to thermal drying, two processes occur simultaneously: transfer of energy (mostly as heat) from the surrounding environment to evaporate the surface moisture, and transfer of internal moisture to the surface of the solid and its subsequent evaporation due to the first process.

The fundamental mechanism of moisture movement within the solid has received much attention in the literature. There appear to be four major modes of transfer: liquid movement caused by capillary forces, liquid diffusion resulting from concentration gradients, vapor diffusion due to partial pressure gradients, and diffusion in liquid layers adsorbed at solid interfaces. The mechanisms of capillarity and liquid diffusion have received the most detailed treatment. In general, capillarity is most applicable to coarse granular materials, and liquid diffusion rules single-phase solids with colloidal or gel-like structure. In many cases the two mechanisms may be applicable to a single drying operation, that is, capillarity dominating moisture movement in the early stages of drying, and liquid diffusion taking over at lower moisture contents (Brennan *et al.*, 1976).

A very important aspect of drying technology is the mathematical modeling of the drying processes and equipment. Its purpose is to allow design engineers to choose the most suitable operating conditions, and then size the drying equipment and drying chamber accordingly to meet desired operating conditions. The principle of modeling is based on having a set of mathematical equations that can adequately characterize the system. The solution of these equations based only on the initial conditions must allow the prediction of the process parameters as a function of time at any point in the dryer.

7.2 Drying Equipment

Drying equipment may be classified in several ways. One classification is based on the method of transferring heat to the wet solids: direct dryers (convection dryers), indirect dryers (conduction or contact dryers), infrared or radiant dryers, and dielectric dryers.

Another classification is based on the handling characteristics and physical properties of the wet material: batch tray dryers, batch through-circulation dryers, continuous tunnel dryers, rotary dryers, agitated dryers, gravity dryers, direct-heat vibrating conveyer dryers, and dispersion dryers.

The first method of classification reveals differences in dryer design and operation, whereas the second method is most useful in the selection of a group of dryers for preliminary consideration in a given drying problem. In the second classification of dryers, heat for drying is transmitted to the wet solid through a retaining wall. The vaporized liquid is removed independently of the heating medium. The rate of drying depends on how the wet material is placed in contact with hot surfaces. These are also known as conduction or contact dryers.

7.3 Drying Periods

In a drying experiment the moisture content data (on a dry basis) are usually measured versus the drying time, as shown in Figure 7.1a. The curves represent the general case when a wet solid loses moisture first by evaporation from a saturated surface on the solid, followed by evaporation from a saturated surface of gradually decreasing area, and, finally, by a regime where moisture diffusion occurs within the solid. Figure 7.1b shows the changes in drying rate versus moisture content (dry basis) of the solid. In addition, Figure 7.1c shows the variation of drying rate versus drying time.

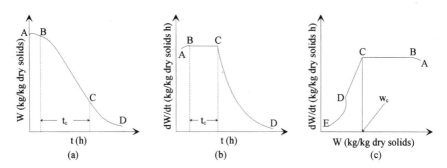

Fig. 7.1. The drying periods for a solid: (a) moisture content versus time; (b) drying rate versus time; (c) drying rate versus moisture content. Drying curves for a set material being dried at a constant temperature and relative humidity.

Drying kinetics refers to the changes of average material moisture content and average temperature with time, unlike drying dynamics which describes changes in the temperature and moisture profiles throughout the drying body. Drying kinetics enables us to calculate the amount of moisture evaporated, drying time, energy consumption, and so on. These depend to a considerable extent on the physicochemical properties of the material. Nevertheless, the changes in material moisture content and temperature are usually controlled by heat and moisture transfer among the body surface, the surroundings, and the internal structure of the drying material. The change in moisture content with time is influenced significantly by the parameters of the drying process, for example, temperature, humidity (pressure), relative velocity of air, or total pressure.

7.4 Basic Heat and Moisture Transfer Analysis

During the drying of a wet solid in heated air, the air supplies the necessary sensible and latent heat of evaporation to the moisture and also acts as a carrier gas for the removal of the water vapor formed from the vicinity of the evaporating surface. The three stages of heat and moisture transfer shown in Figure 7.1 are as follows.

Stage A–B is a warming-up period when the solid surfaces come into equilibrium with the drying air (equilibrium means the same T and P, in particular $P_s = P_a$). It is often a negligible proportion of the overall drying cycle but in some cases it may be significant.

Stage B–C is the period of drying during which the rate of water removal per unit of drying surface is essentially constant. Point C, where the constant-rate period ends, is known as the critical moisture content. During this period the movement of moisture within the solid is rapid enough to keep a saturated condition at the surface, and the drying rate is controlled by the rate at which heat is transferred to the evaporating surface. The surface of the solid remains saturated with liquid water by virtue of the fact that the movement of water within the solid to the surface takes place at a rate as great as the rate of evaporation from the surface. This stage is controlled by the heat and/or moisture transfer coefficients, the area exposed to the drying medium, and the difference in temperature and relative humidity between the drying air and the wet surface of the solid.

The rate of drying depends on the rate of heat transfer to the drying surface. The rate of moisture transfer can be expressed as

$$\frac{dW}{dt} = -h_m A (P_s - P_a), \tag{7.1}$$

where dW/dt is the drying rate, h_m is the moisture transfer coefficient, A is the drying surface area, P_s is the water vapor pressure at the surface (i.e., the vapor pressure of water at surface temperature), and P_a is the partial pressure of water vapor in air. Equation (7.1) can also be written as

$$\frac{dW}{dt} = -h_m A(H_s - H_a), \tag{7.2}$$

where H_s is the absolute humidity at the surface (i.e., the saturation humidity of the air at surface temperature), and H_a is the absolute humidity of air.

In addition, the rate of heat transfer to the drying surface can be expressed as

$$\frac{dQ}{dt} = hA(T_a - T_s), \tag{7.3}$$

where dQ/dt is the rate of heat transfer, h is the convection heat transfer coefficient during heating, A is the surface area, T_a is the dry-bulb temperature of air, and T_s is the surface temperature of the material. Note that in the situation that is considered here (convection heating only) T_s is the wet-bulb temperature of the air. Because equilibrium exists between the rate of heat transfer to the body and the rate of moisture transfer away from the body, the two rates are proportional,

$$\frac{dW}{dt} L = -\frac{dQ}{dt}, \tag{7.4}$$

where L is the latent heat of evaporation at T_s. Combining Equations (7.3) and (7.4) we obtain

$$\frac{dW}{dt} = -\frac{hA}{L}(T_a - T_s). \tag{7.5}$$

If the drying rate is expressed in terms of the rate of change of moisture content W (dry-weight basis), Equation (7.5) can be written as

$$\frac{dW}{dt} = -\frac{hA_{\text{ef}}}{L}(T_a - T_s), \tag{7.6}$$

where A_{ef} is the effective drying surface per unit mass of dry solids.

Assume there is no shrinkage. For the drying of a tray of wet material of depth d with evaporation only from its upper surface, we can write

$$\frac{dW}{dt} = -\frac{h}{\rho_s L d}(T_a - T_s), \tag{7.7}$$

where ρ_s is the bulk density of the dry material. Consequently, the drying time in the constant rate period can be obtained by the integration of Equation (7.7) as

$$t_{CR} = \frac{(W_i - W_c)\rho_s L d}{h(T_a - T_s)}. \tag{7.8}$$

Here t_{CR} is the constant rate drying time, W_i is the initial moisture content of the solid, and W_c is the moisture content at the end of the constant-rate period. In sum, the rate-controlling factors during the constant-rate period are the drying surface area, the difference in temperature or humidity between the air and the drying surface, and the heat or moisture transfer coefficients.

Note that in estimating drying rates, the use of heat transfer coefficients is considered to be more reliable than the use of moisture transfer coefficients. For many cases, the heat transfer coefficient can be calculated from Nu–Re correlations. Thus the air velocity and system dimensions influence the drying rates during the constant-rate period. Alternative expressions for h are used where the airflow is not parallel to the drying surface, or for through-flow situations. When heat is supplied to the material by radiation and/or conduction in addition to convection, then an overall heat transfer coefficient that takes this into account must be substituted for h in Equation (7.7). Under these circumstances the surface temperature during the constant-rate period of drying remains constant, at some value above the wet-bulb temperature of the air and below the boiling point of water.

Stage C–D starts at the critical moisture content, when the constant-rate period ends (Figure 7.1). From point C onwards the surface temperature begins to rise and continues to do so as drying proceeds, approaching the dry-bulb temperature of the air as the material approaches dryness. When the initial moisture content is above the critical moisture content, the entire drying process occurs under the constant-rate conditions. If it is below the critical moisture content, the entire drying process occurs in the falling-rate period. This period usually consists of two zones: the zone of unsaturated surface drying, and the zone in which the controlling mechanism is the internal movement of moisture. At point E the entire exposed surface becomes completely unsaturated, and marks the start of the drying process during which the rate of internal moisture movement controls the drying rate. In Figure 7.1c, CD is defined as the first falling-rate period, and DE is the second falling-rate period. In the falling-rate periods the rate of drying is influenced mainly by the rate of movement of moisture within the solid and the effects of external factors, in particular air velocity, are reduced, especially in the latter stage. In most cases, the falling-rate periods represent the major portion of the overall drying time.

For systems where a capillary flow mechanism applies, the rate of drying can often be expressed with reasonable accuracy by an equation of the type,

$$\frac{dW}{dt} = -h_m(W - W_e), \qquad (7.9)$$

In general, and for a falling rate period from a critical moisture content (W_c) to the equilibrium moisture content (W_e), the moisture transfer coefficient (h_m) becomes

$$h_m = \frac{dW/dt}{W_c - W_e}, \qquad (7.10)$$

where dW/dt is the rate of drying at time t from the start of the falling-rate period, W is the moisture content of the material at any time (varying from W_c to W_e), and W_e is the equilibrium moisture content of material at air temperature and humidity. After combining Equations (7.7), (7.9), and (7.10), we obtain

$$\frac{dW}{dt} = -\frac{h(T_a - T_s)(W_c - W_e)}{\rho_s L d(W - W_e)}. \qquad (7.11)$$

Integrating the above equation from $t = 0$ to t, and from $W = W_c$ to W, yields the drying time

$$t = \frac{\rho_s L d(W_c - W_e)}{h(T_a - T_s)} \ln\left(\frac{W_c - W_e}{W - W_e}\right). \qquad (7.12)$$

Note that the drying equations [Equations (7.7), (7.8), and (7.11) above] are applicable when drying takes place from one side only. In cases where drying occurs from both sides or surfaces, d will be taken as the half-thickness.

7.5 Wet Material

Materials that are subjected to drying processes usually consist of the bone dry material (skeleton) and an amount of moisture, mainly in a liquid state. So-called "wet materials" have different physical, chemical, structural, mechanical, biochemical, and other properties, which result from the properties of the skeleton and the state of the water within it. Although these parameters can significantly influence the drying process and determine the drying technique and technology, the most important in practice are the structural-mechanical properties, the type of moisture in the solids, and the material–moisture bonding.

The moisture content of the material (W) can be defined in two ways: on a dry basis

$$W = \frac{m_m}{m_s}, \qquad (7.13)$$

where m_m is the mass of moisture and m_s is the mass of dry solid material, and on a wet basis,

$$W = \frac{m_m}{m} = \frac{m_m}{m_m + m_s}, \qquad (7.14)$$

where m is total mass of wet material. The moisture content can also be expressed as a percentage,

$$W = \frac{W^*}{1 - W^*} \quad \text{or} \quad W^* = \frac{W}{1 + W}, \qquad (7.15)$$

where W is the dry-basis content and W^* is the wet-basis content. In addition, the volumetric moisture content (W_V) is defined as

$$W_V = \frac{V_{ml}}{V_s + V_{ml} + V_{mv}}, \qquad (7.16)$$

where V_{ml} is the volume of liquid moisture, V_{mv} is the volume of vapor moisture, and V_s is the volume of dry material. The percentage saturation of wet material (Λ_V) becomes

$$\Lambda_V = \frac{V_{ml}}{V_{ml} + V_{mv}} \times 100. \tag{7.17}$$

Equilibrium moisture content is the moisture of a given material that is in equilibrium with the vapor contained in the drying air under specific conditions of air temperature and humidity. This is also called the "minimum hygroscopic moisture content." It changes with the temperature and humidity of the surrounding air. However, at low drying temperatures (e.g., 15 to 50°C), the equilibrium moisture content becomes independent of temperature, and it becomes zero at zero relative humidity. The equilibrium moisture content also depends strongly on the nature of the material being dried. For nonhygroscopic materials it is essentially zero at all temperatures and humidities. For hygroscopic materials (e.g., wood, food, paper, soap, chemicals) it varies regularly with the temperature and relative humidity over a wide range.

Wet materials are classified into three categories, based on their behavior with respect to drying (see Strumillo and Kudra, 1986):

Typical colloidal materials, which change size but preserve their elastic properties during drying (e.g., gelatin, agar);

Capillary porous materials, which become brittle, shrink slightly, and can easily be ground after drying (e.g., sand, charcoal, coffee). In drying, all wet materials the pore radius of which is smaller than 10^{-5} m can be treated as capillary porous with various pore diameter distributions. Note that the moisture in such bodies is maintained mainly by surface tension forces. If the pore size is greater than 10^{-5} m, then in addition to capillary forces, gravitational forces also play a role. These bodies are called "porous." A common feature of porous systems is the presence of void spaces called capillaries or pores. These can have complex shapes and different geometric dimensions.

Colloidal capillary porous materials, which have the properties of the above two types. The walls of the capillaries are elastic, and they swell during drying (e.g., food, wood, paper, leather). These materials are capillary porous as far as their structure is concerned, and colloidal as far as their properties are concerned.

There is an alternative classification for dried materials taking as a basis the states of moisture in the wet materials (see Figure 7.2). Nonhygroscopic material is the one in which the partial pressure of water in the material is equal to the vapor pressure of water, as defined by Karel (1975). It can contain no bound moisture. Such materials may include nonporous or porous bodies with pore radius more than 10^{-7} m. Note that the constant-rate period exists as long as the partial pressure equals the vapor pressure of pure water at that temperature. This means that as long as there is free water on the surface of

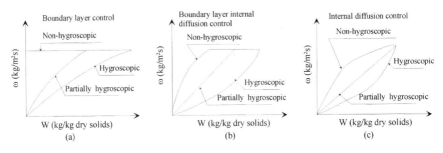

Fig. 7.2. Drying rate curves for various types of materials.

the food, the water will evaporate as if it were evaporating from the surface of a pool of water. Thus the limiting factor is the evaporation rate of water from the surface. For example, for nonhygroscopic foods this occurs as long as the surface water content is greater than zero. For hygroscopic foods, however, this occurs as long as the surface water content is greater than the critical moisture content (W_c). The constant-rate period continues as long as the supply of water to the surface is sufficient to maintain saturation of the surface. It should be noted that the product remains relatively "cool" during this phase of drying, with the surface temperature never exceeding the wet-bulb temperature. However, eventually water cannot get to the surface as quickly and dry patches appear on the surface of the food, and the high rate of evaporation cannot be maintained. Note that W_c is not just product-dependent but also tied to thickness and air conditions. Following the constant-rate period is the falling-rate period. In this case, water removal is limited by diffusion of water to the surface, and diffusion from saturated layers within the food. Water moves to the surface in the gas phase and is removed by the air stream. During one or two falling-rate periods the drying rate slows because there is an increasing resistance to the movement of water to the surface, which is caused by an increasing distance over which the water must move. The end of the falling-rate period occurs when the center of the food is no longer saturated with water in nonhygroscopic food.

Hygroscopic material is the one in which the partial pressure of water becomes less than the vapor pressure of water at some critical level of moisture. This may contain bound moisture. Such materials with bound moisture cover mainly microporous bodies in which liquid exerts a vapor pressure less than that of the pure liquid at the given temperature. If the moisture content in the hygroscopic body exceeds the hygroscopic moisture content, that is, contains unbound water, then up to the moment when this unbound water is removed, it behaves as a nonhygroscopic material.

Partially hygroscopic materials include macroporous bodies which, although they also have bound moisture, exert a vapor pressure that is only slightly lower than that exerted by a free water surface.

In summary, the drying periods are divided into two for nonhygroscopic material as constant-rate and falling-rate periods, and three for hygroscopic material as constant-rate and two falling-rate periods. In hygroscopic materials, however, the partial pressure of water becomes less than the vapor pressure of water at some critical level of moisture. Note that as the moisture content of the material decreases, the transport through capillaries and pores occurs primarily in the vapor phase. Note also that the shape of the drying curve in the falling-rate period depends strongly on the moisture transfer conditions, where the drying behavior is affected by internal and external conditions. No constant drying rate period is commonly observed during the drying of hygroscopic solids under the condition that the drying is controlled internally, which means that the external resistances are negligible.

Several food preservation techniques (e.g., drying) reduce microbial activities by lowering availability of water. Lowering water content also may reduce chemical and physical reactions that limit shelf life. In this regard, it is important to make a distinction between the amount of water and the state of water in the food.

There are several types of moisture in wet materials. *Surface moisture* is liquid that forms as an external film on the material due to surface tension effects. *Free* or *capillary moisture* is *unbound* moisture in a hygroscopic material, in excess of the equilibrium moisture content corresponding to saturation humidity. All water in a nonhygroscopic material is unbound water. All internal moisture in a nonhygroscopic material is unbound. Note that in a hygroscopic material where actual vapor pressure is a function of the saturated state, there is moisture in excess of the equilibrium moisture content corresponding to saturation relative humidity.

Unbound moisture can be categorized into two forms, as shown in Figure 7.3: funicular form, where a continuous liquid state exists within the porous body, and pendular form, where the liquid around and between discrete particles is discontinuous, and moisture is interspersed by air bubbles. In the funicular state liquid movement to the external surface of the material takes place by capillary action. As moisture is removed, the continuity of the liquid phase is interrupted due to suction of air into pores, which leaves isolated pockets of moisture (pendular form). Consequently, capillary flow is possible only on a localized scale. When the material is close to the bone dry (solid) state, the moisture is held as a monolayer of molecules on the pore walls, so it is removed mainly by vapor flow.

Bound, hygroscopic, or *dissolved moisture* is the liquid that exerts a vapor pressure less than that of the pure liquid at the given temperature. Liquid may become bound by retention in small capillaries, by solution in cell or fiber walls, by homogeneous solution throughout the solid, and by chemical or physical adsorption on solid surfaces.

A useful parameter for the classification of material–moisture bonding is the value of the work needed for the removal of 1 mole of moisture from the given material (Rebinder, 1972): chemical moisture bonding (stoichiometric), in which the ionic and molecular bond energy is on the order of 5000 J/mole;

Funicular state   Pendular state

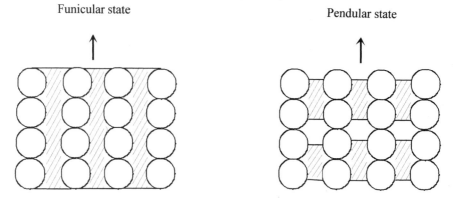

Fig. 7.3. Two forms of unbound moisture: the funicular state is when capillary suction results in air being sucked into the pores; the pendular state is when a continuous film of liquid no longer exists around and between discrete particles, and the flow by capillarity cannot occur. This state succeeds the funicular state.

physicochemical moisture bonding (nonstoichiometric), in which the adsorptive and osmotic bond energy is on the order of 3000 J/mole; and physicomechanical moisture bonding (undetermined proportions), in which the bond energy is about 100 J/mole.

In particular, physicomechanical moisture bonding is of great importance in drying, and classified into: structurally bound moisture (gels with about 1% of solid) that can be removed by evaporation or mechanical dewatering, capillary bound moisture (capillary porous bodies) that can be removed by evaporation from capillaries into ambient air, and unbound moisture (nonporous hydrophilic bodies) that can be removed by surface evaporation into ambient air.

Several parameters are used in the description of the structure of dried materials:

Basis weight (G) is defined as the ratio of the dry mass (m) to its area (A),

$$G = \frac{m}{A}. \tag{7.18}$$

Porosity (ϕ) is defined as the ratio of the total void spaces volume (V_V) to the total volume of the material (V_T), $\phi = V_V/V_T$. See also Section 1.1. An alternative is the absolute porosity $(\bar{\phi})$, which for a wet material (e.g., wood or paper sheet) can be defined as

$$\bar{\phi} = \frac{V_{\text{vo}}}{V_{\text{ap}}} = \frac{V_{\text{ap}} - V_{\text{ce}}}{V_{\text{ap}}}, \tag{7.19}$$

where V_{vo} is the void volume or pores volume, V_{ap} is the apparent or total volume, and the V_{ce} is the volume of cellulose. Drying of industrial pump sheets is a crucial issue, due to the transport of moisture through a sheet during convective drying. Because pulp and paper are made from a dilute aqueous solution of cellulose fibers, the interaction between water and cellulose fibers is a fundamental issue in papermaking. During the removal of the water from a formed sheet moisture gradients will occur. This is due to both process parameters and internal properties of the sheet, such as the diffusion resistance. Shrinkage is closely linked to the distribution of moisture, and the shrinkage affects the strength and curl properties of the sheet. Also other quality parameters and postprocessing steps are affected by the internal moisture distribution. To improve the quality and the production, it is important to improve the understanding of the water distribution and transport mechanisms within the sheet at different stages in the papermaking process. As an example, the absolute porosity is thus determined by knowing the sheet thickness at each level of deformation, and by knowing the oven-dry mass of the sheet,

$$\bar{\phi} = 1 - \frac{m_o}{Ad\rho_c}. \tag{7.20}$$

Here m_o is the oven-dry mass of the sheet, A is the strained area, d is the thickness of the sample, and ρ_c is the density of cellulose ($\sim 1550 \, \text{kg/m}^3$).

Tortuosity (T^*) is defined as the average ratio of path length to the interior of the particle (via the pores) to the distance in a straight line (Section 1.9). As is known, molecular diffusivity of vapor depends on the temperature, moisture, porosity, and tortuosity of food. When considering exclusively the diffusion into a particle with tortuous pores distributed throughout the volume, it is clear that the tortuosity becomes roughly proportional to the square root of the actual pore length to the radius of the particle [see Bilbao *et al.* (2000)].

The pore shape factor (π) is the parameter that characterizes the deviation of the diffusion channel shape when compared to a cylinder. In drying practice, porosity is a function of pore radius. Therefore, a maximum porosity can be introduced. Strumillo and Kudra (1986) indicated that at the maximum value of the radius $(r_{\max})$ the integral curve of pore diameter intersects the line corresponding to the total pore volume in the total body volume $\phi_{\max}$. Therefore, the total pore volume in the total body (e.g., maximum porosity) corresponds to

$$\phi_{\max} = \int_{r_{\min}}^{r_{\max}} \frac{d\phi}{dr} dr. \tag{7.21}$$

Most drying processes involve fluid flow through porous media. In the case of a completely saturated porous medium, the flow is described well by Darcy's law. When the porous medium is only partially saturated with liquid, the flow through the pores is much more complicated. In such cases, the porous material can be regarded as a bundle of capillaries, and the diffusion coefficient can be predicted as a function of moisture content.

Most porous materials have a very intricate solid structure and complex chemical inhomogeneity of the solid, especially at small length scales on the order of the pores. To arrive at a deterministic method of describing moisture transport, the porous material is described with a minimum number of parameters. One parameter is the permeability (K) of the saturated porous material that estimates the flow through the porous material when a pressure gradient dP/dx is applied. Darcy's law can be written as [cf. Equation (1.7)]

$$\frac{\dot{V}}{A} = \frac{K}{\mu}\frac{\Delta P}{d},$$ (7.22)

where $\dot{V}$ is the volumetric flow rate (m^3/s), A is the surface area (m^2), K is the permeability (m^2), μ is the dynamic viscosity at liquid temperature (Pa·s), ΔP is the pressure drop across the body (Pa), and d is the thickness of the body at a given level of deformation (m). The ratio $\dot{V}/A$ is called the superficial liquid velocity (m/s). The permeability is an intrinsic parameter that can be extracted from Equation (7.22), $K = (\mu d/\Delta P)(\dot{V}/A)$. The mechanical properties of porous materials may change during the drying process.

7.6 Types of Moisture Diffusion

Moisture in a drying material can be transferred in both liquid and gaseous phases. Several modes of moisture transport can be distinguished.

Transport by liquid diffusion is governed by a proportionality between the liquid moisture transfer rate and the gradient of moisture concentration inside the material,

$$\omega_\ell = -h_m \frac{\partial(W\rho_1)}{\partial x}.$$ (7.23)

This equation is accepted and applied for the description of liquid moisture movement in materials.

Transport by vapor diffusion is the main mechanism of vapor moisture transfer. It takes place in materials where the characteristic diameter of the free air spaces is greater than 10^{-7} m. The qualitative effect of this transfer can be described by an equation of the Fick type, using instead of the kinematic diffusion coefficient D_{ma}, the effective diffusion coefficient D_{ef} in capillary porous materials,

$$D_{ef} = D_e D_{ma}.$$ (7.24)

Here D_e is the equivalent coefficient of diffusion in the capillary porous material. A general relation for D_e is given by Van Brakel and Heertjes (1974),

$$D_e = \frac{\phi\ell}{\phi_e},$$ (7.25)

where the value of D_e is presented as a function of structural parameters.

Transport by effusion (or *Knudsen-type diffusion*) takes place when a dimension of the space in a capillary porous material is smaller than a characteristic value, which for air is 10^{-7} m. The mass rate of vapor in this case can be obtained from

$$\omega_\ell = -D_{E\mathrm{ef}}\frac{\partial C_m}{\partial x}, \tag{7.26}$$

where $D_{E\mathrm{ef}}$ is the effective effusion which for gel-type materials can be defined as

$$D_{E\mathrm{ef}} = 3^{-1/2}\phi^2 D_E. \tag{7.27}$$

Transport by thermodiffusion can be described by the equation

$$\omega = \rho\frac{D_T}{T}\frac{\partial T}{\partial x} = \rho\beta\frac{D}{T}\frac{\partial T}{\partial x}, \tag{7.28}$$

where D_T is the thermodiffusion coefficient and $\beta \approx D_T/D$ is the thermodiffusion constant. Written for the gas phase, the above equation becomes

$$\omega = \beta\frac{\partial T}{\partial x}. \tag{7.29}$$

Transport by capillary forces occurs when capillaries form interconnected channels. A difference of capillary pressure takes place, and this causes the continuous redistribution of moisture from the large capillaries to the small ones.

Transport by osmotic pressure is another mode. Osmotic pressure is a function of the moisture content in the material. The osmotic moisture transfer can be then described on the basis of liquid diffusion.

Transport due to pressure gradient results from the internal pressure difference due to the local evaporation of liquid or local condensation of vapor. Moisture movement equalizes the pressure in accordance with Darcy's law. In addition to the types of moisture movement listed above, mass transport can also be caused by internal pressure, shrinkage, or external pressure.

7.7 Shrinkage

During drying, the nonuniform temperature and moisture content fields induce thermal and shrinkage stresses inside the dried material. Because drying materials are so diverse, we may expect several mechanisms of interaction between water and solids. We may have unsaturated solutions, saturated solutions with crystals, supersaturated solutions, amorphous hydrophylic solids with limited swelling capacity, porous nondeformable solids with various levels of porosity, porous deformable solids with various levels of porosity, cellular solids, and/or any combination of the above items.

Note that food products undergo a certain degree of shrinkage during drying by all the drying methods, with the possible exception of freeze-drying. Colloidal materials also shrink, for example, meat during cooking (Fowler and Bejan, 1991). In the early stages of drying, at low drying rates, the amount of shrinkage bears a simple relationship to the amount of moisture removed. Toward the end of drying, shrinkage is reduced so that the final size and shape of the material are essentially time-independent before drying is completed.

The bulk density and porosity of dried vegetable pieces depends to a large extent on the drying conditions. At high initial drying rates the outer layers of the pieces become rigid, and their final volume is fixed early in the drying process. As drying proceeds, the tissues split and rupture internally, forming an open structure. The product in this case has a low bulk density and good rehydration characteristics. At low initial drying rates, the pieces shrink inwards and give the food product a high bulk density. Shrinkage of foodstuffs during drying may influence their drying rates because of the changes in the drying surface area and the setting up of pressure gradients within the material. Some work indicates that shrinkage does not affect drying behavior (Brennan *et al.*, 1976). More recent work (Bilbao *et al.*, 2000) suggests that shrinkage cannot always be explained based solely on the amount of moisture evaporated. Shrinkage is specific for each body, depending on the material type, and on the characteristic cell and tissue structure.

The porosity profile of a cylindrical apple as a function of the distance to the interface was studied by Bilbao *et al.* (2000). They found that raw apple exhibits a porosity between 18 and 24%, but as drying proceeds, moisture removal is accompanied by the formation of a more porous structure. A gradual increase in porosity is observed from the inner part of the cylinder, where the tissue is less dry, to the outer regions, but near the interface the porosity decreases as a result of the hardening of the case.

Bilbao *et al.* (2000) also showed that the volume changes in apple tissue during air-drying appear not to be affected by the air temperature. Nevertheless, volume changes are greatly affected by the air velocity, and an equation can be used to predict volume changes in the gas phase as a function of air velocity. The obtained porosity values confirm that food systems could undergo porosity increase as drying proceeds, depending on process variables and system characteristics. Observations of fresh apple tissue usually show the turgid cell walls of parenchyma tissue as bright regions, and small intercellular spaces between cells as dark regions, but drying conditions promote great structural changes. Figure 7.4 shows the appearance of dried apple tissue revealing that cell walls are greatly shrunk, leaving wide spaces between neighboring cells. This can be observed throughout the tissue, except in the first 0.5 mm under the skin, where the tissue is more compact and collapsed.

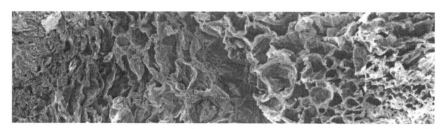

Fig. 7.4. The tissue of dried apple after two hours at 70°C (Bilbao *et al.*, 2000).

Structural changes and shrinkage during drying occur, and phenomena such as shrinkage determine the properties and quality of food products. Quality can be characterized by color, texture, taste, porosity, and other physical properties such as density and specific volume. The quality of dried products changes during drying, depending on the type of drying method and drying conditions. For this reason it is important to be able to anticipate the effect of the drying method and conditions on the final quality. This can be done by constructing a mathematical model to estimate the properties of the product versus material moisture content, and to examine the various features of the model; for example, see Table 7.1.

Krokida and Maroulis (1997) also investigated the effect of the drying method on bulk density, particle density, specific volume, and porosity for five drying methods: freeze, microwave, conventional, vacuum, and osmotic drying. Each method was used to dry apples, carrots, potatoes, and bananas under typical conditions. The approach sketched in Table 7.1 was used to determine the bulk density, particle density, porosity, and specific volume versus material moisture content. The effect of each of the drying methods on the above properties was examined. Equation (T1.1) in Table 7.1 is used for determining the particle density (ρ_p) as a function of moisture content (W). The particle density ranges between the dry solid density (ρ_s) and the

Table 7.1. Mathematical models for property equations

Properties		
ρ_p Particle density	kg/m^3	
ρ_b Bulk density	kg/m^3	
ϕ Porosity		
v Specific volume	m^3/kg db	
Factors		
W Material moisture content (db)	kg moisture/kg dry matter	
Properties equations		
$\rho_p = (1 + W)/(1/\rho_s + W/\rho_w)$	T1.1	
$\rho_b = (1 + W)/(1/\rho_{b0} + \beta' W/\rho_w)$	T1.2	
$\phi = 1 - \rho_b/\rho_p$	T1.3	
$v = 1/\rho_{b0} + \beta' W/\rho_w$	T1.4	
Parameters		
ρ_w Enclosed water density	kg/m^3	
ρ_s Dry solid density	kg/m^3	
ρ_{b0} Dry solid bulk density	kg/m^3	
β' Volume-shrinkage coefficient		
Factors affecting the parameters		
Material		
Drying method		
Drying conditions		

Source: Adapted from Krokida and Maroulis (1997).

density of water (ρ_w). This equation corresponds to a model with two phases in series. Similarly in Equation (T1.2) the bulk density ranges between the bulk density of dry solids (ρ_{b0}) and the density of the enclosed water (ρ_w). Equation (T1.2) comes from the same two-phase structural model. As shown in Equation (T1.3), the total porosity is a function of bulk density and particle density. The definition of specific volume involves three parameters: the bulk density of dry solids (ρ_{b0}), the enclosed water density (ρ_w), and the shrinkage coefficient β', Equation (T1.4). Factors such as material, drying method, and drying conditions affect these parameters and, consequently, the four properties. The shrinkage coefficient can be defined in terms of temperature by an equation of the Arrhenius type,

$$\beta' = C \exp\left(-\frac{E_\mathrm{a}}{RT}\right), \tag{7.30}$$

where C is a constant determined experimentally for each product, E_a is the activation energy of shrinkage, R is the universal gas constant, and T is the absolute drying temperature.

Krokida and Maroulis (1997) determined the values of the required parameters by fitting the proposed model to the experimental data for four types of produce; see Figure 7.5. The porosity of freeze-dried materials is always higher (80 to 90%) when compared with other dehydration processes. The porosities of microwave-dried potato and carrot are next (75%), whereas microwaved apple and banana do not develop high porosities (60% and 25%). Vacuum-dried banana and apple develop high porosities (70%), whereas for vacuum-dried carrot and potato the porosities are lower (50% and 25%). Gabas *et al.* (1999) investigated the effect of drying temperature on the shrinkage of grapes subjected to pretreatment with various substances, as shown in Figure 7.6. They found that the small temperature effect on shrinkage observed in practice may be attributed to the temperature difference of elastic and mechanical properties.

7.8 Modeling of Packed-Bed Drying

The process of drying of grained materials in a packed bed permeated by a gaseous agent is complex because it is time-dependent, and the three fundamental transfer phenomena appear simultaneously: flow of the gaseous phase through layer interstices, heat transfer, and transfer of water vapor from grains to gaseous phase. The drying mechanism, the flow of the gaseous phase through the bed, the heat transfer between the thermal agent and the grained material, and the drying of a monogranular layer are topics of current interest (e.g., Moise and Tudose, 2000).

The packed-bed drying of a granular material takes places in three stages; see Figure 7.7. In the first stage, there are two zones in the layer: zone A, in which the material is kept at the initial humidity, and zone B where the

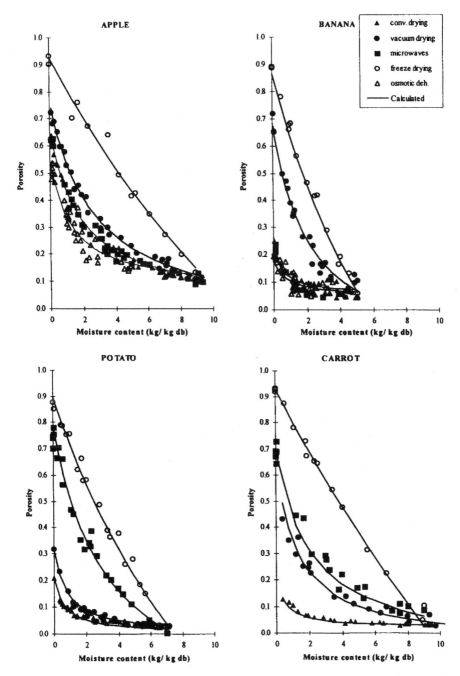

Fig. 7.5. Porosity versus product moisture for various drying methods (reprinted from Krokida and Maroulis (1997) by courtesy of Marcel Dekker, Inc.).

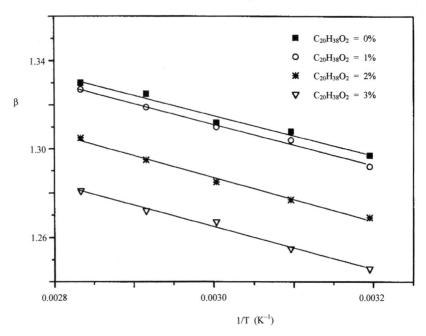

Fig. 7.6. Effect of drying temperature on the shrinkage coefficient of grapes (reprinted from Gabas *et al.* (1999) by courtesy of Marcel Dekker, Inc.).

drying takes place. The height of zone B increases continuously at the expense of zone A. In the drying zone B the material moisture varies in time, and is a function of the axial coordinate.

In the second stage, the new zone C appears. In this zone the material is dried, at the equilibrium humidity. Zone A decreases until it disappears, and the height of zone C increases.

In the third stage, only zones B and C remain in the packed bed. The height of zone C increases until the whole bed reaches the equilibrium humidity. In the case of small beds, or when drying with a gaseous thermal agent with low temperature, the second stage (in which the three zones exist simultaneously) does not appear anymore.

The "plug flow and external transfer model" for packed-bed drying of granular materials was initially extended to time-dependent operation, with specific terms accounting for water evaporation and water vapor transfer (Moise and Tudose, 2000). The model consists of the equations

$$\rho_g \phi \frac{\partial H}{\partial t} = w_v a - w_g \frac{\partial H}{\partial z}, \qquad (7.31)$$

$$\rho_s (1 - \phi) \frac{\partial W}{\partial t} = -F_v a, \qquad (7.32)$$

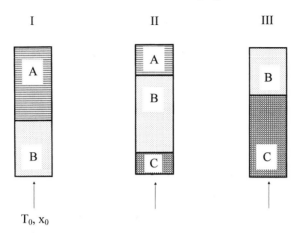

Fig. 7.7. The three stages of packed-bed drying (A: material zone with initial moisture; B: drying zone; and C: dried material zone).

$$\rho_g c_p \phi \frac{\partial T_g}{\partial t} = -F_g c_{p_g} \frac{\partial T_g}{\partial z} - h_g a \left(T_g - T_s \right)$$

$$- F_v a c_{p_v} \left(T_g - T_s \right) - h_w a_w \left(T_g - T_a \right), \qquad (7.33)$$

$$\rho_s c_{p_s} \left(1 - \phi \right) \frac{\partial T_s}{\partial t} = h_g a \left(T_g - T_s \right) - F_v a L, \qquad (7.34)$$

where

$$F_v = \frac{h_{mg}}{R_g T_g} \left(P_s - P_v \right) \frac{1}{1 + Bi \cdot f(w)}, \qquad (7.35)$$

and other new quantities are defined in Nomenclature. The function $f(w)$ is measured experimentally, depending on the material characteristics. The initial conditions are as follows. The absolute humidity of drying air is constant and equal to the atmospheric value $H = H_0$. The moisture content of the bed material is uniformly distributed and equal to the initial moisture $W = W_0$. The air temperature in the bed and in the grain interstices is equal to the environment temperature $T_g = T_s = T_m$. The coordinate z is measured in the direction of flow, where $z = 0$ is the entrance to the packed bed. The conditions at $z = 0$ are: the absolute humidity of drying air is constant $H = H_0$, and the air temperature is constant, $T_g = T_0$. The value of h_g (between gas and particles) generally varies between 46 to $106 \, \text{W/m}^2\text{K}$, and k_g varies between 0.06 and 0.14 [cf. Moise and Tudose (2000)]. Further details may be found in Achenbach (1995). Equations (7.31) to (7.35) can be solved numerically to obtain the unsteady-state moisture content distributions as a function of material properties and experimental drying conditions.

7.9 Diffusion in Porous Media with Low Moisture Content

Although there are many numerical and experimental works focused on the drying of porous materials on higher moisture contents where the liquid phase is mobile and liquid phase transport is dominant, there are a number of applications where drying to low moisture content is important. These include modeling material behavior under fire and combustion conditions, debinding of green ceramic preforms, moisture transport around nuclear waste repositories, and the drying of pharmaceuticals (Plumb et al., 1999).

When the fraction of the pore volume occupied by liquid (the liquid phase saturation) reaches a certain value, the liquid phase is no longer continuous and becomes immobile. This value of liquid phase saturation is called irreducible saturation. Thus moisture cannot be transported to the drying surface or a drying front in the liquid phase as a result of capillary action of pressure gradients in the gas phase. The irreducible saturation for most granular porous materials is 10 to 20% . Below the irreducible saturation the liquid phase is not continuous but exists as liquid islands at the contact points between the individual grains composing the bulk material. Under these conditions diffusion in the vapor phase is the dominant mechanism for drying.

Vapor phase diffusion in porous media is generally quantified on the basis of Fick's law for diffusion in a continuous medium (Plumb et al., 1999):

$$J = D_C \rho_g \nabla W + D_T \nabla T. \tag{7.36}$$

Note that the diffusion effect due to temperature gradients (or Soret effect) is included. For diffusion in a continuum, the Soret effect is generally negligible because the diffusion coefficient (D_T) is small in comparison to the coefficient for diffusion driven by gradients in concentration or partial pressure. For diffusion in a dry porous medium, an effective diffusion coefficient must be defined so that it accounts for the reduced area for diffusion or blockage that results from the presence of the solid phase. This is done by assuming that the effective diffusion coefficient can be obtained by multiplying the diffusion coefficient for free space by the porosity. Furthermore, the increased path length for diffusion is accounted for by including the tortuosity coefficient. The effective diffusion coefficient for a dry porous medium becomes

$$D_{\text{eff}} = \phi D_C / T^* \tag{7.37}$$

The mathematical details for the derivation of the effective diffusion coefficient for a dry porous medium based on volume-averaging can be found in Eidsath et al. (1983).

When immobile liquid is present in the porous medium, the effective diffusion coefficient is further reduced to account for the blockage due to the presence of the liquid phase. This blockage is accounted for by including the gas phase saturation S_g (fraction of pore volume occupied by gas),

$$D_{\text{eff}} = \phi S_g D_C / T^* \tag{7.38}$$

To be consistent with Equation (7.37), for a dry porous medium the tortuosity is also corrected for the presence of the immobile liquid. This line of reasoning leads to (Millington and Quirk, 1961)

$$D_{\text{eff}} = (T^*)^{-10/3} D_C. \tag{7.39}$$

Because diffusion is enhanced as a result of the presence of the liquid phase, an additional correction factor is included in the definition of the effective diffusion coefficient

$$D_{\text{eff}} = \beta \phi S_g D_C / T^* \tag{7.40}$$

where β is an enhancement factor defined as the ratio of the vapor phase diffusion coefficient in a porous medium to that in free space. Although the values of β in liquid phase saturation vary between 0.1 and 0.2, the actual value of the enhancement factor ranges between 1 and 4. For further details, see Plumb *et al.* (1999).

It is important to quantify the effect of the liquid phase saturation on the vapor phase diffusion coefficient. Many of the experimental studies have been indirect, inferring the vapor phase diffusion coefficient by measuring the temperature and using the fact that the effect of saturation on the thermal conductivity is known. Plumb *et al.* (1999) focused on the experimental measurement for quantifying the enhancement factor (β) and its impact on the drying process, and presented the measured values of tortuosities and enhancement factors.

7.10 Modeling of Heterogeneous Diffusion in Wet Solids

Consider now the modeling of heterogeneous diffusion in capillary porous materials during drying. The governing heat and mass transfer equations for liquid as well as vapor flow are based on the work done by Dietl *et al.* (1995). There are two cases: moisture transfer and heat transfer (i.e., conduction). The models and the numerical solutions of the resulting differential equations use as inputs the moisture and the temperature-dependent thermophysical properties of the product. All the equations are given in spherical coordinates, but the numerical calculations are extended to cover other geometries as well.

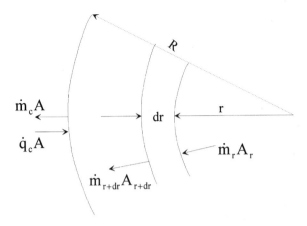

Fig. 7.8. Mass balance in an infinitesimal volume of the food product.

7.10.1 Mass Transfer

With reference to the differential element as shown in Figure 7.8, and taking a mass balance between the incoming and outgoing moisture leads to

$$-\frac{\partial(\dot{m}_r A_r)}{\partial r} dr = \rho_{dr} \frac{\partial W}{\partial t} dV, \tag{7.41}$$

where $\dot{m}_r$ is the total mass flux at the radius r, A_r is the elemental area at the radius r, dr is the infinitesimal thickness, ρ_{dr} is the bone dry density, W is the moisture content (kg/kg) at the time t, and dV is the differential volume corresponding to the radius r. The moisture transport in the vapor phase is described by Fick's law,

$$\dot{m}_v = -\frac{D}{\mu} \frac{1}{R_v T} \frac{\partial P_v}{\partial r}, \tag{7.42}$$

where $\dot{m}_v$ (kg/m² s) is the mass flux of vapor that is being transferred and D(m²/s) is the diffusion coefficient of water vapor in air expressed in m²/s (Kessler, 1981),

$$D = \frac{22.06 \times 10^{-6}}{P} \left(\frac{T_m}{273\,\mathrm{K}}\right)^{1.81}, \tag{7.43}$$

where P is the total pressure expressed in bar, T_m is the mean absolute temperature of product surface and drying air, ranging from $-30°$C to $120°$C, μ is the diffusion resistance factor (dimensionless), R_v is the gas constant for water vapor, and P_v is the partial vapor pressure at point r.

For transport in the sorption (liquid) phase, we have

$$\dot{m}_t = -\rho_{dr} D_{m(W,T)} \frac{\partial W}{\partial r}, \tag{7.44}$$

where $\dot{m}_t$ is the mass flux of liquid that is being transferred, $D_m \, (\mathrm{m^2/s})$ is the moisture diffusivity coefficient, which is a function of moisture content W and temperature T, and M (kg/kg) is the moisture content at point r.

The transport in the vapor phase and the sorption phase are coupled to each other because the vapor pressure in a pore determines the equilibrium sorption moisture content. The equilibrium moisture content is a function of the water activity a_w and the product temperature T. The corresponding saturated vapor pressure P_v^* is related to the product temperature T and the total pressure P,

$$\frac{\partial P_v}{\partial r} = \frac{\partial P_v}{\partial W}\frac{\partial W}{\partial r} = \frac{\partial a_w}{\partial W}P_v^*\frac{\partial W}{\partial r}. \tag{7.45}$$

The total mass flux is then given by

$$\dot{m}_r = \dot{m}_l + \dot{m}_v = -\left(\rho_{dr}K_{(W,T)} + \frac{D}{\mu}\frac{P_v^*}{R_vT}\frac{\partial a_w}{\partial W}\right)\frac{\partial W}{\partial r}. \tag{7.46}$$

By substituting Equation (7.46) into (7.41) we obtain the equation for the mass balance at the infinitesimal level,

$$\frac{\partial}{\partial r}\left[A_r\left(-\rho_{dr}K_{(W,T)} + \frac{D}{\mu}\frac{P_v^*}{R_vT}\frac{\partial a_w}{\partial W}\right)\frac{\partial W}{\partial r}\right]dr = \rho_{dr}\frac{\partial W}{\partial t}\,dV. \tag{7.47}$$

7.10.2 Heat Transfer

For products with high porosity, the drying rate is essentially determined by high mass transfer rates. The heat balance is now made at the boundary of the product for a differential element as follows:

$$A_r = R[\dot{q} - \dot{m}_{r=R}\Delta H_{v(T_o)}] = \rho_{dr}V[c_{pdr} + c_{pw}\bar{W}_{(t)}]\frac{\partial T}{\partial t}, \tag{7.48}$$

where

$$\dot{q} = \dot{q}_c = \frac{h\Delta H_{v(T_o)}}{c_{pv}}\ln\left[1 + \frac{c_{pv}(T_a - T_o)}{\Delta H_{v(T_o)}}\right]. \tag{7.49}$$

In addition, $\dot{q}(=\dot{q}_c)$ is the convection heat flux, h is the convective heat transfer coefficient, $\Delta H_{v(T_o)}$ is the vaporization enthalpy at product surface temperature T_o, c_{pv} is the specific heat capacity of water vapor, and T_a is the drying air temperature. Furthermore,

$$\dot{m}_{r=R} = h_o\frac{P}{R_vT_m}\ln\left(\frac{P - P_{va}}{P - P_{vo}}\right), \tag{7.50}$$

where $\dot{m}_{r=R}$ is the convective mass flux, h_D is the convective mass transfer coefficient, P_{va} is the partial vapor pressure in the drying air, P_{vo} is the partial vapor pressure over the product surface, c_{pdr} is the specific heat capacity of bone-dry product, and

$$\bar{W}_{(t)} = \frac{1}{V_{r=R}} \int_{V_{r=0}}^{V_{r=R}} W_{(r,T)} dV, \qquad (7.51)$$

where $\bar{W}_{(t)}$ is the average moisture content at time t.

For products with low porosity, the drying rate is essentially determined by high energy transfer rates. In this case the heat balance must be made in analogy with the mass balance, that is, at the product element,

$$\frac{\partial}{\partial r} \left\{ A_r \left[\dot{q}_r - \dot{m}_r \Delta H_{v(T_{avg})} \right] \right\} dr = \left[c_{pdr} + c_{pw} W_{(t)} \right] \frac{\partial T}{\partial t} dV, \qquad (7.52)$$

where

$$\dot{q}_r = \frac{q_k}{A} = -k_{(W,T)} \frac{\partial T}{\partial r}. \qquad (7.53)$$

Here $\dot{q}_k$ is the heat flux due to conduction, $k_{(W,T)}$ is the thermal conductivity as a function of moisture content W and temperature T, and

$$\dot{m}_r = \dot{m}_v = -\frac{D}{\mu} \frac{1}{R_v T} \frac{\partial P_v}{\partial r}. \qquad (7.54)$$

7.10.3 Boundary Conditions

The initial conditions for solving the equations are: $W_{(r,t=0)} = W_{in}, T_{(t=0)} = T_{in}$, and $T_{(r,t=0)} = T_{in}$ for all r. There is no moisture gradient at the center of the product, therefore the mass flux is $\dot{m}_{(r=0)} = 0$ for all t. At the surface the vapor pressure must be in equilibrium with the partial vapor pressure of air, therefore the mass flux at the surface is

$$\dot{m}_{r=R} = h_D \frac{P}{R_v T_m} \ln \left(\frac{P - P_{va}}{P - P_{vo}} \right), \qquad P_{v(W,T)} = a_{w(W,T)} P_v^*. \qquad (7.55)$$

7.10.4 Numerical Analysis

The listed equations were solved numerically using the finite-difference procedure. In accordance with Figure 7.9, the product was divided into n_{max} layers of thickness Δr. One can use this approach for the standard geometries: flat plate, cylinder, and sphere. For an element of thickness Δr, one obtains

$$\begin{aligned} V_n &= A\Delta r & \text{plate,} \\ &= \pi (r_n^2 - r_{n-1}^2) h & \text{cylinder,} \\ &= (4\pi/3)(r_n^3 - r_{n-1}^3) & \text{sphere.} \end{aligned} \qquad (7.56)$$

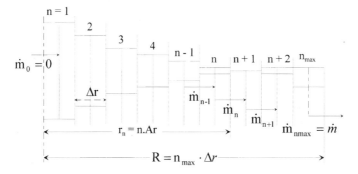

Fig. 7.9. Numerical model for solving the differential equations.

The corresponding position of the surface r_n is

$$A_n = A \qquad \text{plate,}$$
$$= \pi r_n^2 \qquad \text{cylinder,} \tag{7.57}$$
$$= 4\pi r_n^2 \qquad \text{sphere.}$$

At a particular instant in time t, the product has the moisture content W_n in the nth element. The flow of moisture from the nth element to the $(n+1)$th element is $\dot{m}_n A_n$. The mass flux is

$$\dot{m}_n(t) = -\left[\left(\rho_{dr} K_{(W,T)} + \frac{D}{\mu} \frac{P_v^*}{R_v T} \frac{\partial a_w}{\partial W} \right) \frac{W_{n+1} - W_n}{\Delta r} \right] t. \tag{7.58}$$

If the product temperature T and moisture content W_n are known at time t, the moisture content at time $t + \Delta t$ can be calculated for all n elements from the finite-difference equation for the mass balance at the product element,

$$W_{n(t+\Delta t)} = W_{n(t)} - \left(\frac{\dot{m}_n A_n - \dot{m}_{n-1} A_{n-1}}{\rho_{dr} \Delta V_n} \right)_t \Delta t. \tag{7.59}$$

By using the same procedure for the case where heat conduction is not considered, we find that the equation for the product temperature at time $t + \Delta t$ is

$$T_{o(t+\Delta t)} = T_{o(t)} + A_{max} \left[\frac{\dot{q} - \dot{m}_{max} \Delta H_{v(T_o)}}{\rho_{dr} V \left[c_{pdr} + c_{pw} W_{(t)} \right]} \right]_t \Delta t, \tag{7.60}$$

where

$$\bar{W}_{(t)} = \sum_{n=1}^{n=n_{max}} \frac{W_n V_n}{V_n}. \tag{7.61}$$

When heat conduction is taken into account, the corresponding expression is

$$T_{n(t+\Delta t)} = T_{n(t)} + \left[\frac{\dot{q}_n A_n - \dot{q}_{n-1} A_{n-1} - (\dot{m}_n A_n - \dot{m}_{n-1} A_{n-1}) \Delta H_{v(T_{avg})}}{\rho_{dr} \Delta V \left[c_{pdr} + c_{pw} W_{(t)} \right]} \right]_t$$
$$\times \Delta t. \tag{7.62}$$

7.10.5 Heat and Mass Transfer Coefficients

The heat transfer during drying is characterized by a small surface area of the body (temperature T_o) that comes in contact with an air medium, which affects the temperature at the surface layers of the product. The average temperature of the drying air is therefore not much different than its initial temperature. The heat transfer coefficient h can be calculated as follows,

$$h = k\,\mathrm{Nu}/l, \tag{7.63}$$

where Nu is the Nusselt number, k is the thermal conductivity of the air–vapor mixture, and l is a characteristic length of the product. For individual particles in flow, the Nusselt number Nu can be estimated with (Dietl $et\ al.$, 1995)

$$\mathrm{Nu} = \mathrm{Nu_{min}} + (\mathrm{Nu}_{lam}^2 + \mathrm{Nu}_{turb}^2)^{1/2}, \tag{7.64}$$

where

$$\mathrm{Nu}_{lam} = 0.664\,\mathrm{Re}^{0.5}\mathrm{Pr}^{0.33} \qquad \mathrm{Nu}_{turb} = \frac{0.037\,\mathrm{Re}^{0.81}\,\mathrm{Pr}}{1 + 2.443\,\mathrm{Re}^{-0.1}(\mathrm{Pr}^{0.67} - l)}, \tag{7.65}$$

$$\begin{aligned} \mathrm{Nu_{min}} &= 0 && \text{flat plate in parallel flow,} \\ &= 0.3 && \text{cylinder in crossflow,} \\ &= 2 && \text{sphere immersed in free stream.} \end{aligned} \tag{7.66}$$

Here, $\mathrm{Re} = lv/\nu$, v is the velocity of drying air, ν is the kinematic viscosity, $\mathrm{Pr} = \nu/\alpha$ is the Prandtl number, $\alpha(=k/\rho c_p)$ is the thermal diffusivity of drying air, ρ is the density of the air–vapor mixture, and c_p is the specific heat capacity of the air–vapor mixture.

Using the boundary layer analogies (Dincer $et\ al.$, 2000), the mass transfer coefficient h_o is given by

$$h_D = \frac{h}{\rho c_p (\alpha/D)^{1-n}}, \tag{7.67}$$

where n is a positive exponent equal to $1/3$. Note that the values ρ, c_p, α, and D are evaluated at the interface between the product surface and the drying air.

In drying, heat is transferred from hot air to the product, while moisture is transferred from the product to hot air. To account for the coupling between the two fluxes, one defines the corrected transfer coefficients h^* and h_D^*, which are related to the original heat and mass transfer coefficients by

$$\frac{h^*}{h} = \frac{\ln(1+B)^\gamma}{(1+B)^\gamma - 1} \qquad \frac{h_D^*}{h_D} = \frac{\ln(1+B)}{B}, \tag{7.68}$$

where B is known as the driving force,

$$B = \frac{P_{vo} - P_{va}}{P - P_{vo}} \tag{7.69}$$

and

$$\gamma = \frac{c_{pV}\,M_v}{c_p M}\left(\frac{\alpha}{D}\right)^{-(n-1)}, \qquad \bar{c}_p M = \left(1 - \frac{P_{vm}}{P}\right)M_a c_{pa} + \frac{P_{vm}}{P}\,M_v c_{pv}, \tag{7.70}$$

where $\bar{c}_p M$ is the mean specific heat capacity of moist air, $P_{vm} = 0.5(P_{vo} + P_{va})$, M_a is the relative molecular mass of air, c_{pa} is the specific heat capacity of drying air, and M_c is the relative molecular mass of water vapor.

Table 7.2 lists the corresponding equations to calculate fluid properties (e.g., air, vapor, and air–vapor mixture). Because of pure convective heat flow, the heat current q_c and mass flow rate m_c are related by

$$q_c = m_c \Delta H_{v(R_o)}, \tag{7.71}$$

Table 7.2. Fluid properties for numerical calculations[a]

Specific heat (kJ/kg K)	
Air	$c_{pa} = 1.006(1 + 5 \times 10^{-7} T_m^2)$
Vapor	$c_{pv} = 1.88(1 + 2.2 \times 10^{-4} T_m)$
Air–vapor mixture	$c_p = \dfrac{P_{am} M_a c_{pa} + P_{vm} M_v c_{pv}}{P_{am} M_a + P_{vm} M_v}$
	$P_{am} = P - P_{vm}$
Water	$c_{pw} = 4.178 + 9 \times 10^{-6}(T - 35)^2$
Thermal conductivity (W/mK)	
Air	$k_a = 0.02454 \left(\dfrac{T_m}{273.15\,\mathrm{K}}\right)^{0.83}$
Vapor	$k_v = 0.0182 \left(\dfrac{T_m}{273.15\,\mathrm{K}}\right)^{(0.87+0.0010\theta_m)}$
Air–vapor mixture	$k = k_v \dfrac{P_{vm}}{P} + k_a \left(1 - \dfrac{P_{vm}}{P}\right)$
Dynamic viscosity (Pa/s)	
Air	$\eta_a = 17.2 \times 10^{-6} \left(\dfrac{T_m}{273.15\,\mathrm{K}}\right)^{0.7}$
Vapor	$\eta_v = 8.1 \times 10^{-6} \left(\dfrac{T_m}{273.15\,\mathrm{K}}\right)^{1.25}$
Air–vapor mixture	$\eta = \dfrac{\eta_a P_{am} M_a^{1/2} + \eta_v P_{vm} M_v^{1/2}}{P_{am} M_a^{1/2} + P_{vm} M_v^{1/2}}$
Density (kg/m³)	$\rho = \dfrac{P_{am} M_a + P_{vm} M_v}{R_m T_m}$
Latent heat of vaporization (kJ/kg)	$\Delta H_v(T) = 2501 - (c_{pw} - c_{pv})T$
Saturated vapor pressure (Pa)	$P_v^* = 610.7 \times 10^{(7.5T/(237+T))}$
Relative molecular mass (kg/kmol)	$M = M_v \dfrac{P_{vm}}{P} + M_a \left(1 - \dfrac{P_{vm}}{P}\right)$

[a]Properties are expressed in the units $T(\mathrm{K})$ and $P(\mathrm{Pa})$.
Source: Dietl *et al.* (1995).

where the enthalpy of vaporization $\Delta H_{v(T_o)}$ is determined at the temperature of the product surface. Using the above relations, we find that the heat flow q_c can be expressed in terms of the mass flow m_c,

$$q_c = \frac{hA\Delta H_{v(T_o)}}{c_{pv}} \ln \left[1 + \frac{c_{pv}(T_a - T_o)}{\Delta H_{v(T_o)}} \right]. \tag{7.72}$$

The corresponding surface temperature T_o of the product is then given by the expression

$$T_o = T_a - \frac{\Delta H_{v(T_o)}}{c_{pv}}[(1 + B)^\gamma - 1]. \tag{7.73}$$

7.11 Correlation for the Drying of Solids

In spite of numerous theoretical and experimental studies on the determination of drying profiles for various wet materials, limited data on moisture transfer parameters are available in the literature. These data vary greatly because of the complexity of the foods, and because methods of estimation vary from study to study. More recently, a new correlation between Biot number and Dincer number was developed for determining the moisture transfer parameters for wet products subjected to drying:

$$\mathrm{Bi} = 24.85\mathrm{Di}^{-3/8}. \tag{7.74}$$

For the drying process of a wet solid, the Dincer number is defined as the ratio of flow velocity (U) to the drying performance of the solid (SY) and represents the influence of drying air velocity on the drying coefficient of the solid.

The Biot number for moisture diffusion exhibits the magnitude of internal and external resistances to moisture diffusion through the moisture transfer coefficient h_m and moisture diffusivity D. It is therefore defined as

$$\mathrm{Bi} = h_m Y/D. \tag{7.75}$$

Also, knowing the velocity of the drying fluid U, drying coefficient S, and characteristic dimension of the product Y, the Dincer number Di is calculated using (Dincer, 1996)

$$\mathrm{Di} = U/SY. \tag{7.76}$$

The correlation coefficient is greater than 0.8 (as shown in Figure 7.10), and is based on a large number of experimental drying data from the literature. For details, see Dincer et al. (2000). Here we outline the methodology for calculating the drying process and moisture transfer parameters, and present the development of Biot number–lag factor correlation. The experimental moisture content values were nondimensionalized using

$$\Phi = (W - W_e)/(W_t - W_e). \tag{7.77}$$

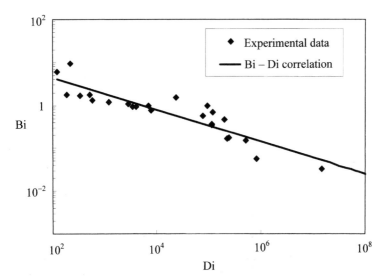

Fig. 7.10. Bi–Di diagram for food products subjected to drying (Dincer and Hussain, 2002).

The dimensionless moisture content values and drying time are regressed using the exponential form

$$\Phi = G\,e^{-St} \tag{7.78}$$

by using the least square curve-fitting method. In this way, the factor G and drying coefficient S are found. The characteristic roots (μ_1s), which appear in the moisture diffusivity relation, are determined using the newly developed expressions as follows (Dincer and Hussain, 2002):

$$\text{Slab: } \mu_1 = -419.24G^4 + 2013.8G^3 - 3615.8G^2 + 2880.3G - 858.94, \tag{7.79}$$

$$\text{Cylinder: } \mu_1 = -3.4775G^4 + 25.285G^3 - 68.43G^2 + 82.468G - 35.638, \tag{7.80}$$

$$\text{Sphere: } \mu_1 = -8.3256G^4 + 54.842G^3 - 134.01G^2 + 145.83G - 58.124. \tag{7.81}$$

The moisture diffusivity is then calculated using

$$D = SY^2/\mu_1^2. \tag{7.82}$$

The Biot number can then be found using Equation (7.74). Finally, the moisture transfer coefficients are calculated using the Biot number equation as $h_m = D\,\text{Bi}/Y$.

For illustration, consider the following example of how to determine the moisture transfer parameters using the existing experimental moisture data of a solid product. The experimental data refer to the drying of prune, okra,

Table 7.3. Thermophysical parameters for drying experiments

	Slab	Cylinder	Sphere
Temperature	60°C	80°C	40°C
Velocity	3 m/s	1.2 m/s	1 m/s
Characteristic			
dimension (Y)	0.0075 m	0.003 m	0.009 m
References	Tsami and Katsioti (2000)	Gogus and Maskan (1999)	McLaughlin and Magee (1999)

Source: Dincer and Hussain (2002).

and potato, as slab, cylinder, and sphere, respectively. The thermophysical parameters of the experiments are given in Table 7.3.

The procedure listed above is employed to determine the moisture transfer parameters and dimensionless moisture distribution. The values of the drying coefficient S, lag factor G, Biot number Bi, root of the characteristic equation μ_1, moisture diffusivity D, and moisture transfer coefficient h_m for the slab, cylindrical, and spherical products were obtained using Equations (7.75) to (7.82). The results are shown in Table 7.4. It is important to emphasize that the moisture content data of these three products were not employed in the development of the correlation (7.74). In fact, the objective was to verify the applicability and accuracy of the present correlation.

By using the data of Table 7.4, we calculate the dimensionless average moisture content profiles for the slab, cylindrical, and spherical products subject to drying at different conditions. Next, we compare the calculated dimensionless moisture content profiles with the experimental profiles. The calculated and experimental profiles are shown in Figures 7.11 to 7.13. The average error between the predicted and measured moisture content values for the products were found to be ±3.38% , ±14.84%, and ±1.31%, respectively. Although the agreement between the calculated values and experimental data for cylindrical products is very good, the agreement for the slab and sphere is excellent.

Table 7.4. Measured drying and moisture transfer parameters for the experimental samples

Process parameters	Slab	Cylinder	Sphere
$S(s^{-1})$	7×10^{-5}	0.0001	0.0009
G	1.0016	1.1981	1.0074
Di	5714286	4000000	123457
Bi	0.0745	0.0851	0.3119
μ_l	0.1407	1.2593	0.2781
$D(\mathrm{m^2 s^{-1}})$	1.989×10^{-7}	5.675×10^{-10}	9.426×10^{-7}
$h_m(\mathrm{ms^{-1}})$	1.976×10^{-6}	1.610×10^{-8}	3.267×10^{-5}

Source: Dincer and Hussain (2002).

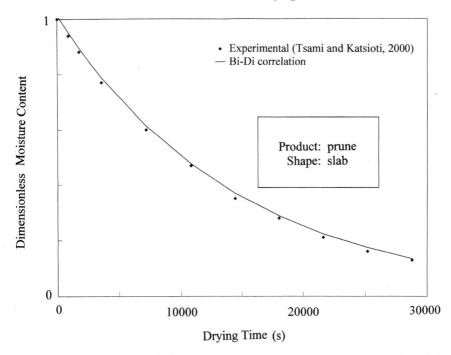

Fig. 7.11. Measured and calculated dimensionless center moisture history for a slab (Dincer and Hussain, 2002).

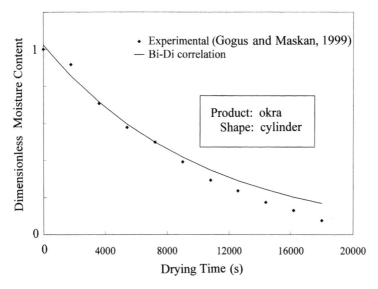

Fig. 7.12. Measured and calculated dimensionless center moisture history for a cylinder (Dincer and Hussain, 2002).

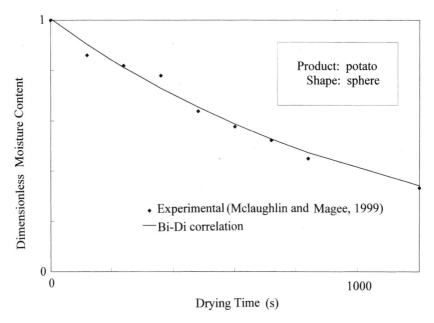

Fig. 7.13. Measured and calculated dimensionless center moisture history for a sphere (Dincer and Hussain, 2002).

In summary, this chapter outlined the fundamental aspects of the drying of solids, the structure and drying characteristics of porous materials (particularly food products), and some key drying equipment. Shrinkage, porosity, and the heat and moisture transfer analyses of drying in porous materials were also discussed. The emphasis was put on the fundamental mechanisms and methods for analyzing and predicting the drying of porous materials.

8

Multidisciplinary Applications

8.1 Walls with Cavities: Insulation and Strength Combined

The void spaces of a porous structure can be interconnected or not. In the latter case the void spaces are disconnected "inclusions" that do not allow a flow to permeate through the porous structure. When the inclusions are large, the flow inside each fluid-filled inclusion may play an important role in the global transport of heat and mass through the structure. This is especially true in coarse cavernous structures encountered in building design.

In this section we show that the internal structure of a cavernous wall can be derived optimally from the competition between the thermal insulation and mechanical strength functions of the wall (Lorente and Bejan, 2002). This combination of two functions, thermal and mechanical, is new in an optimization at such a simple and fundamental level. Previous studies of walls with air enclosures have dealt only with the thermal insulation characteristics of various wall structures.

The design opportunity for varying and optimizing the geometric form of cavities in walls with natural convection was recognized in a few previous studies [e.g., Lorente (2002), Lorente and Lartigue (2002), Lartigue *et al.* (2000), Bejan (1980b), and Frederick (1999)]. The global objective in those studies, however, was the minimization of the thermal resistance of the fluid-filled cavity alone, not the maximization of the insulation capability of an assembly of cavities and separating walls (the present section). For example, it was shown that for two-dimensional natural convection in a vertical cavity with side-to-side heat transfer the global resistance is minimum when the aspect ratio of the cavity has a value of order 1 or smaller, the smaller when the Rayleigh number is larger (Bejan, 1980b; Nield and Bejan, 1999, p. 275). In other words, the heat transfer across the cavity is least impeded when the cavity shape is relatively "round," that is, close to square. The same conclusion was reached in a more recent study by Frederick (1999).

Lorente (2002), Lorente and Lartigue (2002), and Lartigue *et al.* (2000) have documented the global resistance behavior when the side walls of the cavity are deformed (bowed inward) so that the cavity profile resembles a concave lens. Deformations of this kind are commonly found in the air cavities between two glass panes in cold and windy climates. These studies showed that the global resistance to heat transfer increases significantly, in spite of the fact that the narrowing of the cavity midsection suppresses the flow and its natural convection effect.

Even though the few existing studies have dealt with the minimization of the global thermal resistance, they are important because they document the strong relationship that exists between global performance and cavity geometry. This relationship was exploited in a recent paper by Lorente and Bejan (2002), where the objective was the optimization of the wall with internal cavities as an insulation system that must also perform as a strong mechanical structure.

Why should we expect to find an optimal cavity size when we design a cavernous wall as an insulation system? Consider the two-dimensional wall configuration shown in Figure 8.1. Its overall dimensions are fixed: the thickness L, the height H, and the width W perpendicular to the plane of Figure 8.1. There are n vertical air-filled cavities of thickness t_a, which are distributed equidistantly over the wall thickness L. This means that there are $(n + 1)$ slabs of solid wall material (e.g., brick) of individual thickness t_b, which are also distributed equidistantly. We characterize the air and brick (terra-cotta) composite by using the air volume fraction ϕ, which along with the wall volume HLW is a global design parameter,

$$\phi = \frac{nt_a}{L},$$
(8.1)

$$1 - \phi = (n + 1)\frac{t_b}{L}.$$
(8.2)

The overall thermal resistance of this composite is the sum of the resistances of the air and brick layers. If the heat transfer across each air space is by pure conduction, then the thermal resistance posed by each air space is $t_a/(k_a HW)$, where k_a is the thermal conductivity of air. Similarly, the resistance of each layer of brick material is $t_b/(k_b HW)$. The overall resistance is

$$R = \frac{nt_a}{k_a HW} + \frac{(n + 1)t_b}{k_b HW}.$$
(8.3)

or, after using Equations (8.1) and (8.2),

$$R = \frac{\phi L}{k_a HW} + \frac{(1 - \phi)L}{k_b HW}.$$
(8.4)

Equation(8.4) states that the thermal performance of the composite does not depend on the varying geometry, that is, on how many air spaces and slabs

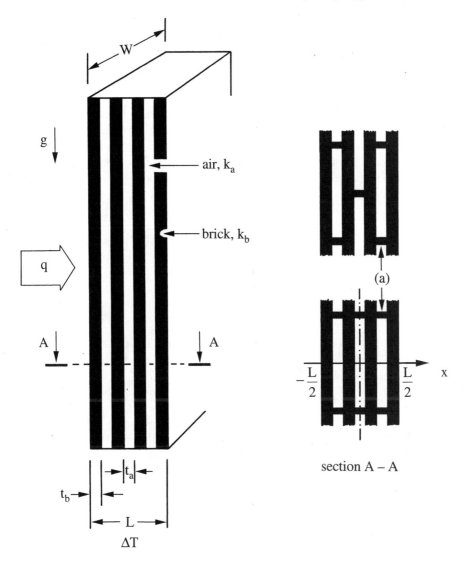

Fig. 8.1. Vertical insulating wall with alternating layers of solid material (brick) and air (Lorente and Bejan, 2002).

of brick we use. This is correct, but only when the air space is ruled by pure conduction, that is, when the thickness t_a is smaller than the thickness of the laminar natural convection boundary layers that would line the vertical walls of each cavity,

$$t_a \lesssim H\mathrm{Ra}_{H,\theta}^{-1/4}. \qquad (8.5)$$

The Rayleigh number $Ra_{H,\theta}$ is based on the height H and temperature difference θ across one air cavity,

$$Ra_{H,\theta} = \frac{g\beta H^3 \theta}{\alpha \nu}. \tag{8.6}$$

The temperature difference θ is smaller than the overall temperature difference ΔT that is maintained across the entire system (Figure 8.1). In the case of air and brick material, the two thermal conductivities are markedly different $(k_b/k_a \approx 20 \gg 1)$, and this means that the overall ΔT is essentially the sum of the temperature differences across all the air cavities,

$$\Delta T \cong n\theta. \tag{8.7}$$

Putting Equations (8.5) to (8.7) together, we see that the insensitivity of R to varying the internal structure n, Equation (8.4), can be expected only when the number of air spaces is sufficiently large so that

$$n^{5/4} \gtrsim \phi \frac{L}{H} Ra_{H,\Delta T}^{1/4}. \tag{8.8}$$

In this inequality, $Ra_{H,\Delta T}$ is based on the overall temperature difference, $Ra_{H,\Delta T} = g\beta H^3 \Delta T/(\alpha \nu)$, and is a known constant because H and ΔT are specified global parameters.

   If the number of air spaces is smaller than in Equation (8.8), the natural convection effect decreases the resistance posed by each air space, and the overall R value is greater than in Equation (8.3). This is why a large enough n, or a small enough t_a, is desirable from a thermal insulation standpoint. On the other hand, the effect of a large n is detrimental to the mechanical stiffness of the wall assembly. When ϕ is prescribed, the stiffest wall is the one where all the solid material is placed in the outermost planes, that is, the wall where two t_b-thin slabs sandwich a single air space. The stiffest wall is the worst thermal insulator, because it contains the thickest air space, which is penetrated by the largest natural convection heat transfer current.

   The optimal internal structure of the wall n results from the competition between thermal performance and mechanical performance. If the mechanical performance is specified, then the wall stiffness serves as a constraint in the process of maximizing thermal performance, from which the optimal geometric form emerges.

   The mechanical strength of the wall, or its resistance to bending and buckling in the plane of Figure 8.1 is controlled by the area moment of inertia of the horizontal wall cross-section [e.g., Beer et al. (2002)]:

$$I_n = \int_{-L/2}^{L/2} x^2 W \, dx. \tag{8.9}$$

The cross-section over which this integral is performed is shown on the right side of Figure 8.1. The area element $W\,dx$ counts only the solid parts of the cross-section, namely, the t_b-thick slabs of brick material. For the sake of simplicity, in this calculation we neglect the transversal ribs [see detail (a) in Figure 8.1] that connect the t_b slabs so that the wall cross-section rotates as a plane during pure bending. It is assumed that the transversal ribs use considerably less material than the t_b slabs. Their role is the same as the role of the central part (the web) of the I profile of an I-beam. In fact, the cross-section of the cavernous wall structure is a conglomerate of I-beam profiles that have been fused solidly over the top and bottom surfaces of the I shape. In practice, the ribs (a) are more commonly arranged in a staggered pattern, as shown in the upper-right corner of Figure 8.1.

In the case of a wall with no cavities ($\phi = 0$) the area moment of inertia is maximum and equal to $L^3W/12$. We use this value as reference in the nondimensionalization of I_n,

$$\tilde{I}_n = \frac{I_n}{L^3W/12}, \qquad (8.10)$$

where the subscript n indicates the number of air gaps. The integral (8.9) can be evaluated case by case, assuming that the cross-section is symmetric about $x = 0$; for example, $\tilde{I}_1 = 1 - \phi^3$ and $\tilde{I}_\infty = 1 - \phi$. These results are displayed in Figure 8.2. The stiffness is larger when n and ϕ are smaller.

An alternative view of this relation is presented in Figure 8.3. When the stiffness is constrained, $\tilde{I}_n$ is constant, and for each geometry n that the designer might contemplate there is one value of ϕ that the wall composite must have. In such cases the ϕ value is larger when the number of air gaps is smaller. Less structural (solid) material is needed when there are fewer air gaps.

When the effect of natural convection cannot be neglected, the overall thermal resistance formula (8.3) has the form

$$R = \frac{nt_a}{k_a HW\mathrm{Nu}} + \frac{(n+1)t_b}{k_b HW}. \qquad (8.11)$$

In the denominator of the first term (the contribution of all the air gaps), Nu is the overall Nusselt number that expresses the relative heat transfer augmentation effect due to natural convection in a single air space,

$$\mathrm{Nu} = \frac{q_{\mathrm{actual}}}{q_{\mathrm{conduction}}}. \qquad (8.12)$$

Several correlations of numerical Nu values have been reported [for a review see Bejan (1995a, p. 234)]: however, they cannot be used in the reported forms because they refer only to the high-Rayleigh number, or the convection-dominated regime (Nu $\gtrsim 2$). More appropriate for the present geometric optimization problem is a Nu function that smoothly covers the entire range of possibilities, from conduction (small t_a) to convection (large t_a).

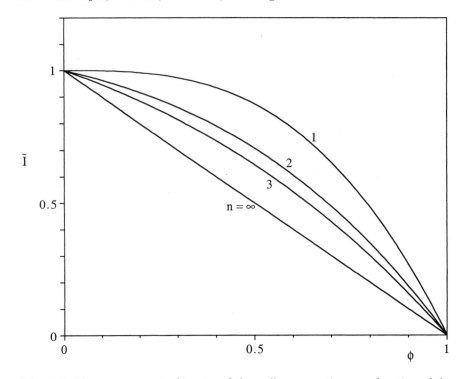

Fig. 8.2. The area moment of inertia of the wall cross-section as a function of the number of air gaps and air volume fraction (Lorente and Bejan, 2002).

The solution we chose is based on the summary presented in Bejan (1995a, Figure 5.8), which shows that the most frequently used high-Ra correlations are well represented by the analytical expression derived based on boundary layer theory,

$$\mathrm{Nu} = 0.364 \frac{t_a}{H} \mathrm{Ra}_{H,\theta}^{1/4} \quad (\text{when } \mathrm{Nu} \gtrsim 2). \tag{8.13}$$

This expression is consistent with the pure conduction criterion (8.5); in other words, Equation (8.13) holds when Equation (8.5) fails. Next, the high-Ra asymptote (8.13) was joined with the pure conduction asymptote (Nu = 1) by using the technique of Churchill and Usagi (1974),

$$\mathrm{Nu} = \left[1 + \left(0.364 \frac{t_a}{H} \mathrm{Ra}_{H,\theta}^{1/4} \right)^m \right]^{1/m} \tag{8.14}$$

with $m = 3$. In summary, the overall resistance formula (8.11) can be nondimensionalized by using as a reference scale the resistance across a completely solid wall $[L/(k_b H W)]$, and converting $\mathrm{Ra}_{H,\theta}$ into $\mathrm{Ra}_{H,\Delta T}$ via Equations (8.6)

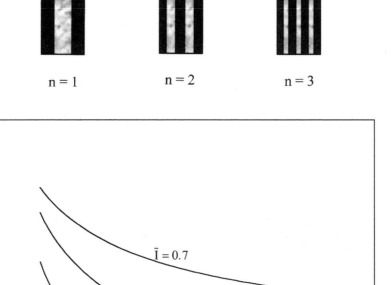

Fig. 8.3. The relation between air volume fraction and number of air gaps when the area moment of inertia of the wall cross-section is constrained (Lorente and Bejan, 2002).

and (8.7):

$$\tilde{R} = \frac{R}{L/(k_b H W)} = \frac{k_b}{k_a}\phi \left[1 + \left(0.364 n^{-5/4}\phi \frac{L}{H} \mathrm{Ra}_{H,\Delta T}^{1/4} \right)^m \right]^{-1/m} + 1 - \phi. \tag{8.15}$$

The overall resistance $\tilde{R}$ emerges as a function of the variable geometric parameters n and ϕ, and the fixed parameters k_b/k_a and the global natural convection parameter

$$b = \frac{L}{H} \text{Ra}_{H,\Delta T}^{1/4}. \tag{8.16}$$

The geometric parameters n and ϕ are related through the global stiffness constraint ($\tilde{I}$ = constant), as shown in Figure 8.3.

In conclusion, when the stiffness constraint is invoked, the global resistance $\tilde{R}$ depends on only one geometric parameter, ϕ or n. This effect is illustrated in Figure 8.4, which shows that $\tilde{R}$ can be maximized with respect to the number of air cavities. The $\tilde{R}$ maximum shifts toward larger n values (more numerous and narrower air gaps) as b increases. The $\tilde{R}$ maximization illustrated in Figure 8.4 was repeated for other $\tilde{I}$ values in the range 0.7 to 0.95.

Let $\tilde{R}_{\max}$ and n_{opt} denote the coordinates of the peak of one of the b = constant curves plotted in Figure 8.4. The maximum resistance $\tilde{R}_{\max}(b, \tilde{I})$ deduced from Figure 8.4 and from similar calculations for other $\tilde{I}$ values, is reported in Figure 8.5. Larger b values represent stronger natural convection, and this is reflected in smaller $\tilde{R}_{\max}$ values. Larger $\tilde{I}$ values represent stiffer walls that use more solid material (Figure 8.3), and, consequently, $\tilde{R}_{\max}$ decreases as $\tilde{I}$ increases.

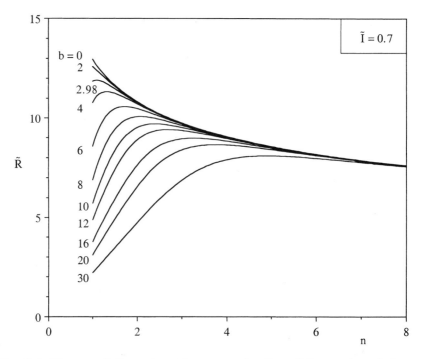

Fig. 8.4. The overall thermal resistance as a function of the number of air gaps when the external parameters b and $\tilde{I}$ are fixed (Lorente and Bejan, 2002).

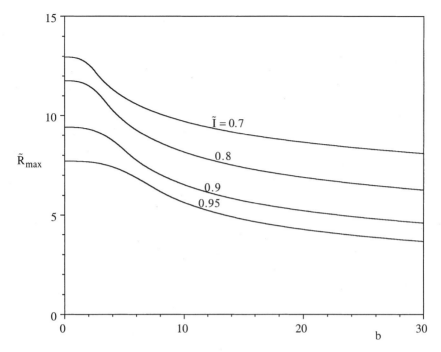

Fig. 8.5. The maximized wall thermal resistance as a function of the global natural convection parameter b and the global stiffness parameter $\tilde{I}$ (Lorente and Bejan, 2002).

The optimal number of air gaps (n_{opt}) that corresponds to the $\tilde{R}_{\mathrm{max}}$ results of Figure 8.5 is reported in Figure 8.6. Fewer air gaps are better when the natural convection effect is weak (small b), and when the required stiffness approaches that of the solid wall ($\tilde{I} = 1$).

In sum, the simultaneous consideration of the thermal and mechanical functions of the complex structure is the defining feature of the idea pursued in this section. The chief conclusion is that the number of air gaps built into the wall can be optimized when the overall stiffness is specified. The optimal number of air gaps increases when the effect of natural convection increases, and when the specified wall stiffness decreases. The maximized wall thermal resistance is larger when the effect of natural convection in the air gaps is weaker, and when the wall stiffness is smaller. The optimal volume fraction occupied by air in the cavernous structure decreases when the natural convection effect becomes stronger, and when the wall stiffness increases.

Future work on the optimization of cavernous wall structures may include additional features, which, for the sake of simplicity, were not used in Figure 8.1. For example, the interplay between air conduction and natural convection in bricks with cavities depends, among other things, on how the

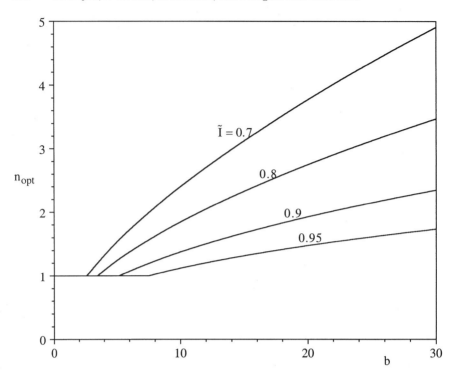

Fig. 8.6. The optimal number of air gaps as a function of the global natural convection parameter b and the global stiffness parameter $\tilde{I}$ (Lorente and Bejan, 2002).

bricks are assembled in the wall. A brick is usually 20 to 25 cm high, and a wall has a height of roughly 250 cm (the height of a floor). This means that the aspect ratio of one continuous (vertical) air space formed by stacking the bricks is 10 times greater than the aspect ratio of the air space of a single brick. Consequently, the global heat transfer across the air space changes. Lorente (2002) reported calculations based on a combined model of conduction, radiation, and natural convection (Lorente et al., 1996). The results reproduced in Figure 8.7 show how the behavior of the global thermal resistance changes when the height of the air space changes. In this model, the total thickness of the bricks remains constant but the number of enclosures (and thus their thickness) could vary. The cavity width was always greater than the air space thickness so that the natural convection pattern was essentially two-dimensional.

Figure 8.7 shows a sample of the results obtained for several values of the temperature difference ($\Delta T = T_h - T_c$) between both sides of the bricks (Lorente et al., 1998; Lorente, 2002). When ΔT is increased, the thermal resistance of the 25-cm high brick decreases by 27%, and the corresponding decrease is less than 10% for the 250-cm high brick. In both cases, the decrease is essentially due to the convective transfer. Indeed, if we analyze the natural

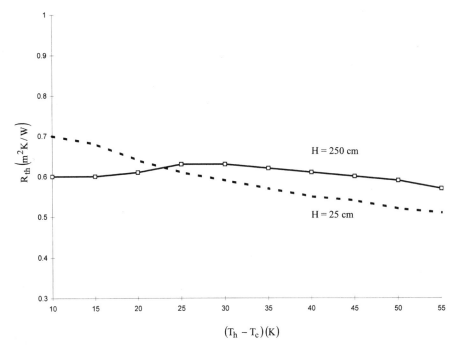

Fig. 8.7. The evolution of the global thermal resistance of bricks with different heights.

convective transfer phenomenon inside enclosures (with constant air thickness and constant temperature difference), the average heat flux density by convection is smaller in a higher cavity. For the range of Rayleigh numbers in Figure 8.7 the flow regime is of the boundary layer type. When the height of the air space increases, boundary layers thicken and temperature gradients decrease. Thus the global heat transfer rate is weaker. This behavior was confirmed in other cavity configurations as well (Lorente *et al.*, 1998).

Another example is the effect of radiation in the air spaces, which can be modeled through the use of surface radiosities [e.g., Lorente *et al.* (1994)]. The shapes of air cavities in terra-cotta walls, and the type of thermal boundary conditions on such cavities also play important roles in the model [e.g., Lorente *et al.* (1996, 1998)]. Furthermore, terra-cotta is a porous material in which thermal diffusion is accompanied by the diffusion of moisture. The latter has an important effect on heat transfer, as documented by Vasile *et al.* (1998). Terra-cotta elements are being contemplated in the modular design and construction of regenerators for the glass industry (Boussant-Roux *et al.*, 2000).

The combined thermal and mechanical optimization can be pursued from alternative points of view, depending on the objectives of the greater system to which the cavernous wall belongs. The combined "flow and strength" geometric optimization method illustrated in this section can be applied in other fields where mechanical structures must carry loads while posing minimum or maximum resistance to internal and external flows [e.g., Gosselin *et al.* (2004)]. This combination may be carried further into the design of structural elements for vehicles, which, like the wall of Figure 8.1, could be conceptualized and "morphed" into geometric forms with more than one function.

8.2 Fibers Coated with Phase-Change Material

It has been shown that polyethylene glycols (polyols) can be bonded stably on fibrous materials. The resulting composite materials—the "thermally active" fabrics—exhibit reversible and reproducible energy storage and release properties (Vigo and Bruno, 1987, 1989; Bruno and Vigo, 1987). The energy storage and release is due to the high latent heat of melting and crystallization of the polyols affixed to the fibers. Prior to this technological breakthrough, which simplifies the manufacturing, weaving, packing, and washing of the fibers, phase-change materials had been used as fillers in hollow fibers (Vigo and Frost, 1982, 1983).

Thermally active fibers have several other attractive properties, for example, improved resistance to oily soiling, static charge, wear and piling, and significantly greater resiliency and hydrophilicity (Vigo and Bruno, 1989). They are projected for use in a wide variety of applications in the apparel, insulation, air conditioning, aerospace, materials, and chemical industries. Yet the work that has been done until now is developmental in nature. It proved the feasibility of these applications, and unveiled the basic properties of the polyol-coated fibers surrounded by air. To proceed from this level to industrial applications of many shapes and sizes (e.g., thermal control in buildings), one must possess a model with which to anticipate the performance of the material in the large-scale system. In this section we outline the model developed by Lim *et al.* (1993).

In this model the fibers and the phase-change material (polyol, liquid or solid) constitute the matrix, and air is the fluid that flows through the interstitial spaces. Of primary interest is the relation between the time of complete melting or solidification of the polyol coatings, and the various dimensions and the external parameters (e.g., flow configuration) of the space filled with fibers.

This model differs fundamentally from the one used in earlier studies of melting and solidification in saturated porous media [e.g., Kazmierczak *et al.* (1986, 1987, 1988), Beckermann and Viskanta (1988), Kazmierczak and Poulikakos (1988), and Jany and Bejan (1988)]. In such studies the melted phase-change material was the only fluid present in the pores and, consequently, there was no flow through regions saturated with solid phase-change material. In the porous medium model for thermally active fibers, the fluid (air) flows through the entire matrix regardless of whether the polyol coatings are liquid or solid.

When the layer of polyol-coated fibers is sufficiently thick and the air flow through the fibers sufficiently slow, the layer may be modeled as a homogeneous porous medium in which the air and the matrix (fibers and polyol) are in local thermal equilibrium. The pores are saturated with air, and the fibers are coated uniformly with polyol (solid or liquid). The effective properties of the equivalent homogeneous porous medium emerge after the volume-averaging of the structure (air, fibers, polyol) and the conservation equations for mass, momentum, and energy. Because of local thermal equilibrium, T represents the local temperature of the fiber and the air and polyol that immediately surrounds it. The composition of the homogeneous porous medium is described by the porosity ϕ (about 80%), and the fraction of the matrix occupied by polyol ε (about 20%). This means that a unit volume is distributed in the following proportions: $\phi = $ air, $(1 - \phi) = $ matrix (fibers and polyol), $(1 - \phi)\varepsilon = $ polyol, and $(1 - \phi)(1 - \varepsilon) = $ fibers. The aggregate heat capacity of the porous medium is then

$$(\rho c)_m = \phi(\rho c_P)_a + (1 - \phi)[\varepsilon(\rho c)_p + (1 - \varepsilon)(\rho c)_f], \qquad (8.17)$$

in which the subscripts m, a, p, and f refer to the averaged porous medium, air, polyol, and fibers.

The air flow through such fabrics is weak, and because of this the effect of thermal dispersion is neglected. The effective thermal conductivity k of thermally active fibers must be determined through direct measurement. It is further assumed that k is a constant, therefore k is independent of T and whether the polyol is solid or liquid.

The volume-averaged energy conservation equation that corresponds to these assumptions is

$$\sigma \frac{\partial T}{\partial t} + u \frac{\partial T}{\partial x} + v \frac{\partial T}{\partial y} = \alpha \left(\frac{\partial^2 T}{\partial x^2} + \frac{\partial^2 T}{\partial y^2} \right), \qquad (8.18)$$

where (u, v) are the volume-averaged air velocity components, and σ and α are the heat capacity ratio and the thermal diffusivity of the porous medium,

$$\sigma = \frac{(\rho c)_m}{(\rho c_P)_a}, \qquad \alpha = \frac{k}{(\rho c_P)_a}. \qquad (8.19)$$

The simplest configuration in which the polyol has an effect on the thermal behavior of the medium is melting by one-dimensional conduction, shown in Figure 8.8. The pores of the one-dimensional layer of thickness L are filled with stagnant air. The initial temperature of the layer is uniform T_i and lower than the polyol melting point T_m. Beginning with the time $t = 0$, the temperature of the left side is maintained at a temperature T_h above the melting point. The right side is insulated. The heating that is administered through the left wall raises the temperature of the medium, and causes the gradual melting of the polyol. The melting front location, or the instantaneous thickness of the sublayer in which the polyol has melted, is $X(t)$. Of primary interest is

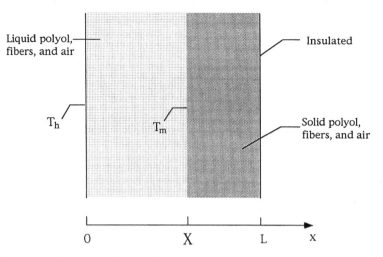

Fig. 8.8. One-dimensional layer of polyol-coated fibers and air (Lim *et al.*, 1993).

the time (t_{end}) when all the polyol of the L-thick layer has melted: this time interval is defined by

$$X(t_{\text{end}}) = L. \tag{8.20}$$

The conservation of energy across the melting front requires

$$-k \left. \frac{\partial T}{\partial x} \right|_{x=X_-} = k \left. \frac{\partial T}{\partial x} \right|_{x=X_+} = (1 - \phi)\varepsilon \rho_p \lambda \frac{dX}{dt}, \tag{8.21}$$

in which λ is the polyol latent heat of melting. The group $(1 - \phi)\varepsilon \rho_p$ represents the polyol density in the porous medium, that is, kilograms of polyol per unit volume of porous medium. The melting front temperature is $T[X(t)] = T_m$. It is important to note that because the polyol coating is thin, and because of the local thermal equilibrium assumption (one local T value for air, polyol, and fiber), in this model we neglect the melting on each fiber, that is, how the melting front propagates through the coating, into the fiber.

The description of this phenomenon can be placed in dimensionless form by using the variables

$$\tilde{x} = \frac{x}{L}, \qquad \theta = \frac{T - T_i}{T_h - T_i}, \qquad \tau = \frac{\alpha t}{\sigma L^2}$$

$$M = \frac{X}{L}, \qquad S = \frac{(1 - \phi)\varepsilon \rho_p \lambda}{(\rho c)_m (T_h - T_i)}. \tag{8.22}$$

The dimensionless group S plays a role similar to that of the inverse Stefan number in a medium consisting of only solid matrix and phase-change material. In the case of polyol-coated fibers, S falls in the range 0.01 to 0.1.

For example, if we consider the properties of cotton–polyester fibers coated with PEG 1000 (i.e., polyol with molecular weight 1000), we use $T_h = 35°C$, $T_i = 0°C$, $\lambda = 10.9$ kJ/kg, $\rho_p = 1130$ kg/m^3, $\phi = 0.8$, $\varepsilon = 0.2$, $(\rho c_P)_a = 1.184$ kJ/m^{3}K, $(\rho c)_p = 2.5 \times 10^6$ J/m^{3}K, $(\rho c)_f \sim (\rho c)_p$, and based on Equations (8.17) and (8.22) we obtain $(\rho c)_m \sim 0.45 \times 10^6$ J/m^{3}K, and $S \approx 0.03$. The corresponding calculation based on the properties of fiberglass coated with the polyol PEG 1450 yields $S \approx 0.023$.

The time of complete melting, τ_{end} is defined by $M(\tau_{end}) = 1$. The dimensionless parameter θ_m measures the position of the melting point T_m relative to the initial (low) and left-side (high) temperatures,

$$\theta_m = \frac{T_m - T_i}{T_h - T_i}. \tag{8.23}$$

The dimensionless complete melting time τ_{end} is a function of two parameters, S and θ_m. This problem can be studied analytically based on classical methods, or numerically.

Figure 8.9 shows the principal results of the numerical solution (Lim et al., 1993), namely, the time needed to melt all the polyol in the L-thick layer. The dimensionless time τ_{end} increases with both θ_m and S; however, the effect of the latter is weak in the range $S = 0.01$ to 0.1. The weakness of the

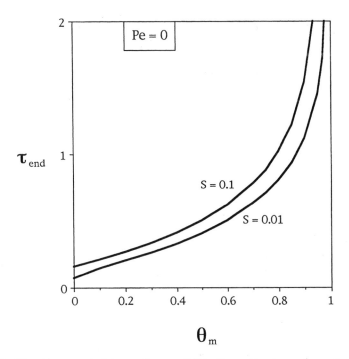

Fig. 8.9. The time needed to melt or solidify the entire layer (one-dimensional conduction) (Lim et al., 1993).

effect of S on τ_{end} makes the calculation of the actual (dimensional) melting time t_{end} easier, because the S group accounts for several of the more complicated features of the fiber-polyol-air porous medium model. In other words, as a first approximation we can evaluate τ_{end} by first estimating θ_m, and then calculating the actual time $t_{\text{end}} = \tau_{\text{end}} \sigma L^2 / \alpha$. For example, in an application where $T_h = 35°\text{C}$, $T_i = 0°\text{C}$, and $T_m = 28°\text{C}$, we find that $\theta_m = 0.8$, and from Figure 8.9 that $\tau_{\text{end}} \sim 0.8$. Furthermore, if $k \sim 0.06$ W/m · K, $(\rho c_P)_a = 1.184$ kJ/m^{3}K, $(\rho c)_m \sim 450$ kJ/m^{3}K, and $L = 10$ cm, we obtain $\sigma \sim 400$, $\alpha \sim 0.5$ cm^2/s and, finally, $t_{\text{end}} \sim 17$ h.

The analogous problem of solidification in a one-dimensional layer can be reduced to the same dimensionless foundation as above, by replacing θ and θ_m with the new definitions

$$\theta = \frac{T_i - T}{T_i - T_c}, \qquad \theta_m = \frac{T_i - T_m}{T_i - T_c}. \tag{8.24}$$

In these definitions T_c is the temperature of the cooled wall ($T_c < T_m < T_i$), that is, the left-hand wall in Figure 8.8. All the fibers are initially coated with liquid polyol at T_i, and the right-hand wall is insulated throughout the cooling and solidification process. The time t_{end} needed by the solidification front to traverse the entire layer can be estimated based on Figure 8.9 and the τ_{end} and S definitions (8.22).

The melting and solidification times are considerably shorter when the boundaries of the porous layer are permeable and air is forced to flow across the layer. In Figure 8.10 we show this by considering the melting under the influence of warm (T_h) air of velocity u that penetrates the layer. The constant velocity u is the actual value outside the layer. If the changes experienced by the air density inside the layer are negligible, this value is the volume-averaged velocity of air in the porous medium (the average velocity in the pores is u/ϕ).

The new feature relative to the flow configurations discussed above is the $u \partial T / \partial x$ term, which is retained in the energy equation (8.18), so that the dimensionless energy equation is

$$\frac{\partial \theta}{\partial \tau} + \text{Pe} \frac{\partial \theta}{\partial \tilde{x}} = \frac{\partial^2 \theta}{\partial \tilde{x}^2}. \tag{8.25}$$

The Peclet number is $\text{Pe} = uL/\alpha$. For example, if $u \sim 1$ cm/s, $L \sim 10$ cm, and $\alpha \sim 0.5$ cm^2/s, the Peclet number is of order 20. The air flow does not have an effect on the two-phase front condition (8.21) because the air stream is single phase and its temperature varies continuously across the front. The adiabatic plane condition $\partial \theta / \partial \tilde{x} = 0$ at $\tilde{x} = 1$ becomes more appropriate as the Peclet number increases, and is consistent with the assumption that the $\tilde{x} = 1$ wall is adiabatic in the absence of the transverse flow (Figure 8.8).

The numerical results are presented in Figure 8.11 as a sequence of three cases according to Peclet number. Figure 8.9 represents the Pe = 0 case in the same sequence. The range covered by τ_{end} in each frame shows that the melting

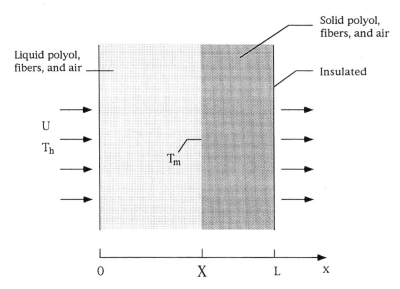

Fig. 8.10. Forced convection perpendicular to the one-dimensional layer (Lim et al., 1993).

time decreases appreciably as the transverse air velocity increases. For example, by repeating the numerical calculation presented above Equation (8.24), we find that when Pe ~ 10 the melting time reduces to $t_{\text{end}} \sim 2$ hours (compared with 17 hours in the case of no flow).

The effect of transverse forced convection on the overall solidification time of the layer can also be estimated using Figure 8.11. In this case θ and θ_m are defined by Equations (8.24), as the cold air stream that crosses the $\tilde{x} = 0$ plane has the temperature T_c. The L-thick layer is originally at the temperature T_i, and all the fibers are coated with liquid polyol.

Natural convection may be an important heat transfer mechanism in large-scale applications of polyol coated fibers, for example, in the walls of buildings. An important question is to what extent the time of complete melting is influenced by natural convection. The answer comes from considering the mass and momentum equations for the volume-averaged air flow, which are accounted for by the single equation

$$\frac{\partial^2 \psi}{\partial x^2} + \frac{\partial^2 \psi}{\partial y^2} = -\frac{Kg\beta}{\nu}\frac{\partial T}{\partial x}, \qquad (8.26)$$

where ψ is the streamfunction ($u = \partial\psi/\partial y, v = -\partial\psi/\partial x$), and ($u$, v) are the volume-averaged velocity components of the air flow. Equation (8.26) is based on the Oberbeck–Boussinesq approximation and the Darcy flow model. The impermeable rectangular boundary indicated in Figure 8.12 means that $\psi = 0$

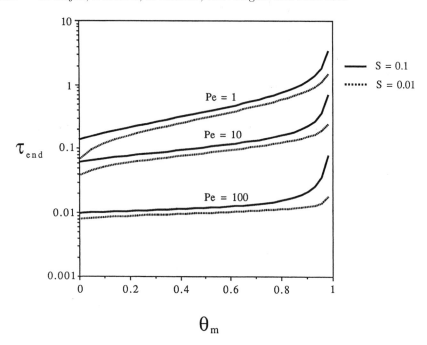

Fig. 8.11. The effect of transverse forced convection on the time needed to melt or solidify the entire layer (Lim *et al.*, 1993).

at $x = 0$, L and $y = 0$, H. The permeability K is treated as a constant; its value is not affected by the phase-change process because the polyol coating is thin and the porosity is high. The K value is controlled by the mix of air and fibers, and the size and shape of the fibers.

In the beginning the fiber coatings are solidified completely and uniformly at the temperature T_i, which is below the polyol melting point T_m. The temperature of the left side is raised suddenly to the level T_h, which is higher than T_h. The remaining three boundaries are well insulated.

As the fiber coating melts, the melting front $X(y, t)$ migrates to the right as the time increases. The air flows both between fibers with liquid coatings and fibers with solid coatings. The only coupling between the melting front position and the flow field is the temperature gradient that appears on the right-hand side of Equation (8.26). The conservation of energy continues to be represented by (8.18). Replacing (8.21) is a new equation for the conservation of energy at the two-dimensional melting front,

$$
-k\left.\frac{\partial T}{\partial y}\right|_{x=X_-} + k\left.\frac{\partial T}{\partial x}\right|_{x=X_+} + \frac{\partial X}{\partial y}\left(k\left.\frac{\partial T}{\partial y}\right|_{x=X_-} - k\left.\frac{\partial T}{\partial y}\right|_{x=X_+}\right)
$$
$$
= (1-\phi)\varepsilon\rho_p\lambda\frac{\partial X}{\partial t}. \tag{8.27}
$$

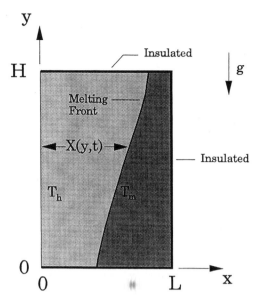

Fig. 8.12. Two-dimensional layer with natural convection due to heating from the side (Lim *et al.*, 1993).

The air flow spreads throughout the $L \times H$ porous medium, regardless of the instantaneous position of the melting front. This feature distinguishes the present phenomenon from the natural convection melting of a porous medium saturated with nothing but a phase-change material [e.g., Kazmierczak *et al.* (1986), Beckermann and Viskanta (1988), and Jany and Bejan (1988)]. In the latter, the flow of liquid phase-change material is confined to only one side of the melting front. Another distinguishing feature of the present phenomenon is that the melting phase-change material (i.e., liquid polyol coatings) does not flow. The flow and the associated natural convection heat transfer are due solely to the air that fills the spaces between the coated fibers.

Numerical solutions were developed by Lim *et al.* (1993) by using the dimensionless variables

$$\tilde{y} = \frac{y}{H}, \qquad \tau_H = \frac{\alpha t}{\sigma H^2}, \qquad \tilde{\psi} = \frac{\psi}{Kg\beta(T_h - T_i)H/\nu}, \qquad (8.28)$$

and $\tilde{x}$, θ, M, and S defined earlier. The new dimensionless time τ_H is based on H, so that the relation between it and τ of Equation (8.22) is $\tau_H = \tau/(H/L)^2$.

The dimensionless groups that appear in the dimensionless governing equations are the geometric aspect ratio and the Darcy modified Rayleigh number for air:

$$R = \frac{H}{L}, \qquad \mathrm{Ra} = \frac{Kg\beta(T_h - T_i)H}{\alpha\nu}. \qquad (8.29)$$

The boundary conditions are: (i) flow, $\tilde{\psi} = 0$ at $\tilde{x} = 0$, 1 and $\tilde{y} = 0$, 1; (ii) temperature to the left of the melting front, $\theta = 1$ at $\tilde{x} = 0$, $\theta = \theta_m$ at $\tilde{x} = M$, $\partial\theta/\partial\tilde{y} = 0$ at $\tilde{y} = 0$, 1; and (iii) temperature to the right of the melting front, $\theta = \theta_m$ at $\tilde{x} = M$, $\partial\theta/\partial\tilde{x} = 0$ at $\tilde{x} = 1$, $\partial\theta/\partial\tilde{y} = 0$ at $\tilde{y} = 0$, 1. The dimensionless melting front temperature θ_m is defined in Equation (8.23). The initial conditions are $\tilde{\psi} = 0$ and $\theta = 0$ at $\tau_H = 0$.

The dimensionless version of Equation (8.26) was solved by using a combined finite-difference and spectral method. Details of the method are given in Lim *et al.* (1993). The time-dependent flow and heat transfer depend on four independent groups, Ra, H/L, S, and θ_m. This set shows that the phenomenon is more complex than natural convection in a porous medium saturated with a phase-change material, where there are only two groups, Ra and H/L, when the Stefan number is small [e.g., Jany and Bejan (1988)].

The effect of each of the four groups was investigated systematically by focusing on the time of complete melting, $\tau_{H,\text{end}}$ defined in Figure 8.13. The figure shows the evolution of the melting front in a square rectangular system, at a moderate Rayleigh number, and with a coating melting point halfway between the initial temperature and the temperature of the heated wall. The time of complete melting $\tau_{H,\text{end}}$ is defined as the moment when the melting front first touches the insulated right-hand side of the system. In the particular case illustrated in Figure 8.13, the melting front is more deformed and tilted than in a porous medium saturated with a phase-change material at a similar Rayleigh number [compare this with Figure 10.3 in Nield and Bejan (1999)].

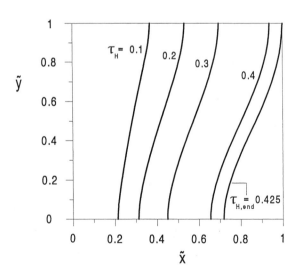

Fig. 8.13. The time evolution of the melting front and the definition of the time of complete melting $\tau_{H,\text{end}}$ (Ra = 10, $H/L = 1$, $S = 0.1$, $\theta_m = 0.5$) (Lim *et al.*, 1993).

Figures 8.14 to 8.16 illustrate the history of the average position of the melting front,

$$\bar{M}(\tau_H) = \int_0^1 M(\tilde{y}, \tau_H)d\tilde{y}. \tag{8.30}$$

In each figure, the time of complete melting $\tau_H = \tau_{H,\mathrm{end}}$ is the point on the abscissa in line with the end of each $M(\tau_H)$ curve. The inflection (S shape) of each curve is considerably more pronounced than in a porous medium saturated with a phase-change material [compare this with Figure 10.6 in Nield and Bejan (1999)].

The effect of the latent heat parameter S is documented in Figure 8.14. A larger latent heat means a larger S value and a longer time until the coating melts on the fibers located the farthest from the heated wall. The time of complete melting decreases sensibly as the Rayleigh number becomes greater than approximately 5.

Figure 8.15 shows how $\bar{M}(\tau_H)$ and $\tau_{H,\mathrm{end}}$ respond to changes in the dimensionless melting temperature θ_m. Three bundles of curves are shown, one bundle for each θ_m value. Each bundle contains four curves according to the

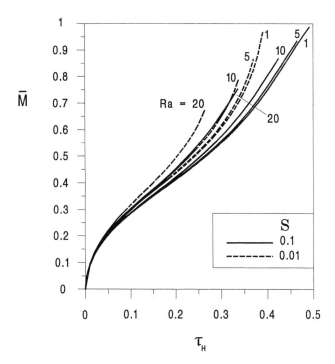

Fig. 8.14. The effect of the latent heat parameter S on the average melting front position and the time of complete melting ($H/L = 1, \theta_m = 0.5$) (Lim $et\ al.$, 1993).

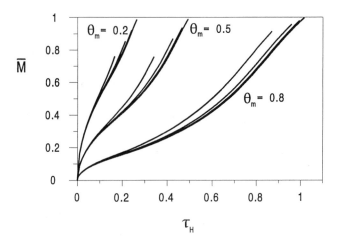

Fig. 8.15. The effect of the melting point parameter θ_m on the average melting front position and the time of complete melting ($H/L = 1, S = 0.1$) (Lim *et al.*, 1993).

Ra value (namely, 20, 10, 5, and 1, counting from the left). The θ_m effect is significant: the melting time increases steadily as θ_m increases, that is, as the temperature difference $T_h - T_m$ becomes small. This trend agrees with what we saw in Figures 8.9 and 8.11.

Figure 8.16 shows the effect of the geometric aspect ratio H/L. Three bundles of $\bar{M}(\tau_H)$ curves are shown, one bundle for each aspect ratio H/L. A fourth bundle, the one for $H/L = 1$, has been plotted in Figure 8.14. Each bundle contains four curves that document the increase in the Rayleigh number from 1 to 20.

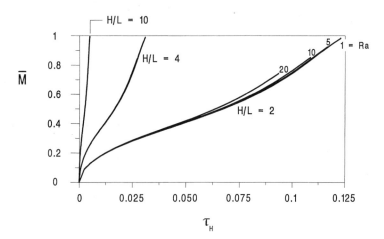

Fig. 8.16. The apparent effect of the geometric aspect ratio H/L, when $\bar{M}$ is plotted against τ_H (Lim *et al.*, 1993).

Expressed in terms of τ_H (i.e., in units of $\sigma H^2/\alpha$), the melting time $\tau_{H,\text{end}}$ decreases significantly as the aspect ratio H/L increases. This strong H/L effect is deceiving, for if $\tau = \alpha t/\sigma L^2$ is used instead of τ_H on the abscissa the three H/L bundles of curves fall almost on top of one another. In conclusion, at Rayleigh numbers smaller than 20 and aspect ratios greater than 1, the H/L effect is insignificant if $\bar{M}$ is plotted against τ. The universal bundle of curves that would result is nearly identical to the $H/L = 1$ bundle ($S = 1$, i.e., the solid lines) plotted in Figure 8.14, where the abscissa parameter τ_H happens to be equal to τ.

The reason for the insignificant H/L effect on the melting time in increasingly taller spaces is that when Ra is small or moderate the heat transfer process becomes dominated by conduction in the horizontal direction. In the $H/L \to \infty$ and finite-Ra limit even the Ra effect becomes insignificant (note the tightening of the bundles as H/L increases in Figure 8.16), and all the melting time results approach very closely the pure conduction results developed in Figure 8.9. This effect is also visible in convection without phase change, Figure 2.6.

There is a perfect analogy between the problem of two-dimensional melting by natural convection (Figure 8.12) and the problem of two-dimensional solidification by natural convection. In the latter, the $H \times L$ medium is initially isothermal and all the fibers are coated with liquid polyol, $T_i > T_m$. The temperature of one of the side walls is lowered suddenly to T_c, which is below the melting point. The remaining walls are insulated. The air flow inhabits the entire $H \times L$ space, that is, both sides of the solidification front. The movement of the solidification front is similar to that of Figure 8.13; however, the shape of the front is the mirror image of the shape shown in Figure 8.13, where the mirror is one of the horizontal boundaries. The time of complete solidification $\tau_{H,\text{end}}$ is defined in the same way, that is, as the moment when the polyol coatings begin to solidify in the plane of the opposite (insulated) side wall. Figures 8.14 to 8.16 can be used to estimate the solidification time, provided θ_m is replaced with the definition (8.24), and the air Rayleigh number Ra is defined by $\text{Ra} = Kg\beta(T_i - T_c)H/(\alpha\nu)$, where $T_i > T_c$.

The existence of the analogy between melting and solidification is one more feature that distinguishes the present phenomena from the corresponding phenomena in a porous medium saturated with a phase-change material. In the latter, such an analogy does not exist because the flow occurs on only one side of the phase-change interface [see, e.g., Oosthuizen (1988)].

8.3 Methane Hydrate Sediments: Gas Formation and Convection

In this section we draw attention to an important new area of fundamental and applied research on convection in porous media: the extraction of methane gas from clathrate hydrates (Rocha et al., 2001; Bejan et al., 2002).

Vast deposits of methane hydrates have been found all over the globe, under the oceanic floor, and under permafrost. Clathrate hydrates are solid crystals of water and methane, which form and exist at sufficiently high pressures and low temperatures; see Figure 8.17. "Clathrate" comes from the Latin verb *clathrare*, which means to endow with a lattice. In chemistry, this terminology refers to a mixture in which the molecules of one substance, such as methane, are completely entrapped in the crystal lattice or cage-like structure of another substance, such as water (Sloan, 1990; de Deugd *et al.*, 2001).

Several gases or gas mixtures form hydrates of three different structures that have repetitive crystal units composed of "cages" (polyhedra) of hydrogen-bounded water molecules (polyhedron's vertices). Each cage (void) contains one guest molecule that is held in by Van der Waals forces. Today, there is a growing interest in understanding the formation and behavior of methane clathrates, because they represent a potentially huge source of gaseous fuel. Several methods are contemplated for capturing methane gas from hydrates: heating the hydrate, depressurizing the hydrate, mining the hydrate, and destabilizing the hydrate by using in situ combustion. The work reviewed in this section refers to the fundamentals of methane gas generation via depressurization; see Rocha *et al.* (2001). In addition, these phenomena are relevant to a wide spectrum of technologies (Sloan, 1990), as clathrate hydrate processes are involved in the plugging of pipelines (Behar, 1994; Lysne, 1994), the sequestration of CO_2 in the ocean (Fontana and Mussumeci, 1994; Makogon, 1981), the endangering of the stability of foundations for offshore oil

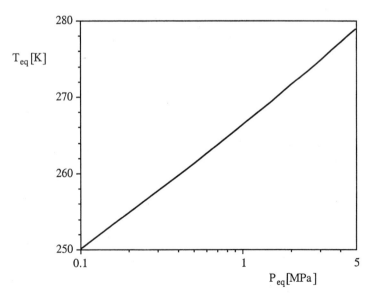

Fig. 8.17. The equilibrium pressure-temperature curve for clathrate hydrates of methane in water.

wells (Briaud and Chaouch, 1997), and the low-temperature storage of energy via clathrate formation (Holder *et al.*, 1994).

The fundamentals of the gas generation process have been studied under two scenarios, the heating of the sediment filled with solid hydrate (Briaud and Chaouch, 1997; Cherskii and Bondarev, 1972; Selim and Sloan, 1989; Islam, 1994), and the depressurization of the sediment (Yousif *et al.*, 1990). Rocha *et al.* (2001) considered the depressurization technique shown in Figure 8.18. The porous layer, which is filled with solid hydrate, lies immediately above sediment containing free gas that has not formed hydrate because the sediment temperature at that depth is too high. In the upper layer the solid hydrate occupies the pore spaces of a permeable porous medium. Here the temperature is sufficiently low and the pressure high such that the solid hydrate is stable. In the lower layer the generated gas flows through the pores of the same or another porous medium. The interface between the two layers marks the level where the initial temperature T_i and pressure P_i correspond to equilibrium (dissociation, phase change, T_{eq} and P_{eq}).

When a well is drilled vertically through the two layers the pressure can be lowered substantially on the lower side of the interface. The hydrate becomes unstable and dissociates into methane gas and liquid water: a dissociated sublayer grows in the lower part of the original upper layer. In one possible scenario, the dissociated region becomes a "lens" that effects the transition between the region with solid hydrate in the pores and the lower layer of gas-saturated porous medium. The gas generated through the dissociation process

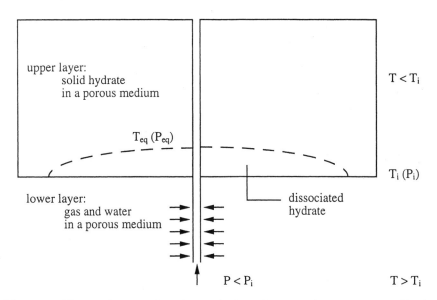

Fig. 8.18. The production of gas from a hydrate layer via depressurization (Rocha *et al.*, 2001).

is driven through the lens, and through the lower porous medium, into the well, where it is captured.

The basic scales of the process characterize the rate at which the dissociated lens advances into the porous medium filled with solid hydrate, and the rate at which gas is being produced and captured in an unsteady, time-dependent fashion. There are two problems—two phenomena—that join hands during this process. One problem is the dissociation and advancement of the phase-change front into the hydrate-filled medium (Figure 8.19a). This problem can be studied first as a one-dimensional, time-dependent heat and fluid flow phenomenon. The generated gas is driven downward through the dissociated layer of thickness $X(t)$. The interface between the original two layers $(x = 0)$ also serves as an interface between the two problems. The second problem is the nearly radial gas flow through the lower porous layer and into the well (Figure 8.19b). The gas enters vertically through the $x = 0$ interface, and flows radially towards the well.

The one-dimensional model of Figure 8.19a is shown in greater detail in Figure 8.20. The dissociated sublayer, region 1, contains a porous medium filled with a combination of gas and liquid water. Region 2 is the original porous medium in which the pores are filled completely or incompletely with a solid hydrate. In the beginning, $t = 0$, region 1 is absent $(X = 0)$, and region 2 is initially isothermal at the equilibrium temperature $T_i(P_i)$ that characterizes the near-interface regions of Figure 8.18.

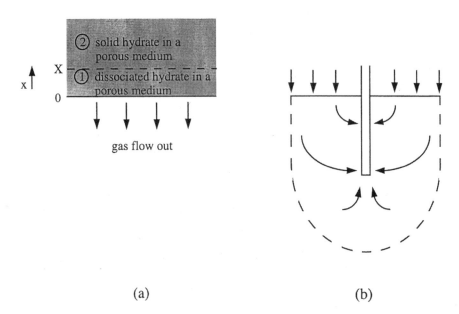

(a)   (b)

Fig. 8.19. The two heat and fluid flow fields separated by the interface between the two layers of Figure 8.18 (Rocha *et al.*, 2001).

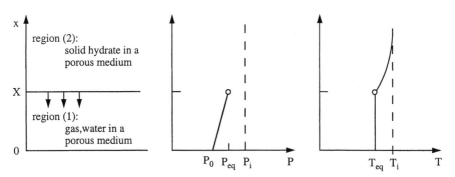

Fig. 8.20. One-dimensional model for phase change, heat transfer, and fluid flow through the upper layer of Figure 8.18 (Rocha *et al.*, 2001).

The dissociation starts when the well begins flowing. This event is modeled as the sudden lowering of the pressure at the $x = 0$ interface: the new pressure P_0 is considerably lower than P_i. Dissociation advances into region 2, on a front $(x = X)$ of pressure P_{eq} and corresponding equilibrium $T_{eq}(< T_i)$. The excess pressure $(P_{eq} - P_0)$ that forms across region 1 drives the generated gas through the $x = 0$ interface. For simplicity it is assumed that the only phase that flows through region 1 is the gas; in other words, the liquid water remains stationary inside the pores. It is also assumed that the gas flow through region 1 is sufficiently slow to conform to the Darcy regime,

$$v = -\frac{K_1}{\mu}\frac{\partial P}{\partial x},$$

(8.31)

where K_1 is the permeability of the porous medium filled partially with liquid water in region 1, and μ is the gas viscosity. Fluid motion was ruled out in region 2, on the assumption that the solid hydrate is sufficiently plentiful to seal the pores. These simplifying assumptions have also been made in earlier studies of one-dimensional dissociations (Cherskii and Bondarev, 1972; Selim and Sloan, 1989), where it was assumed that the dissociation is driven by the heating applied to the underside of the $x = 0$ interface. In the present model, the dissociation is driven by the lower pressure imposed at $x = 0$. This problem is more closely related to the movement of the dissociation front studied by Yousif *et al.* (1990), where the dissociation was driven by depressurization, and where the temperature field on both sides of the front was assumed isothermal. The local description of the gas flow through region 1 continues with the ideal gas model for the gas density ρ,

$$\rho = \frac{PM}{ZRT_1},$$

(8.32)

where T_1 is the local temperature and P is the local pressure in region 1. Storativity and other compressibility effects are neglected because in the present application the pressure variations are small compared to the absolute pressure. The same is true about temperature variations. Temperature and pressure changes tend to offset each other's effect on gas volume in this system. For the conditions at which methane hydrates exist, the compressibility factor Z is on the order of 0.8 to 0.9 for methane gas. In the following analysis Z is treated as a constant, and the numerical results are based on $Z = 1$. In the mass conservation equation for the gas, Equation (1.5) with $u = 0$, it is assumed that the gas motion is quasistationary so that the $\partial \rho / \partial t$ term is negligible; that is,

$$\frac{\partial}{\partial x}(\rho v) = 0. \tag{8.33}$$

The energy absorbed at the dissociation front is being supplied by conduction from the solid region 2. The energy equation and the initial and boundary conditions for this phenomenon are given by

$$\frac{\partial T_2}{\partial t} = \alpha_2 \frac{\partial^2 T_2}{\partial x^2} \tag{8.34}$$

$$T_2 = T_i \quad \text{at } X \leq x < \infty, \quad t = 0, \tag{8.35}$$

$$T_2 = T_{eq} \quad \text{at } x = X, \quad t > 0, \tag{8.36}$$

$$T_2 = T_i \quad \text{as } x \to \infty, \quad t > 0, \tag{8.37}$$

where α_2 is the thermal diffusivity of the porous region filled with solid hydrate. It is necessary to invoke the adiabatic condition on the $x = 0$ plane in order to effect the separation of the phenomenon into the two problems of Figure 8.19. This adiabatic condition is an approximation that becomes better as the time increases, that is, as the distance between the dissociation front $(x = X)$ and the adiabatic plane $(x = 0)$ increases. Thermal diffusion is expected to play a role in the immediate vicinity of the dissociation front. The temperature in region 1 is a record of the dissociation front, which sweeps upward with the temperature $T_{eq}(P_{eq})$, starting with $T_{eq}(P_0)$ at $x = 0$. Rocha et al. (2001) began with the assumption that the temperature throughout $0 \leq x \leq X$ is nearly uniform and equal to the front temperature,

$$T_1(x, t) \cong T_{eq}(t). \tag{8.38}$$

This assumption eliminates conduction through region 1; it was relaxed and tested numerically later in Bejan et al. (2002). The conservation of energy at the dissociation front is a balance between the rate of dissociation and the conduction heat flux supplied by region 2,

$$\phi \rho_H \Delta H \frac{dX}{dt} = -k_2 \frac{\partial T_2}{\partial x} \quad \text{at } x = X(t). \tag{8.39}$$

In this equation ϕ, ρ_H, and ΔH are the volume fraction occupied by hydrate in region 2, the hydrate density, and the hydrate latent heat of dissociation. If hydrate fills 100% of the pore space, then ϕ is also the porosity of the medium.

This equation assumes no kinetic limitation on the rate of dissociation. On the right side of the same equation, k_2 is the thermal conductivity of the medium of region 2. The conservation of species (e.g., methane) at the dissociation front requires

$$\omega\phi\rho_H\frac{dX}{dt} + \rho v = 0 \quad \text{at } x = X(t), \tag{8.40}$$

where ω is the mass fraction of methane in the solid hydrate, for example, $\omega = 0.127$ (Selim and Sloan, 1989). In Equation (8.40) both v and dX/dt are measured relative to the x coordinate. This equation is based on the highly appropriate assumption that the amount of gas left in the space just swept by the front is negligible in comparison with the gas ejected as ρv. Furthermore, the ρv term in (8.40) has the same value throughout the dissociated sublayer [cf. Equation (8.33)], however, this value may vary with time. The problem statement is completed by the relation between pressure and temperature at the dissociation front (Kamath and Holder, 1987; Lundgaard and Mollerup, 1992),

$$P_{eq} = c\exp(a - b/T_{eq}) \quad \text{at } x = X(t), \tag{8.41}$$

where (a, b, c) are three constants listed in Table 8.1. Equation (8.41) is presented in Figure 8.17.

The solution to this problem was developed by Rocha *et al.* (2001) in closed form, by introducing the similarity variables

$$\eta = \frac{x}{(4\alpha_2 t)^{1/2}} \quad \text{and} \quad \xi = \frac{X}{(4\alpha_2 t)^{1/2}} \tag{8.42}$$

such that $\eta = \xi$ represents the position of the dissociation front. Integrating Equation (8.33) once and combining it with (8.31) and (8.32), we obtain

Table 8.1. Physical properties of methane hydrates in subsea sediments

$a = 49.3185$	$T_i = 300\,\mathrm{K}$
$b = 9349\,\mathrm{K}$	$T_{eq}(P_{eq} = P_0) = 280\,\mathrm{K}$
$c = 1\,\mathrm{Pa}$	$Z = 1$
$c_p = 2162\,\mathrm{J/kg\,K}$	$\alpha_1 = 2.9 \times 10^{-6}\,\mathrm{m^2/s}$
$c_1 = 2500\,\mathrm{J/kg\,K}$	$\alpha_2 = 7 \times 10^{-7}\,\mathrm{m^2/s}$
$c_2 = 2500\,\mathrm{J/kg\,K}$	$\Delta H = 4.1 \times 10^{-5}\,\mathrm{J/kg}$
$k_1 = 5.6\,\mathrm{W/m\,K}$	$\mu = 10^{-5}\,\mathrm{kg/sm}$
$k_2 = 2.7\,\mathrm{W/m\,K}$	$\rho_1 = 1000\,\mathrm{kg/m^3}$
$K_1 = 1.4 \times 10^{-13}\,\mathrm{m^2}$	$\rho_2 = 1000\,\mathrm{kg/m^3}$
$M = 16\,\mathrm{kg/kmol}$	$\rho_H = 913\,\mathrm{kg/m^3}$
$P_0 = 5600\,\mathrm{kPa}$	$\phi = 0.3$
$R = 8314\,\mathrm{J/kmol\,K}$	$\omega = 0.127\,\mathrm{kg\ methane/kg\ hydrate}$

Source: Rocha *et al.* (2001).

$(MP/ZRT_1)dP/dx = $ constant, or

$$\frac{M}{ZRT_1}\frac{P}{2(\alpha_2)^{1/2}}\frac{dP}{d\eta} = \text{constant.} \tag{8.43}$$

This constant is supplied by Equation (8.40): constant $= \rho v = -\omega\phi\rho_H\partial X/\partial t$, which also shows the origin of ξ in the subsequent equations [cf. the second of Equations (8.42)]. Integrating Equation (8.43) from $P = P_0$ (at $\eta = 0$) to $P(\eta)$, and using (8.38) leads to

$$P^2 = P_0^2 + \frac{4\eta\xi T_{\text{eq}}\alpha_2\omega\phi\rho_H\mu ZR}{K_1 M}. \tag{8.44}$$

Finally, $P = P_{\text{eq}}$ at $\eta = \xi$ and the equilibrium condition (8.41) yields the relation between T_{eq} and ξ:

$$c^2\left[\exp\left(a - \frac{b}{T_{\text{eq}}}\right)\right]^2 = P_0^2 + \frac{4\xi^2 T_{\text{eq}}\alpha_2\omega\phi\rho_H\mu ZR}{K_1 M}. \tag{8.45}$$

A second relation is supplied by the solution to the transient conduction problem in region 2. By using the classical method of dimensionless unidirectional time-dependent conduction in similarity formulation (Carslaw and Jaeger, 1959; Bejan, 1993), it can be shown that the similarity solution to the problem stated in Equations (8.34) to (8.37) is given by

$$\frac{T_2(x,t) - T_i}{T_{\text{eq}} - T_i} = \frac{\text{erfc}(\eta)}{\text{erfc}(\xi)}. \tag{8.46}$$

Combining this expression with Equations (8.39) and (8.42) yields a second relation between T_{eq} and ξ:

$$(T_i - T_{\text{eq}})\frac{\exp(-\xi^2)}{\xi\text{erfc}(\xi)} = \frac{\pi^{1/2}\phi\rho_H\Delta H\alpha_2}{k_2}. \tag{8.47}$$

Equations (8.45) and (8.47) determine T_{eq} and ξ uniquely, as functions of the imposed boundary pressure P_0, the initial temperature T_i, and the thermophysical properties of regions 1 and 2. These equations establish ξ and T_{eq}/b as functions of five parameters, namely, a, T_i/b, P_0/c, G_2, and G_1, where G_1 and G_2 are dimensionless groups revealed by Equations (8.45) and (8.47), respectively:

$$G_1 = \frac{b\alpha_2\omega\phi\rho_H\mu ZR}{c^2 K_1 M}, \tag{8.48}$$

$$G_2 = \frac{\phi\rho_H\Delta H\alpha_2}{k_2 b}. \tag{8.49}$$

The five parameters are evaluated by using the representative values of thermophysical properties listed in Table 8.1. Not every parameter is a constant.

The least likely to vary is a, and the most likely are G_1 and G_2. Specifically, in a field with methane hydrate sediments the group G_2 may vary (or may be difficult to be described by a single value) because of variations or uncertainties in ϕ. The group G_1 may vary on account of changes in both K_1 and ϕ. Changes in ϕ can also affect α_2 and k_2 if they are large enough, but α_2 and k_2 have been assumed to be constant here.

Figures 8.21 and 8.22 document the sensitivity of the solution (ξ, T_{eq}) with respect to possible variations in several physical parameters. The changes in G_1 and G_2 are attributed to variations in permeability and hydrate saturation, and the remaining properties have the values listed in Table 8.1. Figure 8.21a shows that the equilibrium temperature at the interface is relatively insensitive to changes in permeability and hydrate saturation in the range $10^{-14}\,\mathrm{m}^2 < K_1 < 10^{-11}\,\mathrm{m}^2$ and $0.1 < \phi < 0.3$. The hydrate saturation begins to show an effect in the limit of small K_1 and small ϕ. The corresponding chart for the dimensionless thickness of the dissociated layer (Figure 8.21b) shows that ξ decreases as the hydrate saturation increases. The effect of the permeability on ξ is insignificant.

The effect of variations in the initial temperature and the imposed low pressure is reported in Figure 8.22. The variation, or uncertainty, in the T_i value may be attributed to field conditions such as geographical position and geothermal gradient. The effect of the geothermal gradient is documented later in Bejan $et\ al.$ (2002). Figure 8.22a shows that T_{eq} is relatively insensitive to the value of T_i in the expected range. The effect of the imposed low pressure P_0 is more noticeable: note that T_{eq} increases by about $10\,\mathrm{K}$ whereas P_0 increases

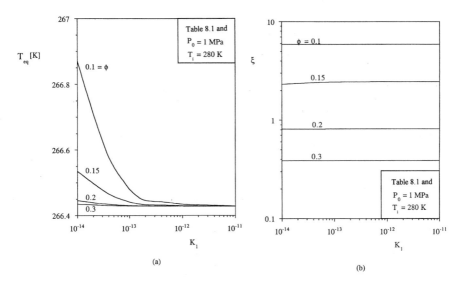

Fig. 8.21. The effect of permeability and hydrate saturation on the temperature and position of the dissociation front (Rocha $et\ al.$, 2001).

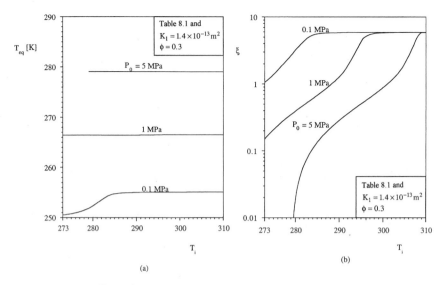

(a)

(b)

Fig. 8.22. The effect of initial temperature and imposed low pressure on the temperature and position of the dissociation front (Rocha *et al.*, 2001).

by a factor of 10. Figure 8.22b shows that the thickness of the dissociated front decreases significantly as P_0 increases, and as T_i decreases, because in both cases the temperature difference $(T_i - T_{eq})$ that drives heat to the dissociation front becomes smaller.

The rate of hydrate dissociation is obtained from Equation (8.40) as follows,

$$\rho(-v) = \omega\phi\rho_H\xi\left(\frac{\alpha_2}{t}\right)^{1/2}. \tag{8.50}$$

Because the group ξ is time-dependent, the flow rate in Equation (8.50) decreases in time as $t^{-1/2}$. This flow rate has the same value at any value of x in the dissociated zone (region 1) [cf. Equation (8.33)]. The rate of gas formation increases not only with ϕ, ω, and ξ, but also with changes in the remaining parameters that lead to increases in ξ, for example, Figures 8.21 and 8.22.

The hydrate dissociation rate—its level and behavior in time—is the central question in the modeling of the phase-change and depressurization process. With this information, one can assess the exergy-production potential of a known deposit, and its lifetime. The decaying production rate is, in this sense, similar to the rate of exergy extraction from hot dry rock deposits (Lim *et al.*, 1992b). The depressurization process is considerably more complicated because of the interaction between region 1 and the well embedded in the lowest porous region.

The similarity solution (8.50) was verified based on a more general numerical simulation in Rocha *et al.* (2002). The same numerical work documented the effect of the vertical geothermal gradient on the phase-change and gas convection process. The geothermal gradient represents the rate of increase of the temperature in the earth with depth. It is generated by the continuous flow of heat outward through the crust of the earth, and its value varies from place to place depending on the heat flow in the region and on the thermal conductivity of the rock.

The geothermal gradient is one of the important parameters that control the thickness and stability of the hydrate zone in a marine environment (Roadifer et al., 1987; Hanumantha Rao et al., 1998; Singh and Singh, 1999; Holder et al., 1987). The depth, extent, and stability of the hydrate zone are governed by the phase diagram for mixtures of methane and hydrate, and are determined by the ambient pressure and temperature. At sea depths greater than about 300 m, the pressure is high enough and the temperature is low enough for hydrates to occur at the sea floor. The base of the hydrate zone is a phase boundary between the solid hydrate and free gas and water and its depth is determined principally by the value of the geothermal gradient (Roadifer et al., 1987; Willoughby and Edwards, 1997). The most favorable conditions under which gas hydrates are likely to occur are the normal range of geothermal gradients, which are below 60 K/km (Subrahmanyam et al., 1998). The geothermal gradient has also been reported to vary in the range 20 to 50 K/km (Briaud and Chaouch, 1997; Win and Rik, 1999; Gering et al., 2000).

The porosity of sediments under the sea is not uniform: for example, the porosity varies in the vertical direction because of sedimentation and compacting in time. According to the Athy model the porosity decreases with depth exponentially (Athy, 1930; Rubey and Hubbert, 1959; Shi and Wang, 1986; Magara, 1971; Hart et al., 1995; Yuan et al., 1994),

$$\phi(x) = \phi_0 \exp\left(-\frac{L-x}{\lambda}\right), \tag{8.51}$$

where λ is an empirical constant, for example, $\lambda = 500$ m, and ϕ_0 is the porosity at the sea bottom. The corresponding permeability decreases with depth as well. The effect of nonuniform sediment porosity is documented numerically in Rocha et al. (2002). The variations in porosity and permeability have a significant effect on global performance indicators, such as the methane flow rate ρv. The group ρv decreases in time as t^{-n}, where n decreases monotonically in time. The decrease in ϕ and K with the depth has the effect of decreasing the gas flow rate.

Nomenclature

a	chemical activity, $\text{mol} \cdot \text{m}^{-3}$; constant modifying hydraulic permeability; specific surface (area/volume), m^2/m^3
a, b	dimensions, m; segments, m, Figure 4.39
a, b, c	constants, Table 8.1
a_S, a_W, a_P, a_E, a_N	finite-difference coefficients
a, a_{sf}	contact area per unit volume, Equation (1.37), m^{-1}
a_p	pore cross-sectional area, m^2
a_s	specific surface area of the porous matrix, $\text{m}^2 \, \text{g}^{-1}$
a_w	water activity
a_M	cross-sectional area of the porous matrix, m^2
A	area, m^2; single particle surface area; area perpendicular to flow of heat or mass, m^2
A_{bed}	particle area per unit volume, m^{-1}
$A_{\dot{s}}$	internal surface area of porous matrix, m^2
$\mathcal{A}$	activity, Bq
ADP	adenosine diphosphate
AFC	alkaline fuel cell
ATP	adenosine triphosphate
ATES	aquifer thermal energy storage
b	dimensionless temperature gradient, Equation (2.48); empirical constant, Equation (1.12); global natural convection parameter, Equation (8.16); plate height, m, Figure 5.8a; source term
B	dimensionless group, Equation (2.56); dimensionless radiation group, Equation (5.44); dimensionless pressure drop, Equation (5.33); driving force
Be	Bejan number, Equation (5.24)
Bi	Biot number, Equation (7.75)
c	concentration, particles m^{-3}
c, c_v, c_P	specific heat, $\text{J} \, \text{kg}^{-1} \, \text{K}^{-1}$

c_F	empirical constant, Equation (1.11)
$\bar{c}_w$	averaged macroscale or coarse-scale aerosol concentration, particles m^{-3}
$\bar{c}_{PM}$	mean specific heat capacity of moist air, J kg^{-1} K^{-1}
C	molar concentration, mol m^{-3}; species concentration, kg m^{-3}; constant, Equation (4.20)
C_p	molar specific heat, J mol^{-1} K^{-1}
C_D	drag coefficient
C_Q, C_r	coefficients, Equations (5.35) and (5.34)
$C_{m,B}$	bound ion concentration per unit of mass of dry material, mol $\cdot$ kg^{-1}
CHP	combined heat and power
CTES	cold thermal energy storage
d	diameter, thickness, m
d_f	fractal dimension
d_h	channel hydraulic diameter, m
D	bulk diffusion coefficient, Equation (6.37), m^2 s^{-1}; diameter, thickness, m; mass diffusivity, m^2 s^{-1}
D_c	binary diffusion coefficient for water vapor in air, m^2 s^{-1}
D_{CT}	Soret coefficient, kg m^{-1}s^{-1} K^{-1}, Equation (1.47)
D_e	effective diffusion coefficient of one species in a porous medium, m^2 s^{-1}
D_h	hydraulic diameter, m
D_K	Knudsen diffusivity, m^2 s^{-1}
D_m	moisture diffusivity coefficient, m^2 s^{-1}
D_M	solute diffusivity, or solute diffusion coefficient, m^2 s^{-1}, Equation (6.38)
D_T	Soret coefficient for thermal diffusion, kg m^{-1} s^{-1} K^{-1}
D_{TC}	Dufour coefficient, m^5 K s^{-1} kg^{-1}, Equation (1.46),
D^*	macroscale or coarse-scale aerosol dispersivity, m^2 s^{-1}
Di	Dincer number, Equation (7.76)
DMFC	direct methanol fuel cell
E	energy, J; equilibrium membrane potential, V; open circuit potential, V
$\dot{E}_Q$	exergy transfer rate, W
e_x	specific flow exergy, J kg^{-1}
E_x	flow exergy, J
$\dot{E}_x$	exergy flow rate, W
EGM	entropy generation minimization
f	flow resistance factor, Equation (4.115); friction factor, Equation (5.9); pressure drop factor, Equation (5.1)
f_{dm}	coefficient accounting for the decrease in solute mobility
f_{pt}	solute partition coefficient
f_K	friction factor, Equation (1.9)

F	Faraday constant, $9.648 \times 10^4\,\mathrm{JV^{-1}mol^{-1}}$
F	force, N; similarity stream function, Equation (4.20)
g	gravitational acceleration, $\mathrm{ms^{-2}}$
G	lag factor; minimum energy required to pump solutes, Equation (6.18), J
Ga	Galileo number, Table 4.1
h	characteristic filtration length, m; height m, Equation (7.56); specific enthalpy, $\mathrm{J\,kg^{-1}}$; heat transfer coefficient, $\mathrm{Wm^{-2}\,K^{-1}}$
h_m	moisture transfer coefficient, $\mathrm{ms^{-1}}$; $\mathrm{kg\,m^{-2}\,s^{-1}}$
h_D	convection mass transfer coefficient, $\mathrm{ms^{-1}}$
h_D^*	corrected convection mass transfer coefficient, $\mathrm{ms^{-1}}$
$\bar{h}$	average or effective heat transfer coefficient, $\mathrm{Wm^{-2}\,K^{-1}}$
H	enthalpy, J; filter thickness, m; height, m; absolute humidity or humidity ratio, kg water vapor/kg dry air
HTF	heat transfer fluid
i	current density, $\mathrm{A\,m^{-2}}$; level of construct, or level of pairing or bifurcation; time step
i_0	exchange current density, $\mathrm{A\,m^{-2}}$
i_l	mass limited current density, $\mathrm{A\,m^{-2}}$
I	current intensity, A
I_n	area moment of inertia, $\mathrm{m^4}$
I_r	interception parameter, r_p/r_c
j	diffusive flux across a unit area of solution, $\mathrm{mol\,m^{-2}\,s^{-1}}$; flux, $\mathrm{mol\,s^{-1}}$
j_e	effective flux across a unit of material, $\mathrm{mol\,m^{-2}\,s^{-1}}$
$\mathbf{j_s}$	entropy flux, $\mathrm{Wm^{-2}\,K^{-1}}$
$\mathbf{j_u}$	internal energy flux, $\mathrm{Wm^{-2}}$
$\mathbf{j_D}$	diffusive mass flux, $\mathrm{kg\,m^{-2}\,s^{-1}}$
J_{sx}	total transfer of solute across the membrane in the x-direction, $\mathrm{mol\,m^{-2}\,s^{-1}}$
$\mathbf{J}_s$	total transfer of solute across the membrane, $\mathrm{mol\,m^{-2}\,s^{-1}}$
k	thermal conductivity, $\mathrm{Wm^{-1}\,K^{-1}}$
k_s	coefficient, Equation (2.81)
k_z	empirical constant, Equation (5.2)
K	permeability, $\mathrm{m^2}$
K_0	initial (particle free) filter permeability, $\mathrm{m^2}$
K_{Bt}	Boltzmann constant, $\mathrm{JK^{-1}}$
K_1	function, Equation (2.66)
Kn	Knudsen number, Table 4.1
Ku	Kuwabara hydrodynamic factor, Table 4.1
l	phenomenological coefficient, Equation (4.78); distance, m; segment, m, Figure 4.39; thickness of electrolyte, Equation (3.42)

L	bed height, m; latent heat of vaporization, $J\,kg^{-1}$; length, m
L_0	elemental length, m
L_r	length scale of one roll, m
L_{ik}	Onsager coefficients
$L_{x,y}$	penetration distances, m
$L_{qq}, L_{qm}, L_{mq}, L_{mm}$	Onsager coefficients
Le	Lewis number, α/D
m	exponent; mass, kg
$\dot{m}$	mass flow rate, $kg\,s^{-1}$; mass flux, $kg\,m^{-2}\,s^{-1}$
$\dot{m}'$	mass flow rate per unit length, $kg\,s^{-1}m^{-1}$
$\dot{m}''$	mass flow rate per unit area, $kg\,s^{-1}m^{-2}$
$\dot{m}'''$	species generation rate per unit volume, $kg\,s^{-1}\,m^{-3}$
M	dimensionless mass addition parameter, Equation (5.53); dimensionless melting zone volume; mass, kg; molecular weight, kmol/kg; moisture content, kg/kg; relative molecular mass, kg/kmol; molecular mass $kg\,kmol^{-1}$
$\bar{M}$	average dimensionless position of the melting front
M_a	relative molecular mass of air, kg/kmol
M_p	mass collected per filter area, $kg\,m^{-2}$
M_0	molecular weight, $kg\,mol^{-1}$
M_v	molecular mass of vapor, $kg\,mol^{-1}$
Mol	solution molarity, moles/liter
MCFC	molten carbonate fuel cell
n	electrons per mole oxidized or reduced; number
n^{K^+}	number of potassium ions pumped into the cell
n^{Na^+}	number of sodium ions pumped out of the cell
N_A	Avogadro constant, $6.023 \times 10^{23} mol^{-1}$
N	number of particles caught per unit of filter area, particles m^{-2}; number
N_0	number of heat loss units, Equation (4.113)
$Nu, Nu_\theta, Nu_D, Nu_y$	Nusselt numbers, Equations (2.13), (2.23), (2.26), and (2.40)
p, P	pressure, Pa
$\tilde{P}$	dimensionless pressure drop, Equation (5.28)
P_a	partial pressure of air, Pa; $kg\,m^{-1}\,s^{-2}$
P_{am}	mean of partial pressures of air over the product surface and in drying air, Pa
P_s	saturation pressure, Pa
P_v	partial pressure of vapor, Pa
P_{va}	partial pressure of vapor in drying air, Pa
P_{vm}	mean of partial vapor pressures of vapor over the product surface and in drying air, Pa
P_{vo}	vapor pressure over the product surface, Pa

P_M	solute permeability, D_M/H, $\mathrm{m\,s^{-1}}$
Pr_p	porous medium Prandtl number, Equation (2.76)
P_v^*	saturated vapor pressure, Pa
$\mathrm{Pe}, \mathrm{Pe}_D$	Peclet numbers, Table 4.1, Equations (2.13) and (2.26)
Pr	Prandtl number, ν/α
PAFC	phosphoric acid fuel cell
PCM	phase-change material
PEFC, PEMFC	proton exchange membrane (polymer) electrolyte fuel cell
PEM	proton exchange membrane
PTFE	polytetrafluoroethylene
q	heat transfer rate, W
q'	heat transfer rate per unit length, $\mathrm{W\,m^{-1}}$
q''	heat transfer rate per unit area, $\mathrm{W\,m^{-2}}$
q'''	volumetric heat generation rate, $\mathrm{W\,m^{-3}}$
$\dot{q}$	heat flux, $\mathrm{W\,m^{-2}}$
Q	dimensionless heat transfer rate per unit volume, Equation (5.31); heat transfer, J
$\dot{Q}$	heat transfer rate, W
r	area specific resistance, $\Omega\,\mathrm{m^{-2}}$; particle coordinate, m; distance normal to the flow of heat/mass, m; radial position, m; radius, m
r_l	pore size characteristic of the lower limit of the self-similar region, m
r_m	molecular radius, m
r_u	pore size characteristic of the upper limit of the self-similar region, m
r_0	radius, m
R	aspect ratio, H/L; radius of a single particle, m; thermal resistance, $\mathrm{K\,W^{-1}}$; universal gas constant, $8.314\ \mathrm{J\,K^{-1}\,mol^{-1}}$
R_v	gas constant for water vapor, $\mathrm{J\,kg^{-1}\,K^{-1}}$
R_M	membrane resistance, Ω
$\mathbf{R^s}$	second order tensor, $\mathrm{m^3\,s^{-1}}$
$R_{\mathrm{lam}}, R_{\mathrm{turb}}$	flow resistances, Equations (4.106) and (4.107)
$\mathrm{Ra}_y, \mathrm{Ra}_{*y}, \mathrm{Ra}_{*H}, \mathrm{Ra}_{\infty,y}, \mathrm{Ra}_{\infty,y}^*, \mathrm{Ra}_H^*, \mathrm{Ra}_I$	Rayleigh numbers, Equations (2.39), (2.44), (2.47), (2.50), (2.51), (2.69) and (2.82)
Ra	Rayleigh number based on source strength, Equation (4.11)
$\mathrm{Ra}_{H,L}$	Rayleigh numbers based on temperature difference, Equations (4.27) and (4.31)
Re_c	Reynolds number, $2\rho r_c u/\mu$
Re_D	Reynolds number, $U_\infty D/\nu$, $V_{\max} D/\nu$
Re_K	Reynolds number, Equation (1.10)

Re_L	Reynolds number, $U_0 L/\nu$
REV	representative elementary volume, Figure 1.1
s	direction, length, m; specific entropy, $\mathrm{J\,kg^{-1}\,K^{-1}}$;
s_v	entropy per unit volume, $\mathrm{J\,K^{-1}\,m^{-3}}$
S	current collecting rib width, m; dimensionless group, Equation (8.22); drying coefficient, $\mathrm{s^{-1}}$; entropy, $\mathrm{J\,K^{-1}}$; spacing, m
S_g	gas phase saturation
$\dot{S}_{\mathrm{gen}}$	entropy generation rate, $\mathrm{W\,K^{-1}}$
$\dot{S}'''_{\mathrm{gen}}$	entropy generation rate per unit volume, $\mathrm{W\,K^{-1}\,m^{-3}}$
St	Stokes number, Table 4.1
SOFC	solid oxide fuel cell
t	time, s; thickness, m
T	temperature, K
T_m	mean temperatures of product surface and drying air, K
T_{ma}	mean absolute temperatures of product surface and drying air, K
T_o	surface temperature, K
T^*	tortuosity
$T_{1/2}$	half-life period, s
TES	thermal energy storage
u	superficial velocity, $\mathrm{m\,s^{-1}}$
u, v, w	volume-averaged velocity components, $\mathrm{m\,s^{-1}}$
u_v	internal energy per unit volume, $\mathrm{J\,m^{-3}}$
U	mean velocity, $\mathrm{m\,s^{-1}}$; overall heat transfer coefficient, $\mathrm{W\,m^{-2}\,K^{-1}}$; reactant crossover bulk velocity, $\mathrm{m\,s^{-1}}$
U^*	macroscale, or coarse-scale aerosol velocity, $\mathrm{m\,s^{-1}}$
U_0	elemental mean velocity, $\mathrm{m\,s^{-1}}$;
v	specific volume, $\mathrm{m^3\,kg^{-1}}$; volumetric flux, $\mathrm{m^3\,m^{-2}\,s^{-1}}$, or $\mathrm{m\,s^{-1}}$
v_{grid}	grid deformation velocity
v, V	velocity, $\mathrm{m\,s^{-1}}$
V	potential, V; voltage, V; volume, $\mathrm{m^3}$
$\dot{V}$	volumetric flow rate, $\mathrm{m^3\,s^{-1}}$
w_v	mechanical energy per unit volume, $\mathrm{J\,m^{-3}}$
$\dot{w}_{v,d}$	rate of mechanical energy destruction per unit volume, $\mathrm{W\,m^{-3}}$
W	absolute humidity, kg/kg; moisture content (dry basis), kg moisture/kg dry matter; membrane thickness, m; width, m
$\dot{W}$	power, W
x	coarse-scale coordinate, m; conductance allocation fraction; dimensionless thickness, Equations (4.118) and (4.119); distance, m; mass fraction; mole fraction
x_i	molar fraction of the ith component in a mixture

X	thickness of dissociated zone, m; thickness of melted zone, m
X_T	thermal entrance length, m
x, y, z	Cartesian coordinates, m
y	dimensionless group, Equation (3.11)
Y	characteristic dimension, m
z	axial coordinate, m; charge number; elevation, m; ion number
Z	compressibility factor; electric charge, C

Greek Symbols

α	angle, rad; charge transfer coefficient; constant; filter solidity, $(1-\phi)$; thermal diffusivity, $m^2\,s^{-1}$
α_o	initial (particle free) filter solidity
α_L, α_T	longitudinal and transversal dispersivities, m
α_p	packing fraction of particles
β	angle, rad; enhancement factor; coefficient of volumetric thermal expansion, Equation (1.38), K^{-1}; volume-shrinkage coefficient
β_C	concentration expansion coefficient, Equation (1.45), $m^3\,kg^{-1}$
γ	activity coefficient, Equation (4.77); angle, rad; parameter; temperature gradient, $K\,m^{-1}$
δ	thickness, m
δ_T	thermal layer thickness, m
Δt	time step, s
$\Delta x, \Delta y, \Delta t$	space and time increments
ΔH	hydrate latent heat of dissociation, $J\,kg^{-1}$
ΔH_v	latent heat of vaporization, $J\,kg^{-1}$
ΔP	pressure difference, Pa
ΔT	temperature difference, K
ΔV_M	difference between the electrical potential across the membrane and the equilibrium membrane potential, Equation (6.68), V
ε	total hemispherical emissivity; volume fraction of phase-change material in the solid matrix
ζ_M	membrane capacitance, F, or $C\,V^{-1}$, or $C^2\,J^{-1}$
η	collector element (single element) efficiency; dynamic viscosity, $Pa\,s^{-1}$; fuel cell electrical energy conversion efficiency, or polarization; similarity variable
η_o	efficiency for a collector element free of particles
η_I	first law efficiency
η_{II}	second law efficiency
θ	angular position, rad; dimensionless temperature, Equations (5.8), (8.22), (8.23) and (8.24); dimensionless time, Equation (3.11); temperature difference, K
Θ	humidity; probability that a carrier is occupied, Equation (6.16)

λ	Lagrange multiplier; latent heat of melting and solidification, $J\,kg^{-1}$; mean free path of gas molecules, m
λ_a	solute-carrier association rate constant, $m^3\,s^{-1}$, or $m^3\,mol^{-1}\,s^{-1}$
λ_d	solute-carrier complex dissociation rate constant, s^{-1}
λ_{drag}	electroosmotic drag coefficient, mol/mol H^+ transferred through electrolyte
μ	chemical potential, $J\,mol^{-1}$; diffusion resistance factor; root of the transcendental characteristic equation; viscosity, $kg\,s^{-1}\,m^{-1}$
$\tilde{\mu}$	modified viscosity coefficient, Equation (1.14), $kg\,s^{-1}\,m^{-1}$
ν	kinematic viscosity, $m^2\,s^{-1}$
ν'	bulk viscosity, $m^2\,s^{-1}$
ξ	Cunningham slip correction factor, Table 4.1; similarity variable, specific nonflow exergy, $J\,kg^{-1}$; stoichiometric flow ratio
Π	osmotic pressure, Pa; penetration, Equation (4.36)
ρ	density, $kg\,m^{-3}$
ρ_{dr}	bone dry density, $kg\,m^{-3}$
ρ_H	hydrate density, $kg\,m^{-3}$
σ	Boltzmann constant, $5.6693 \times 10^{-12}\,W\,cm^{-2}\,K^{-4}$; capacity ratio, Equation (1.29); Staverman reflection coefficient
Ξ	nonflow exergy, J
τ	constrictivity, Equation (4.82); dimensionless temperature difference, Equation (3.13); dimensionless time, Equation (8.22); shear stress, Pa
υ	mobility, $m^2\,mol\,J^{-1}\,s^{-1}$
ϕ	angle, rad; porosity; relative humidity, %; volume fraction occupied by hydrate
φ	filter efficiency, Equation (4.35); specific potential energy, $m^2\,s^{-2}$
Φ	aggregate function, Equation (4.120); dimensionless moisture content; potential, Equation (1.52), $J\,kg^{-1}$; viscous dissipation function, Equation (1.22), s^{-2}
χ	pressure drop factor, Equation (5.1), Figure 5.1
ψ	applied electric field, N/C; electrical potential across the membrane, V; function, Equation (1.17), Pa; stream function, $m^2\,s^{-1}$
ω	mass fraction; mass fraction of methane; wall thickness parameter, Equation (2.54); moisture flux (moisture velocity), $kg\,m^{-2}\,s^{-1}$
ω^*	macroscale, or coarse-scale aerosol deposition-rate coefficient, s^{-1}
Ω	field; filter performance

Subscripts

a	activation; anode; adiabatic; air; environment; inlet flow during charging; surroundings
amb	ambient

avg	average
b	bath; brick; bulk
backing	backing layer
c	carrier; cathode; centerline; cold; collector; convection; critical
d	downstream
D	diffusion
e	effective; equilibrium; equivalent; exit
ef, eff	effective
el	electrostatic
eq	equilibrium
E	east node
Eef	effective effusion
f	fibers; filter; final; fluid
g	gas
G	gravity
h	high; hot
hot	hot spot
H	hydrate
i	initial; ith constituent; inlet; inner; internal; ionic species
I	inertial impaction
j	zone j
k	conduction; number of zones
lam	laminar
ℓ	low
ℓ, liq	liquid
m	averaged porous medium property; environment; mass transport; melting
max	maximum
min	minimum
M	membrane
n	north face
net	net
N	north node
o	initial state; outer; external
oc	open circuit
opt	optimum
out	outlet
p	particle; polyol; pore; pump
r	interception; radial; resistive
rev	reversible
s	solid; solute; storage fluid; south face; surface
S	south node
sat	saturation conditions
st	fuel cell stack; storage (overall)
sv	solute with respect to solvent

t	threshold; top; liquid state; turbine
turb	turbulent
T	total
u	upstream
v	solvent; vapor
w	wall; water; wet basis; west face; working fluid
W	west node
z	longitudinal
0	elemental (smallest, fixed) size; environment; nozzle; reference; surroundings
1	first construct, first root
2	second construct
∞	ambient; free stream; hot-gas inlet temperature
$*$	dimensionless, Equation (4.12)

Superscripts

$(^-)$	averaged
$(^\sim)$	dimensionless
$(^\wedge)$	dimensionless

References

AASHTO (1983) *Standard Method of Test for Rapid Determination of the Chloride Permeability of Concrete*. American Association of State Highway and Transportation Officials, Washington, DC.

Abhat, A. (1983) Low temperature latent heat thermal energy storage: heat storage materials. *Solar Energy* **30**(4), 313–332.

Achenbach, E. (1995) Heat and flow characteristics of packed beds. *Experimental Thermal Fluid Sci.* **10**(1), 17–27.

Adebiyi, G. A., Hodge, B. K., Steele, W. G., Jalalzadeh-Aza, A., and Nsofor, E. C. (1996) Computer simulation of a high temperature thermal energy storage system employing multiple families of phase-change storage materials. *J. Energy Resources Technol.* **118**, 102–111.

Ait Taleb, L., Boussehain, R., and Feidt, M. (2002) Optimisation du transfert conductif dans un milieu poreux bi ou tridimensionnel. *Énergie, Environment, Économie et Thermodynamique Colloque Franco-Roumain (COFRET)*, Bucharest, April 25–27, 291–294.

Alazmi, B. and Vafai, K. (2002) Constant wall heat flux boundary conditions in porous media under local thermal non-equilibrium conditions. *Int. J. Heat Mass Transfer* **45**, 3071–3087.

Alexiades, V. and Solomon, A. D. (1993) *Mathematical Modeling of Melting and Freezing Processes*, Hemisphere, Washington, DC.

Anand, N. K., Kim, S. H., and Fletcher, L. S. (1992) The effect of plate spacing on free convection between heated parallel plates. *J. Heat Transfer* **114**, 515–518.

Anderson, J. L. and Quinn, J. A. (1974) Restricted transport in small pores. *Biophys. J.* **14**, 130–149.

Anderson, K., Torstenfelt, B., and Allard, B. (1983). *Sorption and Diffusion Studies of Cs and I in Concrete*, Svensk Kärnbränslehantering B. (Swedish Nuclear Fuel and Waste Management) Teknisk Rapport, SKB-KBS-TR83.13, Stockholm.

Andrade, C. (1993) Calculation of chloride diffusion coefficients in concrete from ionic migration measurements. *Cement Concrete Res.* **23**, 724–742.

Andrade, C., Diez, J. M., Alaman, A., and Alonso, C. (1995) Mathematical modeling of electrochemical chloride extraction from concrete. *Cement Concrete Res.* **25**, 727–740.

Andrade Jr., J. S., Alencar, A. M., Almeida, M. P., Filho, J. M., Buldyrev, S. V., Zapperi, S., Stanley, H. E., and Suki, B. (1998) Asymmetric flow in symmetric branched structures. *Phys. Rev. Lett.* **81**, 926–929.

Angirasa, D. (2002a) Experimental investigation of forced convection heat transfer augmentation with metallic fibrous materials. *Int. J. Heat Mass Transfer* **45**, 919–922.

Angirasa, D. (2002b) Forced convective heat transfer in metallic fibrous materials. *J. Heat Transfer* **124**, 739–745.

Antohe, B. V. and Lage, J. L. (1997) A general two-equation macroscopic turbulence model for incompressible flow in porous media. *Int. J. Heat Mass Transfer* **40**, 3013–3024.

Asbik, M., Sadki, H., Hajar, M., Zeghmati, B., and Khmou, A. (2002) Numerical study of laminar mixed convection in a vertical saturated porous enclosure: The combined effect of double diffusion and evaporation. *Numer. Heat Transfer A* **41**, 403–420.

Asgharian, B. and Cheng, Y. S. (2002) The filtration of fibrous aerosols. *Aerosol Sci. Technol.* **36**, 10–17.

Athy, L. F. (1930) Density, porosity, and compaction of sedimentary rocks. *Am. Assoc. Pet. Geol. Bull.* **14**, 1–24.

Atkins, P. W. (1998) *Physical Chemistry.* Oxford University Press, Oxford.

Atkinson, A., Nickerson, A. K., and Valentine, T. M. (1984) The mechanism of leaching from some cement based wasteforms. *Radioactive Waste Management Nuclear Fuel Cycle* **4**, 357–378.

Baerlocher, C., Meier, W. M., and Olson, D. H. (2001) *Atlas of Zeolite Framework Types*, 5th ed. Elsevier, Amsterdam.

Bansod, V. J., Singh, P., and Rathish Kumar, B. V. (2000) Heat and mass transfer by natural convection from a horizontal surface in a Darcian fluid. *J. Energy Heat Mass Transfer* **22**, 89–95.

Bansod, V. J., Singh, P., and Rathish Kumar, B. V. (2002) Heat and mass transfer from a vertical surface to a stratified Darcian fluid. *J. Porous Media* **5**, 57–66.

Banu, N. and Rees, D. A. S. (2002) Onset of Darcy–Bénard convection using a thermal nonequilibrium model. *Int. J. Heat Mass Transfer* **45**, 2221–2228.

Bar Cohen, A. and Rohsenow, W. M. (1984) Thermally optimum spacing of vertical, natural convection cooled, parallel plates. *J. Heat Transfer* **106**, 116–123.

Bard, A. J. and Faulkner, L. R. (1980) *Electrochemical Methods.* Wiley, New York.

Baron, J. and Ollivier, J. P. (1992) *La Durabilité des Bétons.* Presses des Ponts et Chaussées, Paris.

Bau, H. H. and Torrance, K. E. (1982) Low Rayleigh number thermal convection in a vertical cylinder filled with porous materials and heated from below. *J. Heat Transfer* **104**, 166–172.

Baumgartner, H. (1987) Elektretfaserschichten für Aerosolfiltration—Unterschungen zum Faserladungszustand und zur Abscheidecharakteristik. *VID Forstschr. Ber.* **3**, VDI-Verlag, Düsseldorf.

Baytas, A. C. and Pop, I. (2000) Entropy generation due to free convection from a trapezoidal cavity filled with a porous medium. *Recent Advances in Transport Phenomena*, I. Dincer and M. F. Yardim, eds., July 16–20, Istanbul, Turkey, 237–240.

Bean, C. P. (1972) The physics of porous membranes—neutral pores. *Membranes* **1**, 1–54.

Bear, J. (1972) *Dynamics of Fluids in Porous Media.* American Elsevier, New York.

Bear, J. and Bachmat, Y. (1990) *Introduction to Modeling of Transport Phenomena in Porous Media.* Kluwer Academic, Dordrecht, The Netherlands.

Beasley, D. E., Ramanarayanan, C., and Torab, H. (1989) Thermal response of a packed bed of spheres containing a phase-change material. *Int. J. Energy Res.* **13**, 253–265.

Beaudoin, J. J., Ramachandran, V. S., and Feldman, R. F. (1990) Interaction of chloride and CSH. *Cement Concrete Res.* **20**, 875–883.

Beavers, G. S. and Joseph, D. D. (1967) Boundary conditions at a naturally permeable wall. *J. Fluid Mech.* **30**, 197–207.

Beavers, G. S., Sparrow, E. M., and Rodenz, D. E. (1973) Influence of bed size on the flow characteristics and porosity of randomly packed beds of spheres. *J. Appl. Mech.* **40**, 655–660.

Beck, R. E. and Schultz, J. S. (1970) Hindered diffusion in microporous membranes with known pore geometry. *Science* **170**, 1302–1305.

Beck, J. S., Vartuli, J. C., Roth W. J., Leonowitz, M. E., Kresge, C. T., Schmitt, K. D., Chu, C. T.-W., Olson, D. H., Sheppard, E. W., McCullen, S. B., Higgins, J. B., and Schlenker, J. L. (1992) A new family of mesoporous molecular sieves prepared with liquid crystal templates. *J. Am. Chem. Soc.* **114**, 10834–10843.

Beckermann, C. and Viskanta, R. (1988) Natural convection solid/liquid phase change in porous media. *Int. J. Heat Mass Transfer* **31**, 35–46.

Beer, F. P., Johnston, Jr., E. R., and DeWolf, J. T. (2002) *Mechanics of Materials*, 3rd ed. McGraw-Hill, Boston.

Behar, E. (1994) Plugging control of production facilities by hydrates. *International Conference on Natural Gas Hydrates, Annals New York Academy of Sciences* **715**, 94–105.

Bejan, A. (1979a) On the boundary layer regime in a vertical enclosure filled with a porous medium. *Lett. Heat Mass Transfer* **6**, 93–102.

Bejan, A. (1979b) A general variational principle for thermal insulation system design. *Int. J. Heat Mass Transfer* **22**, 219–228.

Bejan, A. (1980a) Natural convection in a vertical cylindrical well filled with porous medium, *Int. J. Heat Mass Transfer* **23**, 726–729.

Bejan, A. (1980b) A synthesis of analytical results for natural convection heat transfer across rectangular enclosures. *Int. J. Heat Mass Transfer* **23**, 723–726.

Bejan, A. (1981) Lateral intrusion of natural convection into a horizontal porous structure. *J. Heat Transfer* **103**, 237–241.

Bejan, A. (1982) *Entropy Generation Through Heat and Fluid Flow*. Wiley, New York.

Bejan, A. (1983a) The boundary layer regime in a porous layer with uniform heat flux from the side. *Int. J. Heat Mass Transfer* **26**, 1339–1346.

Bejan, A. (1983b) Natural convection heat transfer in a porous layer with internal flow obstructions. *Int. J. Heat Mass Transfer* **26**, 815–822.

Bejan, A. (1984) *Convection Heat Transfer*. Wiley, New York.

Bejan, A. (1987) Convective heat transfer in porous media. *Handbook of Single-Phase Convective Heat Transfer*, S. Kakac, R. K. Shah, and W. Aung, eds. Wiley, New York.

Bejan, A. (1988) *Advanced Engineering Thermodynamics*. Wiley, New York.

Bejan, A. (1990a) Theory of heat transfer from a surface covered with hair. *J. Heat Transfer* **112**, 662–667.

Bejan, A. (1990b) Optimum hair strand diameter for minimum free-convection heat transfer from a surface covered with hair. *Int. J. Heat Mass Transfer* **33**, 206–209.

Bejan, A. (1993) *Heat Transfer*. Wiley, New York.

Bejan, A. (1995a) *Convection Heat Transfer*, 2nd ed. Wiley, New York.

Bejan, A. (1995b) Optimal spacings for cylinders in crossflow forced convection. *J. Heat Transfer* **117**, 767–770.

Bejan, A. (1996a) *Entropy Generation Minimization*. CRC Press, Boca Raton, FL.

Bejan, A. (1996b) Entropy generation minimization: The new thermodynamics of finite-size devices and finite-time processes. *J. Appl. Phys.* **79**, 1191–1218.

Bejan, A. (1996c) Street network theory of organization in nature. *J. Adv. Transp.* **30**, 85–107.

Bejan, A. (1997) *Advanced Engineering Thermodynamics*, 2nd ed. Wiley, New York.

Bejan, A. (2000) *Shape and Structure, from Engineering to Nature*. Cambridge University Press, Cambridge, UK.

Bejan, A. (2001) The tree of convective heat streams: Its thermal insulation function and the predicted 3/4-power relation between body heat loss and body size. *Int. J. Heat Mass Transfer* **44**, 699–704.

Bejan, A. (2002) Dendritic constructal heat exchanger with small-scale crossflows and larger-scale counterflows. *Int. J. Heat Mass Transfer* **45**, 4607–4620.

Bejan, A. and Anderson, R. (1981) Heat transfer across a vertical impermeable partition imbedded in a porous medium. *Int. J. Heat Mass Transfer* **24**, 1237–1245.

Bejan, A. and Anderson, R. (1983) Natural convection at the interface between a vertical porous layer and an open space. *J. Heat Transfer* **105**, 124–129.

Bejan, A. and Errera, M. R. (2000) Convective trees of fluid channels for volumetric cooling. *Int. J. Heat Mass Transfer* **43**, 3105–3118.

Bejan, A. and Fautrelle, Y. (2003) Constructal multi-scale structure for maximal heat transfer density. *Acta Mechanica* **163**, 39–49.

Bejan, A. and Kraus, A. D., eds. (2003) *Heat Transfer Handbook*. Wiley, New York.

Bejan, A. and Lage, J. L. (1991) Heat transfer from a surface covered with hair. *Convective Heat and Mass Transfer in Porous Media* S. Kakac, B. Kilkis, F. A. Kulacki, and F. Arinc, eds., The Netherlands, Kluwer Academic, Dordrecht, 823–845.

Bejan, A. and Lorente, S. (2001) Thermodynamic optimization of flow geometry in mechanical and civil engineering. *J. Non-Equilib. Thermodyn.* **26**, 305–354.

Bejan, A. and Lorente, S. (2002) Thermodynamic optimization of flow architecture: Dendritic structures and optimal sizes of components. ASME IMECE-vol. 3, *International Mechanical Engineering Congress and Exposition*, New Orleans, November 17–22.

Bejan, A. and Lorente, S. (2003) Thermodynamic formulation of the constructal law. ASME Paper IMECE2003-41167, *International Mechanical Engineering Congress and Exposition*, Washington, DC, November 16–21.

Bejan, A. and Mamut, E., eds. (1999) *Thermodynamic Optimization of Complex Energy Systems*. Kluwer Academic, Dordrecht, The Netherlands.

Bejan, A. and Morega, A. M. (1993) Optimal arrays of pin fins and plate fins in laminar forced convection. *J. Heat Transfer* **115**, 75–81.

Bejan, A. and Poulikakos, D. (1984) The non-Darcy regime for vertical boundary layer natural convection in a porous medium. *Int. J. Heat Mass Transfer* **27**, 717–722.

Bejan, A. and Sciubba, E. (1992) The optimal spacing of parallel plates cooled by forced convection. *Int. J. Heat Mass Transfer* **35**, 3259–3264.

Bejan, A. and Tien, C. L. (1978) Natural convection in a horizontal porous medium subjected to an end-to-end temperature difference. *J. Heat Transfer* **100**, 191–198; errata (1983) **105**, 683–684.

Bejan, A. and Tien, C. L. (1979) Natural convection in horizontal space bounded by two concentric cylinders with different end temperatures. *Int. J. Heat Mass Transfer* **22**, 919–927.

Bejan, A., Fowler, A. J., and Stanescu, G. (1995) The optimal spacing between horizontal cylinders in a fixed volume cooled by natural convection. *Int. J. Heat Mass Transfer* **38**, 2047–2055.

Bejan, A., Rocha, L. A. O., and Cherry, R. S. (2002) Methane hydrates in porous layers: Gas formation and convection. *Transport Phenomena in Porous Media II*, D. B. Ingham and I. Pop, eds., Pergamon, Amsterdam.

Bejan, A., Rocha, L. A. O., and Lorente, S. (2000) Thermodynamic optimization of geometry: T- and Y-shaped constructs of fluid streams. *Int. J. Thermal Sci.* **39**, 949–960.

Bejan, A., Tsatsaronis, G., and Moran, M. (1996) *Thermal Design and Optimization.* Wiley, New York.

Bejan, A., Vadasz, P., and Kröger, D. G., eds. (1999) *Energy and the Environment.* Kluwer Academic, Dordrecht, The Netherlands.

Belghazi, M., Bontemps, A., and Marvillet, C. (2002) Condensation heat transfer on enhanced surface tubes: Experimental results and predictive theory. *J. Heat Transfer* **124**, 754–761.

Benhadji, K. and Vasseur, P. (2001) Double diffusive convection in a shallow porous cavity filled with a non-Newtonian fluid. *Int. Comm. Heat Mass Transfer* **28**, 763–772.

Bennacer, R., Beji, H., and Mohamad, A. A. (2003) Double diffusive convection in a vertical enclosure inserted with two saturated porous layers confining a fluid layer. *Int. J. Thermal Sci.* **42**, 141–151.

Bera, P. and Khalili, A. (2002) Double-diffusive natural convection in an anisotropic porous cavity with opposing buoyancy forces: Multi-solutions and oscillations. *Int. J. Heat Mass Transfer* **45**, 3205–3222.

Bergman, W., Taylor, R. D., Miller, H. H., Biermann, A. H., Hebard, H. D., da Roza, R. A., and Lum, B. Y. (1978) Enhanced filtration program at LLL—a progress report. *Fifteenth DOE Nuclear Air Cleaning Conference*, M. First, ed., New York, 1058–1097.

Bernstein, O. and Shapiro, M. (1994) Direct determination of the orientation distribution of cylindrical particles immersed in laminar and turbulent shear flows. *J. Aerosol Sci.* **25**, 113–136.

Bhattacharjee, S. and Grosshandler, W. L. (1988) The formation of a wall jet near a high temperature wall under microgravity environment. *ASME HTD* **96**, 711–716.

Bhattacharya, A., Calmidi, V. V., and Mahajan, R. L. (2002) Thermophysical properties of high porosity metal foams. *Int. J. Heat Mass Transfer* **45**, 1017–1031.

Bigas, J. P. (1994) La diffusion des ions chlore dans les mortiers. PhD Thesis, Institut National des Sciences Appliquées, Toulouse.

Bilbao, C., Albors, A., Gras, M., Andres, A., and Fito, P. (2000) Shrinkage during apple tissue air-drying: Macro and microstructural changes. *Proceedings of the Twelfth International Drying Symposium*, P. J. A. M. Kerkhof, W. J. Coumans, and G. D. Mooweer, eds., Paper No: 330, August 28–31.

364   A. Bejan, I. Dincer, S. Lorente, A.F. Miguel and A.H. Reis

Bilgen, E. and Mbaye, M. (2001) Bénard cells in fluid-saturated porous enclosures with lateral cooling. *Int. J. Heat Fluid Flow* **22**, 561–570.

Billard, F., Madeleine, G., and Pradel J. (1963) Variation de l'efficacité des filters en fonction de leur colmatage par divers types d'aérossol. *Colloque sur la Pollution Radioactive des Milieux Gazeux* (edited by Presses Universitaires de France), Paris.

Bird, R. B., Steward, W. E., and Lightfoot, E. N. (1960) *Transport Phenomena*, Wiley, New York.

Bockris, J. O'M. and Reddy, A. K. N. (2001) *Modern Electrochemistry*. Kluwer Academic, Dordrecht, The Netherlands.

Boomsma, K. and Poulikakos, D. (2001) On the effective thermal conductivity of a three-dimensionally structured fluid-saturated metal foam. *Int. J. Heat Mass Transfer* **44**, 827–836.

Boomsma, K. and Poulikakos, D. (2002) The effects of compression and pore size variations on the liquid flow characteristics of metal foams. *J. Fluids Eng.* **124**, 263–272.

Boomsma, K., Poulikakos, D., and Zwick, F. (2003) Metal foams as compact high performance heat exchangers. *Mechanics of Materials* **35**, 1161–1176.

Borrvall, T. and Petersson, J. (2002) Topology optimization of fluids in Stokes flow. *Int. J. Numer. Methods Fluids* **41**, 77–107.

Borrvall, T., Klarbring, A., Petersson, J., and Torstenfelt, B. (2002) Topology optimization in fluid mechanics. *Fifth World Congress on Computational Mechanics*, H. A. Mang, F. G. Rammerstorfer, and J. Eberhardsteiner, eds., Vienna, Austria, July 7–12.

Boussant-Roux, Y., Chitti, O., Miller, M., and Chaudourne, S. (2000) Experimental and industrial evaluation of a new fused cast modular checker design for decreasing regenerator plugging. *Glastech. Ber. Glass Sci. Technol.* **73**, 259–269.

Boussinesq, J. (1903) *Théorie Analytique de la Chaleur*, vol. 2. Gauthier-Villars, Paris.

Bradshaw, P. (2001) Shape and structure, from engineering to nature. *AIAA J.* **39** 983.

Brandão, R. M. and Lopes, M. L. (1991) The aeropalinological environment and respiratory allergies in the region of Évora (South of Portugal). *Cadernos de Imuno-Alergologia Pediátrica* **6**, 13–18.

Brennan, J. G., Butters, J. R., Cowell, N. D., and Lilly, A. E. V. (1976) *Food Engineering Operations*, Applied Science, London.

Brenner, H. and Gaydos, L. J. (1977) The constrained Brownian movement of spherical particles in cylindrical pores of comparable radius. *J. Colloid Interface Sci.* **58**, 312–356.

Briaud, J.-L. and Chaouch, A. (1997) Hydrate melting in soil around hot conductor. *J. Geotechn. Geoenviron. Eng.* **123**, 645–653.

Brinkman, H. C. (1947a) A calculation of the viscous force exerted by a flowing fluid on a dense swarm of particles. *Appl. Sci. Res.* A1, 27–34.

Brinkman, H. C. (1947b) On the permeability of media consisting of closely packed porous particles. *Appl. Sci. Res.* A1, 81–86.

Brod, H. (2003) Residence time optimised choice of tube diameters and slit heights in distribution systems for non-Newtonian liquids. *J. Non-Newtonian Fluid Mech.* **111**, 107–125.

Brumleve, T. R. and Buck, R. P. (1978) Numerical solution of the Nernst–Planck and Poisson equation system with application to membrane electrochemistry and solid state physics. *J. Electro-Analytical Chem.* **17**, 1093–1103.

Brunnauer, S., Emmett, P. H., and Teller, E. (1938) Adsorption of gases in multimolecular layers. *J. Am. Chem. Soc.* **60**, 309.

Bruno, J. S. and Vigo, T. L. (1987) Thermally active fabrics containing polyethylene glycols. *J. Coated Fabrics* **16**, 264–274.

Buretta, R. J. and Berman, A. S. (1976) Convective heat transfer in a liquid saturated porous layer. *J. Appl. Mech.* **43**, 249–253.

Burgmayer, P. and Murray, R. W. (1982) An ion gate membrane: Electrochemical control of ion permeability through a membrane with an embedded electrode. *J. Am. Chem. Soc.* **104**, 6139–6140.

Burns, P. J. and Tien, C. L. (1979) Natural convection in porous media bounded by concentric spheres and horizontal cylinders. *Int. J. Heat Mass Transfer* **22**, 929–939.

Butterworth, D. (1978) A model for heat transfer during three-dimensional flow in tube bundles. *Proceedings of the Sixth International Heat Transfer Conference*, Toronto, August 7–11, **4**, 219–224.

Caltagirone, J. P. (1976) Thermoconvective instabilities in a porous medium bounded by two concentric horizontal cylinders. *J. Fluid Mech.* **76**, 337–362.

Carman, P. C. (1937) Flow through a granular bed, *Trans. Inst. Chem. Eng. London* **15**, 150–156.

Carslaw, H. S. and Jaeger, J. C. (1959) *Conduction of Heat in Solids*. Oxford University Press, Oxford, UK.

Chamkha, A. J. and Quadri, M. M. A. (2001) Heat and mass transfer from a permeable cylinder in a porous medium with magnetic field and heat generation/absorption effect. *Numer. Heat Transfer A* **40**, 387–401.

Chan, S. H., Khor, K. A., and Xia, Z. T. (2001) A complete polarization model of a solid oxide fuel cell and its sensitivity to the change of cell component thickness. *J. Power Sources* **93**, 130–140.

Chao, B. H., Cheng, P., and Le, T. (1994) Free convective diffusion flame sheet in porous media. *Combustion Sci. Tech.* **99**, 221–234.

Chao, B. H., Wang, H., and Cheng, P. (1996) Stagnation point flow of a chemically reactive fluid in a catalytic porous bed. *Int. J. Heat Mass Transfer* **39**, 3003–3019.

Chatterji, S. (1998) Colloid electrochemistry of saturated cement paste and some properties of cement based materials. *Adv. Cement Based Materials* **7**, 102–108.

Chatterji, S. and Kawamura, M. (1992) Electrical double layer, ion transport, and reactions in hardened cement paste. *Cement Concrete Res.* **22**, 774–782.

Chaussadent, T. (1999) *État des Lieux et Réflexions sur la Carbonation du Béton Armé*. Laboratoire Central des Ponts et Chaussées, OA29, Paris.

Chella, R., Lasseux, D., and Quintard, M. (1998) Multiphase, multicomponent fluid flow in homogeneous and heterogeneous porous media. *Revue de l'Institut Francais du Pétrole* **53**(3), 335–346.

Chen, Y. and Cheng, P. (2002) Heat transfer and pressure drop in fractal tree-like microchannel nets. *Int. J. Heat Mass Transfer* **45**, 2643–2648.

Chen, Y. and Cheng, P. (2003) Fractal characterisation of wall roughness on pressure drop in microchannels. *Int. Comm. Heat Mass Transfer* **30**, 1–11.

Cheng, P. (1977) Combined free and forced convection flow about inclined surfaces in porous media. *Int. J. Heat Mass Transfer* **20**, 807–814.

Cheng, P. (1978) Heat transfer in geothermal systems. *Adv. Heat Transfer* **14**, 1–105.

Cheng, P. (1982) Mixed convection about a horizontal cylinder and a sphere in a fluid saturated porous medium. *Int. J. Heat Mass Transfer* **25**, 1245–1247.

Cheng, P. and Chang, J. D. (1976) On buoyancy induced flows in a saturated porous medium adjacent to impermeable horizontal surfaces. *Int. J. Heat Mass Transfer* **19**, 1267–1272.

Cheng, P. and Minkowycz, W. J. (1977) Free convection about a vertical flat plate embedded in a saturated porous medium with application to heat transfer from a dike. *J. Geophys. Res.* **82**, 2040–2044.

Cheng, P., Chowdhury, A., and Hsu, C. T. (1991) Forced convection in packed tubes and channels with variable porosity and thermal dispersion effects. *Convective Heat and Mass Transfer in Porous Media*, S. Kakac, B. Kilkis, F. A. Kulacki and F. Arinc, eds. Kluwer Academic, Dordrecht, The Netherlands, 625–653.

Cheng, W. T. and Lin, H. T. (2002) Unsteady forced convection heat transfer on a flat plate embedded in the fluid-saturated porous medium with inertia effect and thermal dispersion. *Int. J. Heat Mass Transfer* **45**, 1563–1569.

Cheng, Y., van den Bleek, C. M., and Coppens, M.-O. (2000) Scale-up and hydrodynamics of fluidized beds with fractal-like injectors. *AIChE Annual Meeting*, Los Angeles, November, 12–17.

Cherskii, N. V. and Bondarev, E. A. (1972) Thermal method of exploiting gas-hydrated strata. *Soviet Physics Doklady* **17**(3), 211–213.

Choi, J. C. and Kim, S. D. (1995) Heat transfer in latent heat-storage system using $MgCl_2 \cdot (6H_2O$ at the melting point. *Energy* **29**, 13–25.

Churchill, S. W. and Usagi, R. 1974. A standardized procedure for the production of correlations in the form of a common empirical equation. *Indust. Eng. Chem. Fund.* **13**, 39–46.

Cizmas, P. G. and Bejan, A. (2001) Optimal placement of cooling flow tubes in a wall heated from the side. *Int. J. Transport Phenomena* **3**, 331–343.

Combarnous, M. A. and Bories, S. A. (1975) Hydrothermal convection in saturated porous media. *Adv. Hydrosci.* **10**, 231–307.

Comer, J. K., Kleinstreuer, V., Hyun, S., and Kim, C. S. (2000) Aerosol transport and deposition in sequentially bifurcating airways. *J. Biomech. Eng.* **122**, 152–158.

Costa, V. A. F. (2003) Unified streamline, heatline and massline methods for the visualization of two-dimensional heat and mass transfer in anisotropic media. *Int. J. Heat Mass Transfer* **46**, 1309–1320.

Costa, V. A. F., Oliveira, L. A., Baliga, B. R. and Sousa, A. C. M. (2004) Simulation of coupled flows in adjacent porous and open domains using a control-volume finite-element method. *Numerical Heat Transfer, Part A* **45**, 1–23.

Crank, J. (1975) *The Mathematics of Diffusion*, 2nd ed. Oxford University Press, Oxford, UK.

Curry, F. E. (1974) A hydrodynamic description of the osmotic reflection coefficients with application to the pore theory exchange. *Microvasc. Res.* **8**, 236–252.

Curry, F. E. and Michel, C. C. (1980) A fiber matrix model of capillary permeability. *Microvasc. Res.* **20**, 96–99.

Daccord, G. and Lenormand, R. (1987) Fractal patterns from chemical dissolution. *Nature* **325**(1), January, 41–43.

Darcy, H. P. G. (1856) *Les Fontaines Publiques de la Ville de Dijon*. Victor Dalmont, Paris.

Daugirdas, J. T., Blake, P. G., and Ing. T. S., eds. (2000) *Handbook of Dialysis*. Lippincott Williams & Wilkins, Philadelphia.

Davies, C. N. (1973) *Air Filtration*. Academic, London.

de Deugd, R. M., Jager, M. D., and de Swaan Arons, J. (2001) Mixed hydrates of methane and water-soluble hydrocarbons modeling of empirical results. *AIChE J.* **47**, 693–704.

Deen, W. M. (1987) Hindered transport of large molecules in liquid-filled pores. *AIChE J.* **33**, 1409–1425.

DePaoli, D. W., Harris, M. T., Morgan, I. L., and Ally, M. R. (1997) Investigation of electrokinetic decontamination of concrete. *Symposium on Separation Science and Technology for Energy Applications* **32**, 387–404.

Dewulf, J., Van Langenhove, H., Mulder, J., van den Berg, M. M. D., van der Kooi, H. J., and de Swaan Arons, J. (2000) Illustrations towards quantifying the sustainability of technology. *Green Chem.* **2**(3), 108–114.

Dhanasekaran, M. R., Das, S. K., and Venkateshan, S. P. (2002) Natural convection in a cylindrical enclosure filled with heat generating anisotropic porous medium. *J. Heat Transfer* **124**, 203–207.

Diamond, S. (1986) Chloride concentration in concrete pore solutions resulting from calcium and sodium chloride admixtures. *Cement Concrete Aggregates* **8**, 97–102.

Dietl, C., George, O. P., and Bansal, N. K. (1995) Modeling of diffusion in capillary porous materials during the drying process. *Drying Technol.* **13**(1–2), 267–293.

Dincer, I. (1996) Development of a new number (the Dincer number) for forced-convection heat transfer in heating and cooling applications. *Int. J. Energy Res.* **20**, 419–422.

Dincer, I. (2002a) Technical, environmental and exergetic aspects of hydrogen energy systems. *Int. J. Hydrogen Energy* **27**(3), 265–285.

Dincer, I. (2002b) The role of exergy in energy policy making. *Energy Policy* **30**(2), 137–149.

Dincer, I. and Dost, S. (1996) A perspective on thermal energy storage systems for solar energy applications. *Int. J. Energy Res.* **20**(6), 547–557.

Dincer, I. and Hussain, M. M. (2002) Development of a new Bi-Di correlation for solids drying. *Int. J. Heat Mass Transfer* **45**, 3065–3069.

Dincer, I. and Rosen, M. A. (2002) *Thermal Energy Storage Systems and Applications*, Wiley, Chichester, UK.

Dincer, I., Dost, S., and Li, X. (1997a) Performance analyses of sensible heat storage systems for thermal applications. *Int. J. Energy Res.* **21**(10), 1157–1171.

Dincer, I., Dost, S., and Li, X. (1997b) Thermal energy storage applications from an energy saving perspective. *Int. J. Global Energy Issues* **9**(4–6) 351–364.

Dincer, I., Sahin, A. Z., Yilbas, B. S., Al-Farayedhi, A. A., and Hussain, M. M. (2000) Exergy and energy analysis of food drying systems. Progress Report 2, KFUPM Project # ME/ENERGY/203.

Donaldson, K., Li, X. Y. and MacNee, W. (1998) Ultrafine (nanometer) particle mediated lung injury. *J. Aerosol Sci.* **29**, 553–560.

Dorman, R. G. (1973) *Dust Control and Air Cleaning*, Pergamon, Oxford.

D'Ottavio, T. and Goren, L. S. (1983) Aerosol capture in granular beds in the impaction dominated regime. *Aerosol Sci. Technol.* **2**, 91–108.

Douglas, R. B. (1989) Occupational lung disease and aerosol. *Aerosols and the Lung*, S. W. Clarke and D. P. Pavis, eds., Butterworths, London, 251–262.

Drazer, G. and Koplik, J. (2001) Tracer dispersion in two-dimensional rough fractures. *Phys. Rev. E* **63**, article 056104, 1–11.

Dullien, F. A. L. (1992) *Porous Media: Fluid Transport and Pore Structure*, 2nd ed. Academic, New York.

Dupouy, M. D. and Camel, D. (1998) Effects of gravity on columnar dendritic growth of metallic alloys: Flow pattern and mass transfer. *J. Crystal Growth* **183**, 469–489.

Dupouy, M. D., Camel, D., and Favier, J. J. (1993) Natural convective effects in directional dendritic solidification of binary metallic allows: Dendritic array morphology. *J. Crystal Growth* **126**, 480–492.

Dupouy, M. D., Drevet, B., and Camel, D. (1997) Influence of convection on the selection of solidification microstructures at low growth rates. *J. Crystal Growth* **181**, 145–159.

Dupuit, A. J. E. J. (1863) *Études Théoriques et Pratiques sur le Mouvement des Eaux dans les Canaux Découverts et a Travers les Terrains Perméables*. Victor Dalmont, Paris.

Eidsath, A., Carbonell, R. G., Whitaker, S., and Hennann, L. R. (1983) Dispersion in pulsed systems-III. Comparison between theory and experiments for packed beds. *Chem. Eng. Sci.* **38**, 1803–1816.

Einstein, A. (1956) *Investigations on the Theory of Brownian Movement*. Dover, New York.

Emi, H., Wang, C. S., and Tien, C. (1982) Transient behavior of aerosol filtration in model filters. *AIChE J.* **28**, 397–405.

Ene, H. I. and Poliševski, D. (1987) *Thermal Flow in Porous Media*. Reidel, Dordrecht, The Netherlands.

Ergun, S. (1952) Fluid flow through packed columns. *Chem. Eng. Prog.* **48**, 89–94.

Errera, M. R. and Bejan, A. (1998) Deterministic tree networks for river drainage basins. *Fractals* **6**, 245–261.

Ewart, H. S. and Klip, A. (1995) Hormonal regulation of the Na^+-K^+-ATPase: Mechanisms underlying rapid and sustained changes in pump activity. *Am. J. Physiol.* **269**, 295–311.

Fand, R. M., Kim, B. Y., Lam, A. C. C., and Phan, R. T. (1987) Resistance to the flow of fluids through simple and complex porous media whose matrices are composed of randomly packed spheres. *J. Fluids Eng.* **109**, 268–274.

Farouk, B. and Guceri, S. (1982) Natural and mixed convection heat transfer around a horizontal cylinder within confining walls. *Numer. Heat Transfer* **35**, 329–341.

Feidt, M. (1987) *Thermodynamique et Optimisation Énergetique des Systèmes et Procedés*. Technique et Documentation, Lavoisier, Paris.

Feidt, M. L. (1998) Thermodynamics and the optimization of reverse cycle machines. *Thermodynamic Optimization of Complex Energy Systems*, A. Bejan and E. Mamut, eds., Kluwer Academic, Dordrecht, The Netherlands, 385–402.

Ferry, J. D. (1936) Statistical evaluation of sieve constants in ultrafiltration. *J. Gen. Physiol.* **20**, 101–179.

Fick, A. (1855) On liquid diffusion. *Philos. Mag.* **4**, 30–39.

Finkelstein, T. and Organ, A. J. (2001) *Air Engines*. Professional Engineering, Bury St. Edmunds, UK.

Fisher, T. S. and Torrance, K. E. (1998) Free convection limits for pin-fin cooling. *J. Heat Transfer* **120**, 633–640.

Flagan, R. C. and Seinfeld, J. H. (1988) *Fundamentals of Air Pollution Engineering*. Prentice-Hall, Englewood Cliffs, NJ.

Fontana, R. L. and Mussumeci, A. (1994) Hydrates offshore Brazil. *International Conference on Natural Gas Hydrates. Annals New York Academy of Sciences* **715**, 106–113.

Forchheimer, P. (1901) Wasserbewegung durch Boden. Z. *Vereines Deutscher Ingenieure* **45**, 1736–1741 and 1781–1788.

Foss, J. M., Frey, M. F., Schamberg, M. R., Peters, J. E., and Leong, K. H. (1989) Collection of uncharged prolate spheroid aerosol particles by spherical collectors—I:(2D) motion. *J. Aerosol Sci.* **20**, 515–532.

Fowler, A. J. and Bejan, A. (1991) The effect of shrinkage on the cooking of meat. *Int. J. Heat Fluid Flow* **12**, 375–383.

Fowler, A. J. and Bejan, A. (1994) Forced convection in banks of inclined cylinders at low Reynolds numbers. *Int. J. Heat Fluid Flow* **15**, 90–99.

Fowler, A. J., Ledezma, G. A., and Bejan, A. (1997) Optimal geometric arrangement of staggered plates in forced convection. *Int. J. Heat Mass Transfer* **40**, 1795–1805.

Francy, O. (1998) Modélisation de la pénetration des ions chlorure dans les mortiers partiellement saturés en eau. PhD Thesis, Université Paul Sabatier, Toulouse.

Frederick, R. L. (1999) On the aspect ratio for which the heat transfer in differentially heated cavities is maximum. *Int. Commun. Heat Mass Transfer* **26**, 549–558.

Freitas, Jr., R. A. (1999) *Nanomedicine, Volume I: Basic Capabilities*. Landes Bioscience, Georgetown.

Friedlander, S. K. (1977) *Smoke, Dust and Haze: Fundamentals of Aerosol Behavior*. Wiley, New York.

Frizon, F. (2003) Décontamination électrocinétique des milieux poreux. Étude expérimentale et modélisation appliquées au césium dans les matériaux cimentaires. PhD Thesis, Laboratoire Matériaux et Durabilité des Constructions Toulouse and Commissariat à l'Energie Atomique, Cadarache, France.

Frizon, F., Lorente, S., Ollivier, J. P., and Thouvenot, P. (2003) Transport model for the nuclear decontamination of cementitious materials. *Comput. Materials Sci.* **27**, 507–516.

Frizon, F., Thouvenot, P., Ollivier, J. P., and Lorente, S. (2002) Description of the radiological decontamination by electro-migration in saturated concrete: A multi-species approach. *Waste Management Conference*, Tucson, AZ.

Fuchs, N. A. (1964) *The Mechanics of Aerosols*. Pergamon, New York.

Furukawa, T. and Yang, W.-J. (2003) Thermal optimization of channel flows with discrete heating sections. *J. Non-Equilib. Thermodyn.* **28**, 299–310.

Gabas, A. L., Menegalli, F. C., and Telis-Romero, J. (1999) Effect of chemical pretreatment on the physical properties of dehydrated grapes. *Drying Technol.* **17**(6), 1215–1226.

Gal, E., Tardos, G. I., and Pfeffer, R. (1985) A study of inertial effects in granular bed filtration. *AIChE J.* **31**, 1093–1104.

Gallily, I., Schiby, D., Cohen, A. H., Holländer, W., Schless, S. D., and Stöber, W. (1986) On inertial separation of nonspherical aerosol particles from laminar flow: I—the cylindrical case. *Aerosol Sci. Technol.* **5**, 267–286.

370   A. Bejan, I. Dincer, S. Lorente, A.F. Miguel and A.H. Reis

Ganderton, D. (1999) Targeted delivery of inhaled drugs: Current challenges and future goals. *J. Aerosol Med.* **12**, S3–S8.

Garror, C. (1999) The frequency of Saharan dust episodes over Tel Aviv Israel. *Atmos. Environ.* **28**, 2867–2871.

Gates, C. M. and Newman, J. (2000) Equilibrium and diffusion of methanol and water in a Nafion 117 membrane. *AIChE J.* **46**, 2076–2085.

Georgiadis, J. G. and Catton, I. (1987) Stochastic modeling of unidirectional fluid transport in uniform and random packed beds. *Phys. Fluids* **30**, 1017–1022.

Gering, K. L., Cherry, R. S., and Weinberg, D. M. (2000) Mechanisms for methane gas accumulation under hydrate deposits in sediment. *Annals New York Academy of Sciences* **912**, 623–632.

Getachew, D., Minkowycz, W. J., and Lage, J. L. (2000) A modified form of the k-ε model for turbulent flows of an incompressible fluid in porous media. *Int. J. Heat Mass Transfer* **43**, 2909–2915.

Gilliland, E. R., Baddour, R. F., Perkinson, G. P., and Sladek, K. J. (1974) Diffusion on surfaces. Effect of concentration on diffusivity of physically adsorbed gases. *Ind. Eng. Chem. Fund.* **13**, 95–100.

Gogus, F. and Maskan, M. (1999) Water adsorption and drying characteristics of okra. *Drying Technol.* **17**(4, 5); 883–894.

Goncalves, L. C. C. and Probert, S. D. 1993 Thermal-energy storage: Dynamic performance characteristics of cans each containing a phase-change material, assembled as a packed-bed. *Appl. Energy* **45**, 117–155.

Gonda, I. (1997) Particle deposition in the human respiratory tract. *The Lungs: Scientific Foundations*, R. G. Crystal, J. B. West, E. R. Weibel, and P. J. Barnes, eds., Lippincott-Raven, Philadelphia, 2289–2294.

Goo, J. and Kim, C. S. (2003) Theoretical analysis of particle deposition in human lungs considering stochastic variations of airway morphology. *J. Aerosol Sci.* **34**, 585–602.

Gosselin, L., Bejan, A. and Lorente, S. (2004) Combined 'heat flow and strength' optimization of geometry: mechanical structures most resistant to thermal attack. *Int. J. Heat Mass Transfer* **47**, in press.

Goyeau, B., Benihaddadene, T., Gobin, D., and Quintard, M. (1999) Numerical calculation of the permeability in a dendritic mushy zone. *Metallurg. Materials Trans. B* **30B**, 613–662.

Gradon, L., Grzybowski, P., and Pilaciski, W. (1988) Analysis of motion and deposition of fibrous particles on a single filter element. *Chem. Eng. Sci.* **43**, 1253–1259.

Greenkorn, R. A. (1983) *Flow Phenomena in Porous Media.* Dekker, New York.

Gregg, S. J. and Sing, K. S. W. (1982) *Adsorption, Surface Area and Porosity.* Academic, London.

Guggenheim, E. A. (1933) *Modern Thermodynamics.* Methuen, London.

Gupta, A., Novick, V., Biswas, P., and Monson, P. R. (1993) Effect of humidity, particle hygroscopicity on mass loading capacity of high efficiency particulate air (HEPA) filters. *Aerosol Sci. Technol.* **19**, 94–107.

Gutch, C. F., Stoner, M. H., and Corea, A. L. (1999) *Review of Hemodialysis for Nurses and Dialysis Personnel,* 6th ed. Mosby, St. Louis.

Gutfinger, C. and Tardos, G. I. (1979) Theoretical and experimental investigation on granular bed dust filters. *Atmos. Envir.* **13**, 853–867.

Guyton, A. C. (1991) *Textbook of Medical Physiology.* Saunders, Philadelphia.

Hanumantha Rao, Y., Reddy, S. I., Khanna, R., Rao, T. G., Thakur, N. K., and Subrahmanyam, C. (1998) Potential distribution of methane hydrates along the Indian continental margins. *Current Sci.* **74**(5), 466–468.

Harris, M. T., DePaoli, D. W., and Ally, M. R. (1997) Modeling the electrokinetic decontamination of concrete. *Symposium on Separation Science and Technology for Energy Applications* **32**, 827–848.

Hart, B. S., Flemings, P. B., and Deshpande, A. (1995) Porosity and pressure: Role of compaction disequilibrium in the development of geopressures in a Gulf Coast pleistocene basin. *Geology* **23**, 45–48.

Hassanizadeh, S. M. and Leijnse, A. (1995) A non-linear theory of high-concentration-gradient dispersion in porous media. *Adv. Water Resources* **18**, 203–215.

Havstad, M. A. and Burns, P. J. (1982) Convective heat transfer in vertical cylindrical annuli filled with a porous medium. *Int. J. Heat Mass Transfer* **25**, 1755–1766.

Heintzenberg, J. (1989) Fine particles in the global troposphere: A review. *Tellus* **41B**, 149–160.

Heintzenberg, J. and Covert, D. S. (1990) On the distribution of physical and chemical particle properties in the atmospheric aerosol. *J. Atmos. Chem.* **10**, 383–397.

Helfferich, F. (1962) *Ion Exchange*. McGraw-Hill, New York.

Hepbasli, A. (1998a) Investigation of the bubble behaviour at the free surface of a large three-dimensional gas fluidized bed. *Int. J. Energy Res.* **22**, 885–909.

Hepbasli, A. (1998b) Estimations of bed expansions in a freely-bubbling three-dimensional gas-fluidized bed. *Int. J. Energy Res.* **22**, 1365–1380.

Hickox, C. E. and Gartling, D. K. (1981) A numerical study of natural convection in a horizontal porous layer subjected to an end-to-end temperature difference. *J. Heat Transfer* **103**, 797–802.

Hinds, W. C. (1999) *Aerosol Technology*. Wiley, New York.

Hirschenhofer, J. H., Stauffer, D. B., Engleman, R. R., and Klett, M. G. (1998) *Fuel Cell Handbook*. Report: DE-AC21-94MC31166, PA, US Department of Energy, Washington, DC.

Hoffmann, M. R. (2000) Fast real space renormalization for two-phase porous media flow. *Computational Methods for Flow and Transport in Porous Media*, J. M. Crolet, ed. Kluwer Academic, Dordrecht, The Netherlands, 83–91.

Hoffmann, M. R. and van der Meer, F. M. (2002) A simple space-time averaged porous media model for flow in densely vegetated channels. *Developments in Water Science*, vol. 2, Hassanizadeh, S. M., Schotting, R. J., Gray, W. G., and Pinder, G. F., eds. (2002) *Computational Methods in Water Resources*, Elsevier, Amsterdam, 1661–1668.

Holder, G. D., Kamath, V. A., and Godbole, S. P. (1994) The potential of natural gas hydrates as an energy resource. *Annals New York Academy of Sciences* **715**, 427–445.

Holder, G. D., Malone, R. D., and Lawson, W. F. (1987) Effects of gas composition and geothermal properties on the thickness and depth of natural-gas-hydrate zones. *J. Petroleum Techn.* **39**(9), 1147–1152.

Horno, J. and Castilla, J. (1994) Application of networks thermodynamics to the computer simulation of non-stationary ionic transport in membranes. *J. Membrane Sci.* **90**, 173–181.

Horton, C. W. and Rogers, F. T. (1945) Convection currents in a porous medium. *J. Appl. Phys.* **16**, 367–370.

Hunter, R. J. (1981) *Zeta Potential in Colloid Science, Principles and Applications.* Academic, Sydney.

ICRP (International Commission on Radiological Protection) (1994) Human respiratory tract model for radiological protection. *Annals of the ICRP 66*, Elsevier, New York.

Ilias, S. and Douglas, P. L. (1989) Inertial impaction of aerosol particles on cylinders at intermediate and high Reynolds numbers. *Chem. Eng. Sci.* **44**, 81–99.

Ingham, D. B. and Pop, I., eds. (1998) *Transport Phenomena in Porous Media.* Elsevier, Amsterdam.

Ingham, D. B. and Pop, I., eds. (2002) *Transport Phenomena in Porous Media II.* Elsevier, Amsterdam.

Islam, M. R. (1994) A new recovery technique for gas production from Alaskan gas hydrates. *J. Petroleum Sci. Eng.* **11**, 267–281.

Ismail, K. A. R. and Stuginsky, R. (1999) A parametric study on possible fixed bed models for pcm and sensible heat storage. *Appl. Thermal Eng.* **19**, 757–788.

IUPAC Recommendations (2001) *Pure Appl. Chem.* **73**, 381–394.

Jaluria, Y. (2001) Fluid flow phenomena in materials processing—the 2000 Freeman Scholar Lecture. *J. Fluids Eng.* **123**, 173–210.

Jankowaska, E., Reponen, T., Willeke, K., Grinshpun, S. A., and Choi, K.-J. (2000) Collection of fungal spores on air filters and spore reentrainment from filters into air. *J. Aerosol Sci.* **31**, 969–978.

Jany, P. and Bejan, A. (1988) Scales of melting in the presence of natural convection in a rectangular cavity filled with a porous medium. *J. Heat Transfer* **110**, 526–529.

Joseph, D. D., Nield, D. A., and Papanicolaou, G. (1982) Nonlinear equation governing flow in a saturated porous medium. *Water Resources Res.* **18**, 1049–1052 and **19**, 591.

Jung, Y. and Tien, C. (1991) New correlations for predicting the effect of deposition on collection efficiency and pressure drop in granular filters. *J. Aerosol Sci.* **22**, 187–200.

Kalla, L., Mamou, M., Vasseur, P., and Robillard, L. (2001) Multiple solutions for double diffusive convection in a shallow porous cavity with vertical fluxes of heat and mass. *Int. J. Heat Mass Transfer* **44**, 4493–4504.

Kamath, V. A. and Holder, G. D. (1987) Dissociation heat transfer characteristics of methane hydrates. *AIChE J.* **33**(2), 347–350.

Karathanos, V. T., Kanellopoulos, N. K., and Belessiotis, V. G. (1996) Development of porous structures during air drying of agricultural plant products. *J. Food Eng.* **29**, 167–183.

Karel, M. (1975) Dehydration of foods. *Physical Principles of Food Preservation*, M. Karel, O. R. Fennema, and D. B. Lund, eds., Dekker, New York.

Karim, F., Farouk, B., and Namer, I. (1986) Natural convection heat transfer from a cylinder between confining adiabatic walls. *J. Heat Transfer* **108**, 291–298.

Katul, G., Ellsworth, D. S., and Lai, C.-T. (2000) Modelling assimilation and intercellular CO_2 from measured conductance: A synthesis of approaches. *Plant Cell Envir.* **23**, 1313–1328.

Katz, A. J. and Thompson, A. H. (1985) Fractal sandstone pores: Implications for pore conductivity and pore formation. *Phys. Rev. Let.* **54**, 1325–1328.

Kaviany, M. (1995) *Principles of Heat Transfer in Porous Media*, 2nd ed. Springer-Verlag, New York.

Kazmierczak, M. and Poulikakos, D. (1988) Melting of an ice surface in a porous medium. *J. Thermophys. Heat Transfer* **2**, 352–358.

Kazmierczak, M., Poulikakos, D., and Pop, I. (1986) Melting from a flat plate embedded in a porous medium in the presence of steady natural convection. *Numer. Heat Transfer* **10**, 571–582.

Kazmierczak, M., Poulikakos, D., and Sadowski, D. (1987) Melting of a vertical plate in porous medium controlled by forced convection of a dissimilar fluid. *Int. Comm. Heat Mass Transfer* **14**, 507–517.

Kazmierczak, M., Sadowski, D., and Poulikakos, D. (1988) Melting of a solid in a porous medium induced by free convection of a warm dissimilar fluid. *J. Heat Transfer* **110**, 520–523.

Kedem, O. and Katchalsky, A. (1958) Thermodynamic analysis of the permeability of biological membranes to nonelectrolytes. *Biochem. Biophys. Acta* **27**, 229–235.

Kessler, H. G. (1981) *Food Engineering and Dairy Technology*. Springer-Verlag, Berlin.

Kim, C. S. and Fisher, D. M. (1999) Deposition characteristics of aerosol particles in sequentially bifurcating airway model. *Aerosol Sci. Tech.* **31**, 198–220.

Kim, C. S., Brown, L. K., Lewars, G. G., and Sackner, M. A. (1983) Deposition of aerosol particles and flow resistance in mathematical and experimental airway models. *J. Appl. Physiol.* **55**, 154–163.

Kim, G. B., Hyun, J. M., and Kwak, H. S. (2001) Buoyant convection in a square cavity partially filled with a heat-generating porous medium. *Numer. Heat Transfer A* **40**, 601–618.

Kim, S. and Kim, M. C. (2002) A scale analysis of turbulent heat transfer driven by buoyancy in a porous layer with homogeneous heat sources. *Int. Commun. Heat Mass Transfer* **29**, 127–134.

Kim, S. H., Anand, N. K., and Fletcher, L. S. (1991) Free convection between series of vertical parallel plates with embedded line heat sources. *J. Heat Transfer* **113**, 108–115.

Kim, S. J. and Lee, S. W., eds. (1996) *Air Cooling Technology for Electronic Equipment*. CRC Press, Boca Raton, FL, 1–46.

Kimura, S., Bejan, A., and Pop, I. (1985) Natural convection near a cold plate facing upward in a porous medium. *J. Heat Transfer* **107**, 819–825.

Kimura, S., Kiwata, T., Okajima, A., and Pop, I. (1997) Conjugate natural convection in porous media. *Adv. Water Resources* **20**, 111–126.

Kimura, S., Okajima, A., and Kiwata, T. (2002) Natural convection heat transfer in an anisotropic porous cavity heated from side (2nd report, experiment using a Hele-Shaw Cell). *Heat Transfer—Asian Res.* **31**, 463–474.

Kimura, S., Schubert, G., and Straus, J. M. (1986) Route to chaos in porous-medium thermal convection. *J. Fluid Mech.* **166**, 305–324.

Kondepudi, D. and Prigogine, I. (1998) *Modern Thermodynamics, From Heat Engines to Dissipative Structures*. Wiley, Chichester, UK.

Kondepudi, S., Somasundaram, S., and Anand, N. K. (1988) A simplified model for the analysis of a phase-change material-based, thermal energy storage system. *Heat Recovery Syst. CHP* **8**(3), 247–254.

Koponen, A., Kandhai, D., Hellén, E., Alava, M., Hoekstra, A., Kataja, M., Niskanen, K., Sloot, P., and Timonen, J. (1998) Permeability of three-dimensional random fiber web. *Phys. Rev. Lett.* **80**, 716–719.

Kotas, T. J. (1995) *The Exergy Method of Thermal Plant Analysis.* Krieger, Melbourne, FL.

Koulich, V., Lage, J. L., Hsia, C. C. W., and Johnson, Jr., R. L. (1999) A porous medium model of alveolar gas diffusion. *J. Porous Media* **2**, 263–275.

Kraus, A. D. and Bar-Cohen, A. (1995) *Design and Analysis of Heat Sinks.* Wiley, New York.

Krokida, M. K. and Maroulis, Z. B. (1997) Effect of drying method on shrinkage and porosity. *Drying Technol.* **15**(10), 2441–2458.

Kulacki, F. A. and Freeman, R. G. (1979) A note on thermal convection in a saturated heat-generating porous layer. *J. Heat Transfer* **101**, 169–171.

Kulish, V. and Lage, J. L. (2001) Fundamentals of alveolar diffusion: A new modeling approach. *Automedica* **20**, 225–268.

Kulish, V., Lage, J. L., Hsia, C. C. W., and Johnson, Jr., R. L. (2002) Three-dimensional, unsteady simulation of alveolar respiration. *J. Biomed. Eng.* **124**, 609–616.

Kuwahara, F., Kameyama, Y., Yamashita, S., and Nakayama, A. (1998) Numerical modeling of turbulent flow in porous media using a spatially periodic array. *J. Porous Media* **1**, 47–55.

Kuznetsov, A. V. (1997) Thermal nonequilibrium, non-Darcian forced convection in a channel filled with a fluid saturated porous medium—a perturbation solution. *Appl. Sci. Res.* **57**, 119–131.

Kuznetsov, A. V. (2002) Forced convection in a heterogeneous parallel-plate channel: Use of the Brinkman–Forchheimer flow model. *Int. J. Transport Phenomena* **4**, 97–108.

Kuznetsov, A. V. and Nield, D. A. (2001) Effects of heterogeneity in forced convection in a porous medium: Triple layer or conjugate problem. *Numer. Heat Transfer A* **40**, 363–385.

Lacey, J. and Dutkiewicz, J. (1994) Bioaerosols and occupational lung disease. *J. Aerosol Sci.* **25**, 1371–1404.

Lacroix, M. (2002) Modelling of latent heat storage systems. *Thermal Energy Storage Systems and Applications*, I. Dincer and M. A. Rosen, eds., Wiley, Chichester, Chapter 7.

Lage, J. L. (1998) The fundamental theory of flow through permeable media from Darcy to turbulence. *Transport Phenomena in Porous Media*, D. B. Ingham and I. Pop, eds., Elsevier, Amsterdam, 1–30.

Lage, J. L., de Lemos, M. J. S., and Nield, D. A. (2002) Modeling turbulence in porous media. *Transport Phenomena in Porous Media*, D. B. Ingham and I. Pop, eds., vol. II. Pergamon, Oxford, 198–230.

Lai, C.-T. and Katul, G. (2000) The dynamic role of root-water uptake in coupling potential to actual transpiration. *Adv. Water Resources* **23**, 427–439.

Lai, C.-T., Katul, G., Oren, R., Ellsworth, D., and Schaäfer, K. (2000) Modelling CO_2 and water vapor turbulent flux distributions within forest canopies. *J. Geophys. Res.* **105**(D21), 26333–26351.

Lane, G. A. (1988) *Solar Heat Storage: Latent Heat Materials, Vol. 1: Background and Scientific Principles*, CRC Press, Boca Raton, FL.

Lapwood, E. R. (1948) Convection of a fluid in a porous medium. *Proc. Cambridge Philos. Soc.* **44**, 508–521.

Lartigue, B., Lorente, S., and Bourret, B. (2000) Multicellular natural convection in a high aspect ratio cavity: Experimental and numerical results. *Int. J. Heat Mass Transfer* **43**, 3157–3170.

Lauffenberger, D. A. and Linderman, J. J. (1993) *Receptors.* Oxford University Press, New York.

Läuger, P. (1970) *Physical Principles of Biological Membranes.* Gordon and Breach, New York.

Läuger, P. (1991) *Electrogenic Ion Pumps.* Sinauer, Sunderland.

Ledezma, G. A. and Bejan, A. (1997) Optimal geometric arrangement of staggered vertical plates in natural convection. *J. Heat Transfer* **119**, 700–708.

Ledezma, G., Morega, A. M., and Bejan, A. (1996) Optimal spacing between pin fins with impinging flow. *J. Heat Transfer* **118**, 570–577.

Lee, K. W. and Liu, B. Y. H. (1982) Theoretical study of aerosol filtration by fibrous filters. *Aerosol Sci. Technol.* **1**, 147–161.

Leers, R. (1957) Die Abscheidung von Schwebstoffen in Faserfiltern. *Staub.* **17**, 402–417.

Lehnert, B. E. (1990) Lung defense mechanisms against deposited dust. *Prob. Res. Care* **3**, 130–162.

Lems, S., van der Kooi, H. J., and de Swaan Arons, J. (2002) The sustainability of resource utilization. *Green Chem.* **4**(4), 308–313.

Leopold, L. B., Wolman, M. G., and Miller, J. P. (1964) *Fluvial Processes in Geomorphology.* Freeman, San Francisco.

Lewins, J. (2003) Bejan's constructal theory of equal potential distribution. *Int. J. Heat Mass Transfer* **46**, 1541–1543.

Li, L. Y. and Page, C. L. (1998) Modelling of electrochemical chloride extraction from concrete: Influence of ionic activity coefficients. *Comput. Materials Sci.* **9**, 303–308.

Li, L. Y. and Page, C. L. (2000) Finite element modeling of chloride removal from concrete by an electrochemical method. *Corrosion Sci.* **42**, 2145–2165.

Lim, J. S., Bejan, A., and Kim, J. H. (1992a) Thermodynamic optimization of phase-change energy storage using two or more materials. *J. Energy Resources Technol.* **114**, 84–90.

Lim, J. S., Bejan, A., and Kim, J. H. (1992b) Thermodynamics of energy extraction from fractured hot dry rock. *Int. J. Heat Fluid Flow* **13**, 71–77.

Lim, J. S., Fowler, A. J., and Bejan, A. (1993) Spaces filled with fluid and fibers coated with a phase-change material. *J. Heat Transfer* **115**, 1044–1050.

Liu, B. Y. H. and Ahn, K. (1987) Particle deposition on semiconductor wafers. *Aerosol Sci. Technol.* **6**, 215–224.

Liu, B. Y. H. and Rubow, K. K. (1990) Efficiency, pressure drop and figure of merit of high efficiency fibrous and membrane filter media. *Fifth World Filtration Congress*, Nice.

Loeffler, R. (1971) Collection of particles by fiber filters. *Air Pollution Control*, W. Strauss, ed., Academic, New York, 223–285.

Lorente, S. (2002) Heat losses through building walls with closed, open and deformable cavities. *Int. J. Energy Res.* **26**, 611–632.

Lorente, S. and Bejan, A. (2002) Combined "flow and strength" geometric optimization: Internal structure in a vertical insulating wall with air cavities and prescribed strength. *Int. J. Heat Mass Transfer* **45**, 3313–3320.

Lorente, S. and Lartigue, B. (2002) Maximization of heat flow through a cavity with natural convection and deformable boundaries. *Int. Commun. Heat Mass Transfer* **29**, 633–642.

Lorente, S., Javelas, R., Petit, M., and N'Guessan, K. (1994) Modélisation simplifiée des écoulements dans une cavité verticale de petites dimensions. *Rev. Gén. Thermique* **33**(388), 273–279.

Lorente, S., Ollivier, J. P., Frizon, F., and Thouvenot, P. (2002b) Porous medium decontamination by electromigration. *First International Conference On Applications of Porous Media*, Djerba, Tunisia.

Lorente, S., Petit, M., and Javelas, R. (1996) Simplified analytical model for thermal transfer in a vertical hollow brick. *Energy Buildings* **24**, 95–103.

Lorente, S., Petit, M., and Javelas, R. (1998) The effects of temperature conditions on the thermal resistance of walls made with different shape vertical hollow bricks. *Energy Buildings* **28**, 237–240.

Lorente, S., Wechsatol, W., and Bejan, A. (2002a) Tree-shaped flow structures designed by minimizing path lengths. *Int. J. Heat Mass Transfer* **45**, 3299–3312.

Lundgaard, L. and Mollerup, J. (1992) Calculation of phase diagrams of gas-hydrates. *Fluid Phase Equilibria* **76**, 141–149.

Lysne, D. (1994) Hydrate plug dissociation by pressure reduction. *International Conference on Natural Gas Hydrates, Annals New York Academy of Sciences* **715**, 514–517.

Ma, L., van der Zanden, J., van der Kooi, J., and Nieuwstadt, F. T. M. (1994) Natural convection around a horizontal circular cylinder in infinite space and within confining plates: A finite element solution. *Numer. Heat Transfer A* **25**, 441–456.

Magara, K. (1971) Permeability considerations in generation of abnormal pressures. *Soc. Pet. Eng.* **11**, 236–242.

Magyari, E., Pop, I., and Keller, B. (2001) Exact dual solutions occurring in the Darcy mixed convection flow. *Int. J. Heat Mass Transfer* **44**, 4563–4566.

Makogon, Y. F. (1981) *Hydrates of Natural Gas*. Pennwell, Tulsa, OK.

Malmberg, P. (1990) Health effects of organic dust exposure in dairy farmers. *Am. J. Industrial Med.* **17**, 7–15.

Mamou, M. (2002) Stability analysis of double-diffusive convection in porous enclosures. *Transport Phenomena in Porous Media*, D. D. Ingham and I. Pop, eds., vol. II, 113–154.

Marafie, A. and Vafai, K. (2001) Analysis of non-Darcian effects on temperature differentials in porous media. *Int. J. Heat Mass Transfer* **44**, 4401–4411.

Marcondes, F., de Medeiros, J. M., and Gurgel, J. M. (2001) Numerical analysis of natural convection in cavities with variable porosity. *Numer. Heat Transfer A* **40**, 403–420.

Marsters, G. F. (1975) Natural convection heat transfer from a horizontal cylinder in the presence of nearby walls. *Canadian J. Chem. Eng.* **35**, 144–149.

Masliyah, J. H. (1994) *Electrokinetic Transport Phenomena*. AOSTRA Technical Publication Series 12, Alberta, Canada.

Masuoka, T. and Takatsu, Y. (1996) Turbulence model for flow through porous media. *Int. J. Heat Mass Transfer* **39**, 2803–2809.

Masuoka, T. and Takatsu, Y. (2002) Turbulence characteristics in porous media. *Transport Phenomena in Porous Media*, D. B. Ingham and I. Pop, eds., vol. II. Pergamon, Oxford, 231–256.

Mat, M. D. and Ilegbusi, O. J. (2002) Application of a hybrid model of mushy zone to macro segregation in alloy solidification. *Int. J. Heat Mass Transfer* **45**, 279–289.

McLaughlin, C. P. and Magee, T. R. A. (1999) The effects of air temperature, sphere diameter, and puffing with CO_2 on the drying of potato spheres. *Drying Technol.* **17**(1–2), 119–136.

Mench, M. M., Wang, C.-Y., and Thynell, S. T. (2001) An introduction to fuel cells and related transport phenomena. *Int. J. Transport Phenomena* **3**(3), 151–176.

Mendes, N., Philippi, P. C., and Lamberts, R. (2002) A new mathematical method to solve highly coupled equations of heat and mass transfer in porous media. *Int. J. Heat Mass Transfer* **45**, 509–518.

Mercier, J.-F., Weisman, C., Firdaouss, M., and Le Quéré, P. (2002) Heat transfer associated to natural convection flow in a partly porous cavity. *J. Heat Transfer* **124**, 130–143.

Mereu, S., Sciubba, E., and Bejan, A. (1993) The optimal cooling of a stack of heat generating boards with fixed pressure drop, flow rate or pumping power. *Int. J. Heat Mass Transfer* **36**, 3677–3686.

Michel, C. C. (1997) Starling: The formulation of his hypothesis of microvascular fluid exchange and its significance after 100 years. *Exp. Physiol.* **82**, 1–30.

Miele, A. (1965) *Theory of Optimum Aerodynamic Shapes*. Academic, New York.

Miguel, A. F. (1998a) Airflow through porous screens: From theory to practical considerations. *Energy Buildings* **28**, 63–69.

Miguel, A. F. (1998b) Transport phenomena through openings and screens. PhD Thesis, Wagenigen Universiteit en Researchcentrum, The Netherlands.

Miguel, A. F. (2000) Contribution to flow characterization through porous media. *Int. J. Heat Mass Transfer* **43**, 2267–2272.

Miguel, A. F. (2003a) Effect of air humidity on the evolution of permeability and performance of a fibrous filter during loading with hygroscopic and non-hygroscopic particles. *J. Aerosol Sci.* **34**, 783–799.

Miguel, A. F. (2003b) Filtration and filters: Classical approaches and new developments. *Emerging Technologies and Techniques in Porous Media*, D. B. Ingham et al., eds., Kluwer Academic, Dordrecht, The Netherlands.

Miguel, A. F. and Reis, A. H. (2004) Transient forced convection in an isothermal fluid-saturated porous medium layer: Effective permeability and boundary layer thickness. *J. Porous Media* (in press).

Miguel, A. F. and Silva, A. M. (2001) Experimental study of mass loading behaviour of fibrous filters. *J. Aerosol Sci.* **32**(S1), 851–852.

Millington, R. J. and Quirk, J. P. (1961) Permeability of porous solids. *Trans. Faraday Soc.* **57**, 1200–1207.

Mohamad, A. A. and Bennacer, R. (2002) Double diffusion, natural convection in an enclosure filled with saturated porous medium subjected to cross gradients; stably stratified fluid. *Int. J. Heat Mass Transfer* **45**, 3725–3740.

Moise, A. and Tudose, R. Z. (2000) A study on the drying of the packed beds circulated by the thermal agent. *Proceedings of the Twelfth International Drying*

Symposium, K. P. J. A. M. Kerkhof, W. J. Coumans, and G. D. Mooweer, eds., Paper No: 34, August 28–31.

Moran, M. J. (1989) *Availability Analysis: A Guide to Efficient Energy Use*, 2nd ed. ASME, New York.

Moran, M. J. and Shapiro, H. N. (1995) *Fundamentals of Engineering Thermodynamics*, 3rd ed. Wiley, New York.

Morega, A. M. and Bejan, A. (1994) Heatline visualization of convection in porous media. *Int. J. Heat Fluid Flow* **15**, 42–47.

Morega, A. M., Bejan, A., and Lee, S. W. (1995) Free stream cooling of a stack of parallel plates. *Int. J. Heat Mass Transfer* **38**, 519–531.

Mullins, L. J. and Noda, K. (1963) The influence of sodium-free solutions on the membrane potential of frog muscle fibers. *J. Gen. Physiol.* **43**, 117–132.

Murray, C. D. (1926) The physiological principle of minimal work, in the vascular system, and the cost of blood-volume. *Proc. Acad. Nat. Sci.* **12**, 207–214.

Muskat, M. (1937) *The Flow of Homogeneous Fluids Through Porous Media*. McGraw-Hill, New York.

Nagataki, S. (1995) Concrete technology in Japan. *Proceedings of CONSEC 95*, Chapman and Hall, London.

Nagataki, S., Otsuki, N., Wee, T. H., and Nakashita, K. (1993) Condensation of chloride ion in hardened cement matrix materials and on embedded steel bars. *ACI Materials J.* **90**, 323–332.

Nakayama, A. (1995) *PC-Aided Numerical Heat Transfer and Convective Flow*. CRC Press, Tokyo.

Nakayama, A. and Kuwahara, F. (1999) A macroscopic turbulence model for flow in a porous medium. *J. Fluids Eng.* **121**, 427–433.

Nakayama, A. and Kuwahara, F. (2000) Diffusion-controlled catalytic reaction on a monolithic catalytic converter. *J. Porous Media* **3**, 115–122.

Nakayama, A., Kuwahara, F., Sugiyama, M., and Xu, G. (2001) A two-energy equation model for conduction and convection in porous media. *Int. J. Heat Mass Transfer* **44**, 4375–4379.

Narasimhan, A. and Lage, J. L. (2001) Forced convection of a fluid with temperature-dependent viscosity flowing through a porous medium channel. *Numer. Heat Transfer A* **40**, 801–820.

Narasimhan, A. and Lage, J. L. (2002) Inlet temperature influence on the departure from Darcy flow of a fluid with variable viscosity. *Int. J. Heat Mass Transfer* **45**, 2419–2422.

Nazaroff, W. and Cass, G. (1991) Protecting museum collections from soiling due to the deposition of airborne particles. *Atmos. Envir.* **25A**, 841–852.

Nelson, Jr. R. A. and Bejan, A. (1998) Constructal optimization of internal flow geometry in convection. *J. Heat Transfer* **120**, 357–364.

Nepf, H. M. (1999) Drag, turbulence, and diffusion in flow through emergent vegetation. *Water Resources Res.* **35**, 479–489.

Newman, J. (1999) *Electrochemical Systems*. Prentice-Hall, New York.

Nield, D. A. (1968) Onset of thermohaline convection in a porous medium. *Water Resources Res.* **4**, 553–560.

Nield, D. A. (1991) Estimation of the stagnant thermal conductivity of saturated porous media. *Int. J. Heat Mass Transfer* **34**, 1575–1576.

Nield, D. A. (1997) Comments on "Turbulence model for flow through porous media." *Int. J. Heat Mass Transfer* **40**, 2499.

Nield, D. A. (2000) Resolution of a paradox involving viscous dissipation and nonlinear drag in a porous medium. *Transport in Porous Media* **41**, 349–357.

Nield, D. A. and Bejan, A. (1999) *Convection in Porous Media*, 2nd ed. Springer Verlag, New York.

Nield, D. A. and Joseph, D. D. (1985) Effects of quadratic drag on convection in a saturated porous medium. *Phys. Fluids* **28**, 995–997.

Nield, D. A. and Kuznetsov, A. V. (2001) The interaction of thermal nonequilibrium and heterogeneous conductivity effects in forced convection in layered porous channels. *Int. J. Heat Mass Transfer* **44**, 4369–4373.

Nield, D. A., Junqueira, S. L. M., and Lage, J. L. (1996) Forced convection in a fluid-saturated porous-medium channel with isothermal or isoflux boundaries. *J. Fluid Mech.* **322**, 201–214.

Nield, D. A., Kuznetsov, A. V., and Xiong, M. (2002) Effect of local thermal non-equilibrium on thermally developing forced convection in a porous medium. *Int. J. Heat Mass Transfer* **45**, 4949–4955.

Nilsson, L. O., Sandberg, P., Poulsen, E., Tang, L., Andersen, A., and Frederiksen, J. M. (1996) *HETEK, Chloride Penetration into Concrete, State of the Art, A System for Estimation of Chloride Ingress into Concrete, Theoretical Background.* Danish Road Directorate, Copenhagen.

Nishizawa, M., Menon, V. P., and Martin, C. R. (1995) Metal nanotubule membranes with electrochemically switchable ion-transport selectivity. *Science* **268**, 700–702.

Nobel Lectures (1973) *Nobel Lectures in Physiology Medicine 1963–1970.* Elsevier, Amsterdam.

Norton, B. (1992) *Solar Energy Thermal Technology.* Springer-Verlag, London.

Nottalle, L. (1993) *Fractal Space Time and Microphysics.* World Scientific, Singapore.

Novick, V. J., Klassen, J. F., Monson, P. R., and Long, T. A. (1993) Predicting mass loadings as a function of pressure difference across prefilter/HEPA in filter systems. *Conference Proceedings of the 22nd DOE/NRC Air Cleaning* (CONF-9020823), 554–572.

Nutting, P. G. (1930) Physical analysis of oil sands. *Bull. Am. Assoc. Petr. Geol.* **14**, 1337–1349.

Oberbeck, A. (1879) Ueber die Wärmeleitung der Flüssigkeiten bei Berücksichtigung der Strömungen infolge von Temperaturdifferenzen. *Ann. Phys. Chem.* **7**, 271–292.

Ogston, A. G., Preston, B. N., and Wells, J. D. (1973) On the transport of compact particles through solutions of chain-polymers. *Proc. R. Soc. Lond.* **A333**, 297–316.

Oldham, M. J., Phalen, R. F., and Heistracher, T. (2000) Computational fluid dynamics and experimental results for particle deposition in an airway model. *Aerosol Sci. Technol.* **32**, 61–71.

Oosthuizen, P. H. (1988) The effects of free convection on steady-state freezing in a porous medium-filled cavity. *ASME HTD* **96**(1), 321–327.

Ordonez, J. C. and Bejan, A. (2003) System-level optimization of the sizes of organs for heat and fluid flow systems. *Int. J. Thermal Sci.* **42**, 335–342.

Ordonez, J. C., Bejan, A., and Cherry, R. S. (2003) Designed porous media: Optimally nonuniform flow structures connecting one point with more points. *Int. J. Therm. Sci.* **42**, 857–870.

Organ, A. J. (1992) *Thermodynamics and Gas Dynamics of the Stirling Cycle Machine.* Cambridge University Press, Cambridge, UK.

Organ, A. J. (1997) *The Regenerator and the Stirling Engine.* Professional Engineering Publishing, Bury St. Edmunds, UK.

Ozkaynak, H. and Thurston, G. D. (1987) Associations between 1980 U.S. mortality rates and alternative measures of airborne particle concentration. *Risk Analysis* **7**, 449–461.

Padet, J. (1991) *Fluides en Écoulement. Méthodes et Modèles.* Masson, Paris.

Paine, P. L. and Scherr, P. (1975) Drag coefficients for the movement of rigid spheres through liquid-filled cylindrical pores. *Biophys. J.* **15**, 1087–1091.

Pannetier, R. (1982) *Vade-mecum du Technicien Nucléaire.* Ed. SCF du Bastet, Massy.

Papathanasiou, T. D. (2001) The hydraulic permeability of periodic arrays of cylinders of varying size. *J. Porous Media* **4**, 323–336.

Pavlík, V. (2000) Water extraction of chloride, hydroxide and other ions from hardened cement pastes. *Cement Concrete Res.* **30**, 895–906.

Payet, S., Boulaud, D., Madelaine, G., and Renoux, A. (1992) Penetration and pressure drop of a HEPA filter during loading with submicron liquid particles. *J. Aerosol Sci.* **23**, 723–735.

Pedras, M. H. J. and de Lemos, M. J. S. (2001a) Macroscopic turbulence modeling for incompressible flow through undeformable porous media. *Int. J. Heat Mass Transfer* **44**, 1081–1093.

Pedras, M. H. J. and de Lemos, M. J. S. (2001b) Simulation of turbulent flow in porous media using a spatially periodic array and a low Re two-equation closure. *Numer. Heat Transfer A* **39**, 35–59.

Pence, D. V. (2002) Reduced pumping power and wall temperature in microchannel heat sinks with fractal-like branching channel networks. *Microscale Thermophys. Eng.* **6**, 319–330.

Pendse, H. and Tien, C. (1982) General correlation of the initial collection efficiency of granular filter bed. *AIChE J.* **28**, 677–686.

Petrescu, S. (1994) Comments on the optimal spacing of parallel plates cooled by forced convection. *Int. J. Heat Mass Transfer* **37**, 1283.

Phanikumar, M. S. and Mahajan, R. L. (2002) Non-Darcy natural convection in high porosity metal foams. *Int. J. Heat Mass Transfer* **45**, 3781–3793.

Pich, J. (1966) Theory of aerosol filtration by fibrous and membrane filters. *Aerosol Science*, C. N. Davies, ed., 223–285, Academic, London.

Plumb, O. A. and Huenefeld, J. S. (1981) Non-Darcy natural convection from heated surfaces in saturated porous media. *Int. J. Heat Mass Transfer* **24**, 765–768.

Plumb, O. A., Gu, L., and Webb, S. W. (1999) Drying of porous materials at low moisture content. *Drying Technol.* **17**(10), 1999–2011.

Pop, I. and Ingham, D. B. (2001) *Convective Heat Transfer: Mathematical and Computational Modeling of Viscous Fluids and Porous Media.* Pergamon, Oxford.

Pop, I., Merkin, J. H., and Ingham, D. B. (2002) Chemically driven convection in porous media. *Transport Phenomena in Porous Media*, D. B. Ingham and I. Pop, eds., vol. II. Pergamon, Oxford, 341–364.

Poulikakos, D. and Bejan, A. (1983a) Natural convection in vertically and horizontally layered porous media heated from the side. *Int. J. Heat Mass Transfer* **26**, 1805–1814.

Poulikakos, D. and Bejan, A. (1983b) Numerical study of transient high Rayleigh number convection in an attic-shaped porous layer. *J. Heat Transfer* **105**, 476–484.

Prasad, V., Kulacki, F. A., and Keyhani, M. (1985) Natural convection in porous media. *J. Fluid Mech.* **150**, 89–119.

Prasetyo, I., Do H. D., and Do, D. D. (2002) Surface diffusion of strong adsorbing vapours on porous carbon. *Chem. Eng. Sci.* **57**, 133–141.

Quintard, M. and Whitaker, S. (1995) Aerosol filtration: An analysis using the method of volume averaging. *J. Aerosol Sci.* **26**, 1227–1255.

Radcenco, V. (1994) *Generalized Thermodynamics*. Editura Tehnica, Bucharest.

Ramachadran, V. S., Seeley, R. C., and Polomark, G. M. (1984) Free and combined chloride in hydrating cement and cement composites. *Materials and Structures* **17**, 285–289.

Rao, N. and Faghri, M. (1988) Computer modeling of aerosol filtration by fibrous filters. *Aerosol Sci. Technol.* **8**, 133–156.

Rathish Kumar, B. V., Singh, P., and Bansod, V. J. (2002) Effect of thermal stratification on double-diffusive natural convection in a vertical porous enclosure. *Numer. Heat Transfer A* **41**, 1–28.

Rebinder, P. A. (1972) *Physical-Chemical Principles of Food Production*. Pishche-promizdizdat, Moscow.

Rees, D. A. S. (2002) The onset of Darcy–Brinkman convection in a porous layer: An asymptotic analysis. *Int. J. Heat Mass Transfer* **45**, 2213–2220.

Rees, D. A. S. and Hossain, M. A. (2001) The effect of inertia on free convective plumes in porous media. *Int. Comm. Heat Mass Transfer* **28**, 1137–1142.

Rees, D. A. S. and Pop, I. (2000) Vertical free convective-layer flow in a porous medium using a two-temperature model. *J. Porous Media* **3**, 31–43.

Reid, R. C., Prausnitz, J. M., and Sherwood, T. K. (1977) *The Properties of Gases and Liquids*, 3rd ed. McGraw-Hill, New York.

Reis, A. H., Miguel, A. F., and Aydin, M. (2004) Constructal theory of the flow architectures of the lungs. *Medical Physics*, in press.

Reis, A. H., Silva, O., and Rosa, R. (1994) Heat and mass transfer in porous materials. *IUPAC Symposium Characterisation of Porous Solids II*, J. Rouquerol, F. Rodriguez-Reinoso, K. Sing, and K. Unger, eds., Elsevier Science, Marseille, France, 207–209.

Ren, X. and Gottesfeld, S. (2001) Electro-osmotic drag of water in poly (perfluorosulfonic acid) membranes. *J. Electrochem. Soc.* **148**(1), A87-A93.

Renkin, E. M. (1954) Filtration, diffusion and molecular sieving through porous cellulose membranes. *J. Gen. Physiol.* **38**, 225–243.

Renkin, E. M. (1988) Transport pathways and processes. *Endothelial Cell Biology*, N. Simionescu and M. Simionescu, eds., Plenum, New York.

Reuss, J. D., Fowler, A. J., Kim, Y. K., and Lewis, A. (2002) Manufacture of thick cross-section composites using a pre-catalyzed fabric technique. *J. Composite Materials* **36**, 1367–1379.

Richet, C. (1992) Etude de la migration des radio elements dans les liants hydrauliques—Influence du vieillissement sur les mécanismes et la cinétique des transferts. PhD dissertation, Université de Paris XI Orsay.

Roadifer, R. D., Godbole, S. P., and Kamath, V. A. (1987) Thermal model for establishing guidelines for drilling in the Arctic in the presence of hydrates. *SPE California Regional Meeting*, Ventura, CA, SPE 16361.

Rocha, L. A. O. and Bejan, A. (2001) Geometric optimization of periodic flow and heat transfer in a volume cooled by parallel tubes. *J. Heat Transfer* **123**, 233–239.

Rocha, L. A. O., Lorente, S., and Bejan, A. (2002) Constructal design for cooling a disc-shaped area by conduction. *Int. J. Heat Mass Transfer* **45**, 1643–1652.

Rocha, L. A. O., Neagu, M., Bejan, A., and Cherry, R. S. (2001) Convection with phase change during gas formation from methane hydrates via depressurization of porous layers. *J. Porous Media* **4**, 283–295.

Rodriguez-Iturbe, I. and Rinaldo, A. (1997) *Fractal River Basins.* Cambridge University Press, Cambridge, UK.

Rohsenow, W. M. and Choi, H. Y. (1961) *Heat, Mass and Momentum Transfer.* Prentice-Hall, Englewood Cliffs, NJ.

Rohsenow, W. M. and Hartnett, J. P. (1973) *Handbook of Heat Transfer.* McGraw-Hill, New York.

Rosen, M. A. and Dincer, I. (2001) Exergy as the confluence of energy, environmental and sustainable development. *Exergy Int. J.* **1**(1), 3–13.

Rosen, M. A. and Dincer, I. (2003) Exergy-cost-energy-mass analysis of thermal systems and processes. *Energy Conversion Management* **44**(10), 1633–1651.

Rouquerol, J., Avnir, D., Everett, D. H., Fairbridge, C., Haynes, M., Pernicone, N., Ramsay, J. D. F., Sing, K. S. W., and Unger, K. K. (1994) Guidelines for the characterization of porous solids. *Studies Surface Sci. Catalysis* **87**, 1–9.

Rubey, W. W. and Hubbert, M. K. (1959) Role of fluid pressure in mechanics of overthrust faulting, II. *Bull. GSA* **70**(2), 167–206.

Rubinstein, I. (1990) *Electro-Diffusion of Ions.* SIAM, Philadelphia.

Rusakov, S. D. A. and Kullmann, D. M. (1998) Geometric and viscous components of the tortuosity of the extracellular space in the brain. *Proc. Nat. Acad. Sci.* **95**, 8975–8980.

Ruthven, D. (1984) *Principles of Adsorption and Adsorption Processes.* Wiley, New York, 135–137.

Saborio-Aceves, S., Nakamura, H., and Reistad, G. M. (1994) Optimum efficiencies and phase change temperature in latent heat storage systems. *Energy Resources Technol.* **116**, 79–86.

Sackmann, E. (1995) Biological membranes architecture and function. *Structure and Dynamics of Membranes*, R. Lipowsky and E. Sackmann, eds., Elsevier, Amsterdam.

Sadeghipour, M. S. and Kazemzadeh, S. H. (1992) Transient natural convection from a horizontal cylinder confined between vertical walls: A finite element solution. *Int. J. Numer. Methods Eng.* **34**, 621–635.

Sadeghipour, M. S. and Razi, Y. P. (2001) Natural convection from a confined horizontal cylinder: The optimum distance between the confining walls. *Int. J. Heat Mass Transfer* **44**, 367–374.

Saetta, A. V., Schrefler, B. A., and Vitaliani, R. V. (1995) 2-D Model for carbonation and moisture/heat flux in porous media. *Cement Concrete Res.* **25**, 1703–1712.

Saghir, M. Z., Comi, P., and Mehrvar, M. (2002) Effects of interaction between Rayleigh and Marangoni convection in superposed fluid and porous layers. *Int. J. Therm. Sci.* **41**, 207–215.

Sahimi, M. (1995) *Flow and Transport in Porous Media and Fractured Rock: From Classical Methods to Modern Approaches.* VCH, Weinheim.

Sahin, A. Z. and Dincer, I. (2000) Analytical modeling of transient phase-change problems. *Int. J. Energy Res.* **24**, 1029–1039.

Samson, E. and Marchand, J. (1999) Numerical solution of the extended Nernst–Planck model. *J. Colloid Interface Sci.* **215**, 1–8.

Sangani, A. S. and Acrivos, A. (1982) Slow flow past periodic arrays of cylinders with application to heat transfer. *Int. J. Multiphase Flow* **8**, 193–206.

Scheidegger, A. E. (1974) *The Physics of Flow Through Porous Media.* University of Toronto Press, Toronto.

Schmidt, E. W. (1978) Filtration of aerosols in granular bed. *J. APCA* **28**, 143–157.

Schmidt-Nielsen, K. (1984) *Scaling (Why is Animal Size So Important?).* Cambridge University Press, Cambridge, UK.

Schumm, S. A., Mosley, M. P., and Weaver, W. E. (1987) *Experimental Fluvial Geomorphology.* Wiley, New York.

Schwartz, J. and Dockery, D. W. (1992) Increased mortality in Philadelphia associated with daily air pollution concentration. *Am. Rev. Respir. Dis.* **145**, 600–604.

Sciubba, E. (1999a) Optimisation of turbomachinery components by constrained minimisation of the local entropy production rate. *Thermodynamic Optimization of Complex Energy Systems*, A. Bejan, and E. Mamut, eds., Kluwer Academic, Dordrecht, The Netherlands.

Sciubba, E. (1999b) Allocation of finite energetic resources via an exergetic costing method. *Thermodynamic Optimization of Complex Energy Systems*, A. Bejan and E. Mamut, eds., Kluwer Academic, Dordrecht, The Netherlands, 151–162.

Sciubba, E. and Melli, R. (1998) *Artificial Intelligence in Thermal Systems Design: Concepts and Applications.* Nova Science, New York.

Selim, M. S. and Sloan, E. D. (1989) Heat and mass transfer during dissociation of hydrates in porous media. *AIChE J.* **35**(16), 1049–1052.

Sellier, A., Bary, B., Aït-Moktar K., Delmi M., Miragliotta R., and Rougeau P. (2000) Etude de la mise en œuvre et de la durabilité d'une barriére ouvragée cimentaire pour le stockage de déchets "B" en formation géologique profonde. Influence de la phase aérée. C RP 0CIB 01–001/A.

Sen, P. N. (1989) Unified model of conductivity and membrane potential of porous media. *Phys. Rev. B* **39**, 9508–9517.

Shackelford, C. D. (1991) Diffusion in saturated soils. I: Background. *J. Geotechnical Eng.* **117**, 467–484.

Shapiro, M. and Brenner, H. (1990) Dispersion/reaction model of aerosol collection by porous filters. *J. Aerosol Sci.* **21**, 97–125.

Shapiro, M., Kettner, I., and Brenner, H. (1991) Transport mechanics and collection of submicrometer particles in fibrous filters. *J. Aerosol Sci.* **6**, 707–722.

Sharma, S. K., Sethi, P. B. S., and Chopra, S. (1992) Kinetic and thermophysical studies of acetamide sodium bromide eutectic for low temperature storage applications. *Energy Conversion Management* **33**(2), 145–150.

Shenoy, A. V. (1993) Darcy–Forchheimer natural, forced and mixed convection heat transfer in non-Newtonian power-law fluid-saturated porous media. *Transport Porous Media* **11**, 219–241.

Shi, Y. and Wang, C. (1986) Pore pressure generation in sedimentary basins: Overloading versus aquathermal. *J. Geophys. Res.* **91**, 2153–2162.

Shimizu, M. and Fujita, K. (1985) Actual efficiencies of thermally stratified thermal storage tanks. *IEA Heat Pump Center Newslett.* **3**(1/2), 20–25.

Shiner, J. S., ed. (1996) *Entropy and Entropy Generation*. Kluwer Academic, Dordrecht, The Netherlands.

Shu, J.-J. and Pop, I. (1997) Inclined wall plumes in porous media. *Fluid Dynamics Res.* **21**, 303–317.

Sieniutycz, S. and Salamon, P., eds. (1990) *Finite-Time Thermodynamics and Thermoeconomics*. Taylor and Francis, New York.

Simpkins, P. G. and Blythe, P. A. (1980) Convection in a porous layer. *Int. J. Heat Mass Transfer* **23**, 881–887.

Singh, A. and Singh, B. D. (1999) Methane gas: An unconventional energy resource. *Current Science* **76**(12), 1546–1551.

Sinkar, K. K. (1975) Transport in packed beds at intermediate Reynolds number. *Ind. Eng. Chem. Funda.* **14**, 73–81.

Siqueira, M., Lai, C.-T. and Katul, G. (2000) Estimating scalar sources, sinks and fluxes in a forest canopy using Lagrangian, Eulerian, and hybrid inverse models. *J. Geophys. Res.* **105**(D24), 29475–29488.

Skalny, J. P. (1987) *Concrete Durability: A Multibillion-Dollar Opportunity*. National Research Council, USA.

Sloan, E. D. (1990) *Clathrate Hydrates of Natural Gases*. Dekker, New York.

Smith, S. J. and Bernstein, J. A. (1996) Therapeutic uses of lung aerosols. *Inhalation Aerosol: Physical and Biological Basis for Therapy*, A. J. Hickey, ed., Dekker, New York, 233–269.

Sparrow, E. M. and Pfeil, D. (1984) Enhancement of natural convection heat transfer from horizontal cylinder due to vertical shrouding surfaces. *J. Heat Transfer* **106**, 124–130.

Spencer, H. (1985) *Pathology of the Lung*. Pergamon, Oxford.

Springer, T. E., Rockward, T., Zawodzinski, T. A., and Gottesfeld, S. (2001) Model for polymer electrolyte fuel cell operation on reformate feed. *J. Electrochem. Soc.* **148**(1), A11–A23.

Spurny, K. (1997) *Faserige Mineralstäube*. Springer-Verlag, Heidelberg.

Stanescu, G., Fowler, A. J., and Bejan, A. (1996) The optimal spacing of cylinders in free-stream cross-flow forced convection. *Int. J. Heat Mass Transfer* **39**, 311–317.

Staverman, A. J. (1951) The theory of measurement of osmotic pressure. *Rec. Trav. Chim.* **70**, 344–349.

Stecco, S. S. and Moran, M. J., eds. (1990) *A Future for Energy*. Pergamon, Oxford, UK.

Stecco, S. S. and Moran, M. J., eds. (1992) *Energy for the Transition Age*. Nova Science, New York.

Strumillo, C. and Kudra, T. (1986) *Drying: Principles, Applications and Design*. Gordon and Breach Science, New York.

Subrahmanyam, C., Reddy, S. I., Thakur, N. K., Gangadhara Rao, T., and Sain, K. (1998) Gas-hydrates—a synoptic view. *J. Geolog. Soc. India* **52**, 497–512.

Suneja, S. K. and Lee, C. H. (1974) Aerosol filtration by fibrous filters at intermediate Reynolds numbers ($\leq$100). *Atmos. Envir.* **8**, 1081–1094.

Tan, A. Y., Prasher, B. D., and Guin, J. A. (1975) Mass transfer in non-uniform packing. *AIChE J.* **21**, 396–701.

Tang, L. (1996) Electrically accelerated methods for determining chloride diffusivity in concrete—current developments. *Mag. Concrete Res.* **48**, 173–179.

Tang, L. and Nilsson, L. O. (1993) Chloride binding capacity and binding isotherms of OPC pastes and mortars. *Cement Concrete Res.* **23**, 247–253.

Tardos, G. I. and Pfeffer, R. (1980) Interceptional and gravitational deposition of inertialess particles on a single sphere in a granular bed. *AIChE J.* **26**, 698–701.

Thomas, S. and Zalbowitz, M. (1999) *Fuel Cells—Green Power.* Los Alamos National Laboratory, NM.

Thompson, A. H., Katz, A. J., and Krohn, C. E. (1987) The microgeometry and transport properties of sedimentary rock. *Adv. Phys.* **36**, 625–694.

Thompson, D'A. W. (1942) *On Growth and Form.* Cambridge University Press, Cambridge, UK.

Tien, C. (1989) *Fundamentals of Granular Filtration of Aerosols and Hydrosols.* Butterworths, Boston.

Tomlinson, J. J. and Kannberg, L. D. (1990) Thermal energy storage. *Mech. Eng.* **112**, 68–72.

Tondeur, D., Luo, L., and D'Ortona, U. (2000) Optimisation des transferts et des materiaux par l'approche constructale. *Entropie* **30**, 32–37.

Torab, H. and Beasley, D. E. (1987) Optimization of a packed bed thermal energy storage unit. *J. Solar Energy Eng.* **109**, 170–175.

Truc, O. (2000) Prediction of chloride penetration into saturated concrete, multi-species approach. PhD Thesis, Chalmers University of Technology, Göteborg and Institut National des Sciences Appliquées, Toulouse.

Truc, O., Ollivier, J. P., and Nilsson, L. O. (2000) Numerical simulation of multispecies transport through saturated concrete during a migration test. MsDiff Code. *Cement Concrete Res.* **10**, 1581–1592.

Tsami, E. and Katsioti, M. (2000) Drying kinetics for some fruits: Predicting of porosity and color during drying. *Drying Technol.* **18**(7), 1559–1581.

Tsong, T. Y. (2002) Na, K-ATPase as a Brownian motor: Electric field-induced conformational fluctuation leads to uphill pumping of cation in absence of ATP. *J. Biol. Phys.* **28**, 309–325.

Tsong, T. Y. and Chang, C.-H. (2003) Ion pump as Brownian motor: Theory of electroconformational coupling and proof of ratchet mechanism for Na, K-ATPase action. *Phys. A* **321**, 124–138.

Turner, J. C. R. (1975) Diffusion coefficients near consolute points. *Chem. Eng. Sci.* **30**, 1304–1305.

Tyvand, P. A. (2002) Onset of Rayleigh–Bénard convection in porous bodies. *Transport Phenomena in Porous Media*, D. B. Ingham and I. Pop, eds., vol. II, 82–112.

Vafai, K., ed. (2000) *Handbook of Porous Media.* Dekker, New York.

Vafai, K. and Tien, C. L. (1981) Boundary and inertia effects on flow and heat transfer in porous media. *Int. J. Heat Mass Transfer* **24**, 195–203.

Valero, A. and Tsatsaronis, G., eds. (1992) *ECOS '92, Proceedings of the International Symposium on Efficiency, Costs, Optimization and Simulation of Energy Systems*, Zaragoza, Spain. ASME, New York.

Van Brakel, J. and Heertjes, P. M. (1974) Analysis of diffusion in macroporous media in terms of a porosity, a tortuosity and a constrictivity factor. *Int. J. Heat Mass Transfer* **17**, 1093–1103.

Vasile, C., Lorente, S., and Perrin, B. (1998) Study of convective phenomena inside cavities coupled with heat and mass transfers through porous media— application to vertical hollow bricks—a first approach. *Energy Buildings* **28**, 229–235.

Vigo, T. L. and Bruno, J. S. (1987) Temperature-adaptable textiles containing durably bound polyethylene glycols, *Textile Res. J.* **57**, 427–429.

Vigo, T. L. and Bruno, J. S. (1989) Improvement of various properties of fiber surfaces containing crosslinked polyethylene glycols. *J. Appl. Polym. Sci.* **37**, 371–379.

Vigo, T. L. and Frost, C. E. (1982) Temperature-adaptable hollow fibers containing inorganic and organic phase change materials. *Thermal Analysis, Proceedings of the Seventh International Conference on Thermal Analysis*, vol. 2, Wiley, New York, 1286–1295.

Vigo, T. L. and Frost, C. E. (1983) Temperature-adaptable hollow fibers containing polyethylene glycols. *J. Coated Fabrics* **12**, 243–254.

Vogel, S. (1988) *Life's Devices*. Princeton University Press, Princeton, NJ.

Voller, V. R. (1997) An overview of numerical methods for solving phase change problems. *Advances in Numerical Heat Transfer*, W. J. Minkowycz and E. M. Sparrow, eds., Taylor and Francis, New York, Chapter 9.

Ward, J. C. (1964) Turbulent flow in porous media. *ASCE J. Hydraul. Div.* **90**(HY5), 1–12.

Weber, J. E. (1975) The boundary layer regime for convection in a vertical porous layer. *Int. J. Heat Mass Transfer* **18**, 569–573.

Wechsatol, W., Lorente, S., and Bejan, A. (2001) Tree-shaped insulated designs for the uniform distribution of hot water over an area. *Int. J. Heat Mass Transfer* **44**, 3111–3123.

Wechsatol, W., Lorente, S., and Bejan, A. (2002a) Development of tree-shaped flows by adding new users to existing networks of hot water pipes. *Int. J. Heat Mass Transfer* **45**, 723–733.

Wechsatol, W., Lorente, S., and Bejan, A. (2002b) Optimal tree-shaped networks for fluid flow in a disc-shaped body. *Int. J. Heat Mass Transfer* **45**, 4911–4924.

Wechsatol, W., Lorente, S., and Bejan, A. (2003) Dendritic convection on a disc. *Int. J. Heat Mass Transfer* **46**, 4381–4391.

Weibel, E. R. (1963) *Morphometry of the Human Lung*. Academic, New York.

Weinbaum, S. and Jiji, L. M. (1985) A new simplified bioheat equation for the effect of blood flow on the local average tissue temperature. *J. Biomech. Eng.* **107**, 131–139.

Whitby, K. T. and Severdrup, G. M. (1980) California aerosols: Their physical and chemical characteristics. *Ad. Envir. Sci. Technol.* **10**, 477–485.

Wiens, B. (2000) *The Future of Fuel Cells*. June 27. Available at http://www.benwiens.com/energy4.html#energy1.9.

Willoughby, E. C. and Edwards, R. N. (1997) On the resource evaluation of marine gas-hydrate deposits using seafloor compliance methods. *Geophys. J. Int.* **131**, 751–766.

Wilson, R. and Spengler, J. D. (1996) *Particles in Our Air: Concentration and Health Effects*. Harvard University Press, Boston.

Win, J. A. M. and Rik, J. J. (1999) Thermal reservoir simulation model of production from naturally occurring gas hydrate accumulations. *1999 SPE Annual Technical Conference*, Houston, TX, SPE 56550.

Wong, P.-Z. (1988) Statistical physics of sedimentary rock. *Phys. Today* **41**(12), 2–10.

Woods, L. C. (1986) *The Thermodynamics of Fluid Systems.* Oxford University Press, Oxford, UK.

Wowra, O. and Setzer, M. J. (2000) About the interaction of chloride and hardened cement paste. *Second International RILEM Workshop*, Paris.

Wylie, D. (1990) Evaluating and selecting thermal energy storage, *Energy Eng.* **87**(6), 6–17.

Xia, Z.-Z., Li, Z.-X., and Guo, Z.-Y. (2002) Heat conduction optimization: High conductivity constructs based on the principle of biological evolution. *Twelfth International Heat Transfer Conference*, Grenoble, France, August 18–23.

Yih, C. S. (1969) *Fluid Mechanics.* McGraw-Hill, New York.

Yousif, M. H., Li, P. M., Selim, M. S., and Sloan, E. D. (1990) Depressurization of natural gas hydrates in Berea sandstone cores. *J. Inclusion Phenomena Molecular Recognition Chem.* **8**, 71–88.

Yu, B. and Cheng, P. (2002) A fractal permeability model for bi-dispersed porous media. *Int. J. Heat Mass Transfer* **45**, 2983–2993.

Yuan, J., Rokni, M., and Sundén, B. (2002) Analysis of fluid flow and heat transfer in proton exchange membrane fuel cell ducts by an extended Darcy model. *Progress in Transport Phenomena*, S. Dost, H. Struchtrup, and I. Dincer, eds., Elsevier, Paris, 837–842.

Yuan, T., Spence, G. D., and Hyndman, R. D. (1994) Seismic velocities and inferred porosities in the accretionary wedge sediments at the Cascadia margin. *J. Geophys. Res.* **99**, 4413–4427.

Zhang, Z. and Bejan, A. (1987) The horizontal spreading of thermal and chemical deposits in a porous medium. *Int. J. Heat Mass Transfer* **37**, 129–138.

Zhang, Z., Kleinstreuer, V., and Kim, C. S. (2002) Aerosol deposition efficiencies and upstream release position for different inhalation modes in an upper bronchial airway model. *Aerosol Sci. Technol.* **36**, 828–844.

Zhao, Z. M., Tardos, G., and Pfeffer, R. (1991) Separation of airborne dust in electrostatically enhanced fibrous filters. *Chem. Eng. Commun.* **108**, 307–332.

Zukauskas, A. (1987) Convective heat transfer in cross flow. *Handbook of Single-Phase Convective Heat Transfer*, S. Kakac, R. K. Shah, and W. Aung, eds., Wiley, New York, Chapter 6.

Index